Chemical, Gas, and Biosensors for Internet of Things and Related Applications

$175.00

(189) J.R Bus

Chemical, Gas, and Biosensors for Internet of Things and Related Applications

Edited by

Kohji Mitsubayashi

Department of Biomedical Devices and Instrumentation, Institute of Biomaterials and Bioengineering, Tokyo Medical and Dental University, Tokyo, Japan

Osamu Niwa

Saitama Institute of Technology, Fukaya, Japan

Yuko Ueno

NTT Basic Research Laboratories, NTT Corporation, Atsugi, Japan

ELSEVIER

Elsevier
Radarweg 29, PO Box 211, 1000 AE Amsterdam, Netherlands
The Boulevard, Langford Lane, Kidlington, Oxford OX5 1GB, United Kingdom
50 Hampshire Street, 5th Floor, Cambridge, MA 02139, United States

British Library Cataloguing-in-Publication Data
A catalogue record for this book is available from the British Library

Library of Congress Cataloging-in-Publication Data
A catalog record for this book is available from the Library of Congress

ISBN: 978-0-12-815409-0

For Information on all Elsevier publications
visit our website at https://www.elsevier.com/books-and-journals

Publisher: Susan Dennis
Acquisition Editor: Kathryn Morrissey
Editorial Project Manager: Vincent Gabriel
Production Project Manager: Prem Kumar Kaliamoorthi
Cover Designer: Greg Harris

Typeset by MPS Limited, Chennai, India

Contents

Part II Flexible, Wearable, and Mobile Sensors and Related Technologies 161

List of Contributors

Takahiro Arakawa Department of Biomedical Devices and Instrumentation, Institute of Biomaterials and Bioengineering, Tokyo Medical and Dental University, Tokyo, Japan

Hongmei Bi College of Science, Heilongjiang Bayi Agricultural University, Daqing, P.R. China

Chao-Min Cheng Institute of Biomedical Engineering, National Tsing Hua University, Hsinchu, Taiwan

Akihiro Chiba NTT Service Evolution Laboratories, Yokosuka, Japan

Shih-Wen Chiu National Tsing Hua University, Hsinchu, Taiwan

Ting-I Chou National Tsing Hua University, Hsinchu, Taiwan

Kana Eguchi NTT Service Evolution Laboratories, Yokosuka, Japan

Shunsuke Furutani Biomedical Research Institute, National Institute of Advanced Industrial Science and Technology (AIST), Osaka, Japan; Advanced Photonics and Biosensing Open Innovation Laboratory (PhotoBIO-OIL), National Institute of AIST, Osaka, Japan

Xiaojun Han State Key Laboratory of Urban Water Resource and Environment, School of Chemistry and Chemical Engineering, Harbin Institute of Technology, Harbin, P.R. China

Kenshi Hayashi Department of Electronics, Graduate School of Information Science and Electrical Engineering, Kyushu University, Fukuoka, Japan

Satoshi Hiyama Research Laboratories, NTT DOCOMO, Inc., Yokosuka, Japan

Jen-Hsuan Hsiao Department of Power Mechanical Engineering, National Tsing Hua University, Hsinchu, Taiwan

Chun-Jen Huang Department of Biomedical Sciences and Engineering, National Central University, Jhong-Li, Taiwan; Department of Chemical and Materials Engineering, National Central University, Jhong-Li, Taiwan; R&D Center for Membrane Technology, Chung Yuan Christian University, Chung-Li City, Taiwan

Kazunori Ikebukuro Department of Biotechnology and Life Science, Graduate School of Engineering, Tokyo University of Agriculture and Technology, Tokyo, Japan

Satoshi Ikeda TANITA Corporation, Tokyo, Japan

Gaku Imamura Center for Functional Sensor & Actuator (CFSN), National Institute for Materials Science (NIMS), Tsukuba, Japan; International Research Center for Materials Nanoarchitectonics (WPI-MANA), National Institute for Materials Science (NIMS), Tsukuba, Japan

Narushi Ito PROVIGATE Inc., Tokyo, Japan

Wataru Iwasaki Advanced Manufacturing Research Institute, National Institute of Advanced Industrial Science and Technology, Tokyo, Japan

Ryohei Kanzaki Research Center for Advanced Science and Technology, The University of Tokyo, Tokyo, Japan

Junichi Kodate NTT Device Innovation Center, NTT Corporation, Atsugi, Japan

Izumi Kubo Graduate School of Bioinformatics, Soka University, Tokyo, Japan

Hisashi Kurasawa NTT Service Evolution Laboratories, Yokosuka, Japan

Ryoji Kurita National Institute of Advanced Industrial Science and Technology, Tsukuba, Japan

Chuanjun Liu Research Laboratory, U.S.E. Co., Ltd, Tokyo, Japan

Yasuko Yamada Maruo Tohoku Institute of Technology, Sendai, Japan

Hiroaki Matsuura Department of Life Science & Green Chemistry, Faculty of Engineering, Saitama Institute of Technology, Fukaya, Japan

Kosuke Minami Center for Functional Sensor & Actuator (CFSN), National Institute for Materials Science (NIMS), Tsukuba, Japan; International Research Center for Materials Nanoarchitectonics (WPI-MANA), National Institute for Materials Science (NIMS), Tsukuba, Japan

Kohji Mitsubayashi Department of Biomedical Devices and Instrumentation, Institute of Biomaterials and Bioengineering, Tokyo Medical and Dental University, Tokyo, Japan

Hidefumi Mitsuno Research Center for Advanced Science and Technology, The University of Tokyo, Tokyo, Japan

Mariko Miyashita TANITA Corporation, Tokyo, Japan

Fumio Mizutani Graduate School of Material Science, University of Hyogo, Kamigori, Japan

Nobutomo Morita Advanced Manufacturing Research Institute, National Institute of Advanced Industrial Science and Technology, Tokyo, Japan

Kuniaki Nagamine Research Center for Organic Electronics (REOL), Yamagata University, Yonezawa, Japan

Maria Portia Briones Nagata Advanced Manufacturing Research Institute, National Institute of Advanced Industrial Science and Technology, Tokyo, Japan

Hiroshi Nakashima NTT Basic Research Laboratories, Nippon Telegraph and Telephone Corporation, Tokyo, Japan

Huynh Thien Ngo Center for Functional Sensor & Actuator (CFSN), National Institute for Materials Science (NIMS), Tsukuba, Japan; International Research Center for Materials Nanoarchitectonics (WPI-MANA), National Institute for Materials Science (NIMS), Tsukuba, Japan

Osamu Niwa Saitama Institute of Technology, Fukaya, Japan

Takeshi Sakurai Department of Agricultural Innovation for Sustainability, Faculty of Agriculture, Tokyo University of Agriculture, Kanagawa, Japan

Fumihiro Sassa Department of Electronics, Graduate School of Information Science and Electrical Engineering, Kyushu University, Fukuoka, Japan

Kota Shiba Center for Functional Sensor & Actuator (CFSN), National Institute for Materials Science (NIMS), Tsukuba, Japan; International Research Center for Materials Nanoarchitectonics (WPI-MANA), National Institute for Materials Science (NIMS), Tsukuba, Japan

Masato Suzuki Graduate School of Material Science, University of Hyogo, Kamigori, Japan

Kea-Tiong Tang National Tsing Hua University, Hsinchu, Taiwan

Shizuo Tokito Research Center for Organic Electronics (REOL), Yamagata University, Yonezawa, Japan

Koji Toma Department of Biomedical Devices and Instrumentation, Institute of Biomaterials and Bioengineering, Tokyo Medical and Dental University, Tokyo, Japan

Yu-Ting Tsao School of Traditional Chinese Medicine, Chang Gung University College of Medicine, Taoyuan, Taiwan

Shingo Tsukada NTT Basic Research Laboratories, Nippon Telegraph and Telephone Corporation, Tokyo, Japan

Kaori Tsukakoshi Department of Biotechnology and Life Science, Graduate School of Engineering, Tokyo University of Agriculture and Technology, Tokyo, Japan

Kinuko Ueno Department of Biotechnology and Life Science, Graduate School of Engineering, Tokyo University of Agriculture and Technology, Tokyo, Japan

Yuko Ueno NTT Basic Research Laboratories, NTT Corporation, Atsugi, Japan

Chung-Yao Yang Hygeia Touch Inc., Taipei, Taiwan

Tomoyuki Yasukawa Graduate School of Material Science, University of Hyogo, Kamigori, Japan

Genki Yoshikawa Center for Functional Sensor & Actuator (CFSN), National Institute for Materials Science (NIMS), Tsukuba, Japan; International Research Center for Materials Nanoarchitectonics (WPI-MANA), National Institute for Materials Science (NIMS), Tsukuba, Japan; Materials Science and Engineering, Graduate School of Pure and Applied Science, University of Tsukuba, Tsukuba, Japan

Preface

The growing interest in medical and environmental sensors can be attributed to the increasing awareness that health is essential for enjoying and keeping a good quality of life. The advent of wearable and/or flexible sensors for healthcare represents such a leap. The demand for quality healthcare has strained the current healthcare system, which faces large workforce shortages and inadequate medical facilities. It is paramount to develop consumer-based devices to augment the current healthcare demands. One emerging technology is sensor networks connected to the Internet of Things (IoT), a human-oriented network of tiny human monitoring devices developed on the body and in the surrounding infrastructure.

Chemicals, gas, and biosensors will be required for preventive medicine and environmental assessment for healthcare demands in the near future. The reason for this stems from the fact that all branches of modern medicine, ranging from prevention to complex intervention, rely on early and accurate diagnosis followed by close monitoring of the results. The research field of gas and bio/chemical sensors has been developed by a group of truly interdisciplinary researchers involving chemists, biologists, physicists, material scientists, and computer engineers to create a range of novel configurations exploiting bio/chemical recognition systems allied with physiochemical transducers. Novel chemical, gas, and biosensors would further progress with the introduction of a range of optical, acoustic, magnetic, thermal, and electrical technologies, coupled with microelectronics and MEMS devices for the future medical and healthcare systems. This book is organized with the following parts and chapters. Part 1 introduces novel bio/chemical sensors with several recognition materials such as enzymes, antibodies, aptamers, receptors, and artificial materials for analytes in body fluids and gas-phase samples. In Part 2, flexible and mobile sensors and related techniques are explained with some wearable and cavitas (body cavity) sensors, and smell detectors. Information and network technologies for the sensors are illustrated with industrial applications in Part 3.

This book is intended for graduate students, academic researchers, and professors who work in the field of medical and environmental research, and also for industry professionals involved in development of devices and systems with IoT for human bio/chemical measurement, medical monitoring, and healthcare services with Internet technologies. We would like to sincerely express our appreciation to the distinguished authors of the chapters whose expertise has certainly contributed significantly to the book. We hope that this book can shed light on various technological aspects related to bio/chemical sensors with IoT and related applications in a healthcare context, and will stimulate further research in this field.

Kohji Mitsubayashi, Osamu Niwa and Yuko Ueno
Editors

Part I

Sensors and Devices for Internet of Things Applications

Portable urine glucose sensor

1

Narushi Ito[1], *Mariko Miyashita*[2] *and Satoshi Ikeda*[2]
[1]PROVIGATE Inc., Tokyo, Japan, [2]TANITA Corporation, Tokyo, Japan

1.1 Introduction

Development of a noninvasive blood glucose monitoring system is based on measurement of near infrared light passing through fingers and arms [1,2], measurement of interstitial fluid collected from the skin surface with a micro glucose sensor [3,4], and measurement of contact lens type tears [5,6], etc. It has been done over the past 30 years and enormous amount of research and development has been published, however, there are still no practical products.

A series of products that have succeeded in development include a continuous blood glucose monitor and a flash glucose monitor that place small needles in the abdomen and the like. These are minimally invasive, and continuous monitoring for 2 weeks is possible. Meanwhile, as a noninvasive measurement, a portable urine glucose meter that makes it possible to quantitatively measure urine glucose levels correlated with blood glucose levels has been put to practical use.

In this section, we describe the principle and structure of the microplanar type urine glucose sensor and examples of application of commercialized urine glucose meter to healthcare.

1.2 Significance of urine glucose measurement

In the urine glucose tests, diabetes screening tests are being conducted to test positive (+) or negative (−) by measuring fasting urine such as common in medical examinations.

Positive (+) is based on 100 mg/dL as the urine glucose concentration. Medically urine glucose positivity is considered as a suspicious indicator of diabetes first. Then further inspections are necessary, because there are transient cases such as renal diabetes, stress, pregnancy, etc., in addition to other findings for diagnosis of diabetes.

Ultimately diabetes is confirmed by the 75 g oral glucose tolerance test. Urine glucose test is regarded as an auxiliary test and a large number of screening tests are still being carried out at present, due to its advantage of noninvasive measurement.

Chemical, Gas, and Biosensors for the Internet of Things and Related Applications.
DOI: https://doi.org/10.1016/B978-0-12-815409-0.00001-2

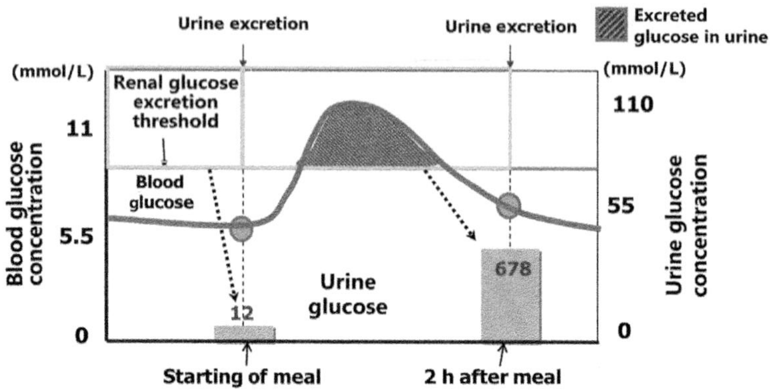

Figure 1.1 Relationship between urine glucose concentration and blood glucose concentration after meal.

Changes in urine glucose concentration after meals are shown in Fig. 1.1. The postprandial urine glucose concentration is linked to blood glucose levels, and it is important that it does not overlook postprandial hyperglycemia occurring after meals. When the urine glucose concentration exceeds 50 mg/dL, it means that the blood glucose level exceeds the glucose excretion threshold of 160–180 mg/dL in the kidney.

Even when blood glucose level rises with the diet, it decreases after 1 hour due to the action of insulin. In other words, the blood glucose level measured at 2 hours after a meal usually returns to the normal range. On the other hand, it is known that the urine glucose concentration 2 hours after a meal reflects the elevated blood glucose level with the meal. As well, it is demonstrated that urine glucose level correlates with the mean blood glucose level.

1.3 Operating principle of urine glucose sensor and laminated structure

1.3.1 *Principle of operation*

Urine glucose sensor is an enzyme electrode method combining glucose oxidase (GOX) and hydrogen peroxide electrode.

The electrode is fabricated by photolithography technology. Glucose is enzymatically converted to hydrogen peroxide (H_2O_2) by GOX, and the yielded H_2O_2 is electrochemically detected by the electrodes.

The enzymatic reaction (1) and the electrode reactions (2) are as follows:

1. Enzymatic reaction

   GOX: Glucose + $O_2 \rightarrow$ gluconolactone + H_2O_2

2. Electrochemical reactions at electrodes

   Working electrode: $H_2O_2 \rightarrow 2H^+ + O_2 + 2e^-$

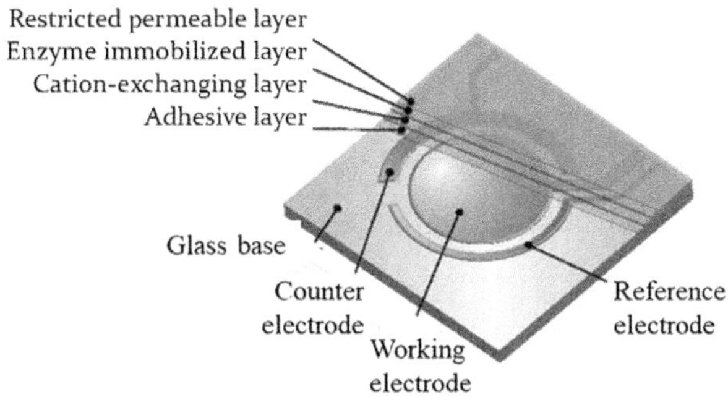

Figure 1.2 Perspective view of the sensor.

Counter electrode: $2H^+ + 1/2O_2 + 2e^- \rightarrow H_2O$
Entire electrode system: $H_2O_2 \rightarrow H_2O + 1/2O_2$

A perspective view of the sensor is shown in Fig. 1.2. Three electrodes, a working electrode, a counter electrode, and a reference electrode, are formed in the hole of the cartridge. The working electrode and the counter electrode are Pt electrodes, and thin film Ag/AgCl electrodes are formed as reference electrode. The reference electrode has the role of stabilizing the potential after immersion in the solution. The outermost layer of the electrode is coated with a thin film of a fluorinated polymer to prevent contamination due to urine components while protecting the electrode system as a whole and stabilizing the operation of the electrodes for more than 1 year in solution.

1.3.2 Laminated structure of urine glucose sensor

To accurately measure postprandial urine, means to eliminate the influence of vitamin C (ascorbic acid) among substances released from foods are required. Ascorbic acid has a reaction that gives electrons to an electrode and another reaction decomposing hydrogen peroxide, and it is typical of an interfering substance of an amperometric sensor.

Furthermore, it is necessary to fabricate a membrane structure so that interfering substances other than ascorbic acid contained in the urine do not affect the measured values.

Fig. 1.3 shows a laminated structure of the urine glucose sensor. This sensor is composed of four layers: a restricted permeable layer, an enzyme immobilized layer, a cation-exchanging layer, and an adhesive layer.

1. The restricted permeable layer has a wide measurement range from 10 to 2000 mg/dL, limiting the diffusion of molecules larger than glucose. It has the role of preventing the influence of adhered substances in urine.
2. The enzyme immobilized layer is the one where enzyme (GOX) and bovine serum albumin are crosslinked and immobilized so as not to inactivate the enzyme; as a result, repetition of the sensor is possible.

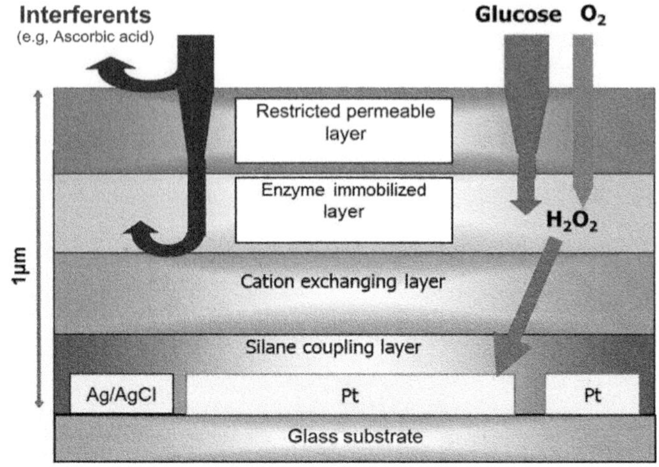

Figure 1.3 Laminated structure of the urine glucose sensor.

3. The cation-exchanging layer has the role of permeating hydrogen peroxide and limiting the diffusion of molecules larger than hydrogen peroxide. Furthermore, it has the function of preventing permeation of ionized molecules.
4. The adhesive layer has the role of covalently bonding the selectively permeable film, which is an organic material, to the surface of the glass substrate or the electrode and stably adhering for a long time in water.

This sensor is formed of a thin film of four layers with a total thickness of 1 μm or less, effectively eliminating the influence of interfering substances in the urine and an early time response [7].

1.4 Development of portable urine glucose meter

1.4.1 Composition of urine glucose meter

This urine glucose meter consists of a body and a sensor. Portability is designed so that the sensor section is folded down to be compact, and at the time of measurement it is extended. Photo 1.1 shows a urine glucose meter in a stored state. Photo 1.2 shows the urine glucose meter extended at the time of measurement. The urine glucose meter sensor section is composed of a preservation solution bottle, which makes the sensor wet. The preservation solution is reserved to hold the sensor, which can cause optimal enzymatic reaction with pH buffer and physiological saline. After the sensor is taken out, it becomes possible to measure instantly.

Photo 1.2 shows a urine glucose meter at the time of measurement. When measured, total length 210 mm of the meter can be directly applied to urine with one hand. The urine glucose sensor at the tip is equipped with a thermistor for detecting the water temperature so that the output can be corrected. The sensor needs to be

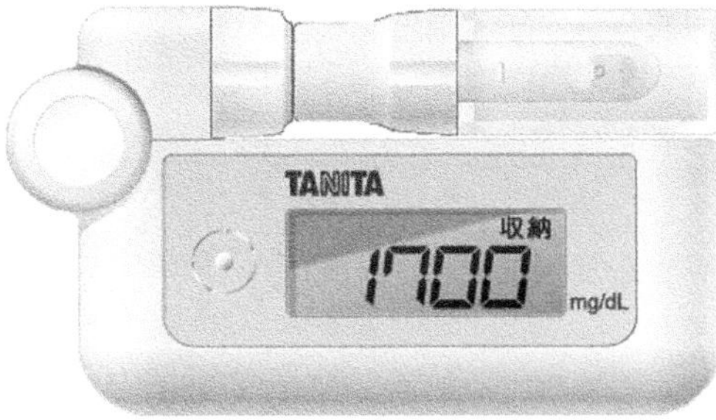

Photo 1.1 Urine glucose meter folded (closed).

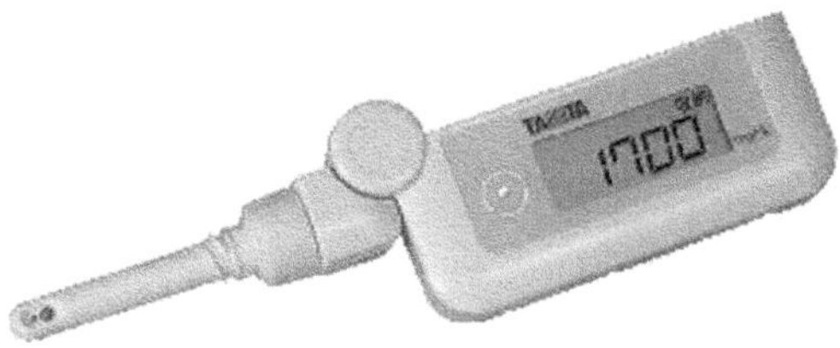

Photo 1.2 Urine glucose meter extended (opened).

replaced either after 200 measurements or within 60 days due to the removable socket structure. The main body measures the minute current and controls the device. It also has a liquid crystal display that indicates urine glucose concentration, and switches for calibration and measurement value recall. It also operates for 8 months with one lithium battery.

1.4.2 Performance evaluation of urine glucose meter

The final test of the urine glucose meter requires performance evaluation by the urine of patients. The results of simultaneous measurement of patients' urine specimens with urine glucose meter and clinical urine glucose analyzer (A&T GA03R) and correlation evaluation are shown in Fig. 1.4. The primary equation obtained by the method of least squares is urine glucose meter ① $Y = 0.925\ X + 53.3$, $R = 0.987$, urine glucose meter ② $Y = 0.939\ X + 62.9$, $R = 0.987$, urine glucose meter ③ $Y = 0.968\ X + 40.4$, $R = 0.99$, showing a high correlation with the conventional clinical urine glucose analyzer. Especially, the deviation of the measured values in the low concentration region is small, although the deviation is medically regarded as a problem. However, Fig. 1.4 demonstrates sufficient results for performance of a compact and simple measuring instrument. As well, those results showed that as a self-measuring tool at home, its portability is a plus, and it can measure urine glucose with high accuracy with a small size of 210 mm in total length [8].

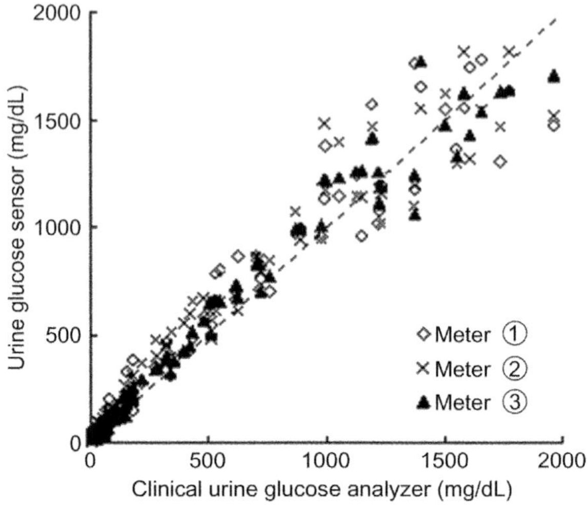

Figure 1.4 Correlation between urine glucose meter and clinical urine glucose analyzer.

1.5 Clinical application of urine glucose meter

1.5.1 Relationship between the amount of boiled rice and urine glucose concentration in impaired glucose tolerance

Meal load test was conducted on subjects judged to have impaired glucose tolerance by 75 g oral glucose tolerance test. In the method, blood glucose level and urine glucose level up to 3 hours after starting a meal were measured for 320 kcal salad and meat, boiled rice with different amount of 100–300 g (145–435 kcal). The blood glucose level was measured using a self-monitoring blood glucose meter (GLUCOCARD: ARKRAY), and the urine glucose concentration was measured with a developed urine glucose meter.

Fig. 1.5 (A) shows changes in blood glucose concentration, and (B) shows changes in urine glucose concentration. The results confirmed an increase in urine glucose concentration with the rise in blood glucose concentration reflecting the difference in the amount of boiled rice. In particular, differences in urine glucose concentrations of 400 and 600 mg/dL are difficult to determine with conventional urine glucose test paper. The above results indicate that the quantitative measurement using the urine glucose meter can accurately capture postprandial hyperglycemia that changes depending on the carbohydrate intake [9].

1.5.2 Results of urine glucose monitoring on impaired glucose tolerance case

The results of urine glucose measurement before and after a meal for 7 days are shown in Fig. 1.6. In this impaired glucose tolerance case, self-monitoring of

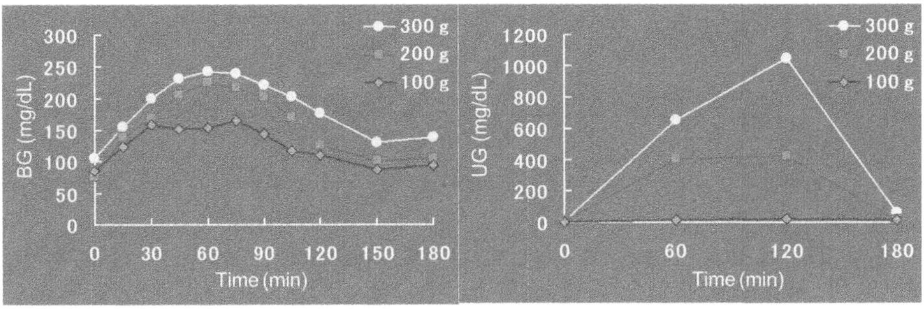

Figure 1.5 Blood glucose concentration (BG) and urine glucose concentration (UG) differ by changing the volume of rice (In case of impaired glucose tolerance). (A) BG changes after meal; (B) UG changes after meal.

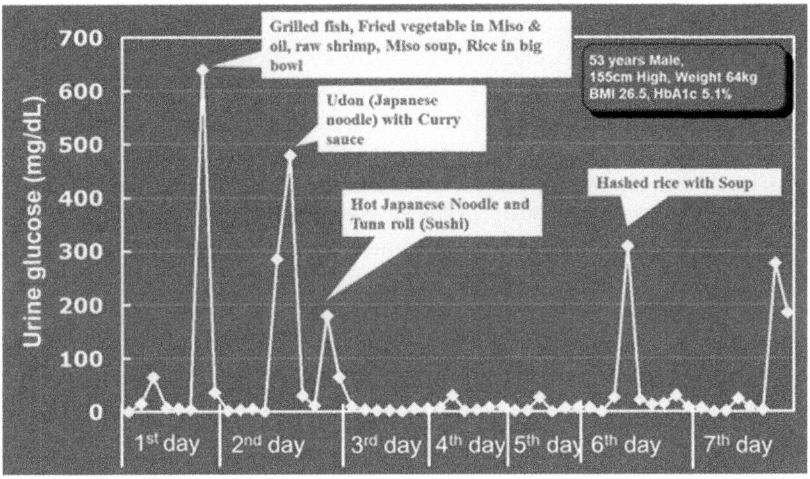

Figure 1.6 Results of urine glucose measurement before and after meal for 7 days (In case of impaired glucose tolerance).

urine glucose (SMUG) was carried out by instructing the user to pay attention to the relationship between urine glucose concentration after meals and meal contents. The meal content ingested was recorded at the same time. As a result, it was revealed that dietary control becomes possible by monitoring the urine glucose concentration after meals. Also, over the next 8 months, as a result of eating meals that did not raise the urine glucose concentration after meals, weight decreased from 63.9 to 59.0 kg, body fat decreased from 20.5% to 13.7%, and it was also effective to reduce body weight. In conclusion, postprandial hyperglycemia that occurs from early stage of diabetes can be controlled by urine glucose meter [9].

1.5.3 Results of a case of self-monitoring of urine glucose in diabetes

SMUG was conducted for 6 months on a voluntary type 2 diabetes patient (woman aged 69 years, height 153.2 cm, weight 51.8 kg, BMI 22.1 kg/m^2). The method was to measure urine glucose concentration 8 times a day (before morning, before breakfast, after breakfast, before lunch, after lunch, before dinner, after dinner, before going to bed). Also, meal contents were recorded at the same time. The doctor monitored the feedback of the relationship between the urine glucose concentration and meal contents after meals. Meanwhile, at a hospital every month, HbA1c and body weight were measured. The results of SMUG are shown in Fig. 1.7

The transition of measured values over 100 mg/dL during 4 months from the start of urine glucose measurement was 9 times in the first 2 weeks after SMUG began, 5 times in the next 2 weeks, 6 times in the next 2 weeks, then 3 times, 1 time, 0 times, 3 times, 0 times, and 3 times, all 2-week periods. HbA1c and body weight change are shown in Fig. 1.8. During the 5 months before SMUG, HbA1c had been in the 8% range, but it decreased from 8.7% (glycemic control status: unacceptable) to 5.8% (glycemic control status: excellent) in about 3 months after starting urine glucose measurement. Therefore, glycemic control improved. As a result, a decrease in HbA1c was observed 1 month after starting measurement of postprandial urine glucose concentration, and it was effective for blood glucose control of type 2 diabetic patients. The finding of the interview after use is that urine glucose measurement is easy to introduce and continue because it is noninvasive and the measured value changes dynamically in the range of 10–2000 mg/dL, so the results show it was easy to understand, and it was thought that the motivation for the patient's blood glucose control was improved.

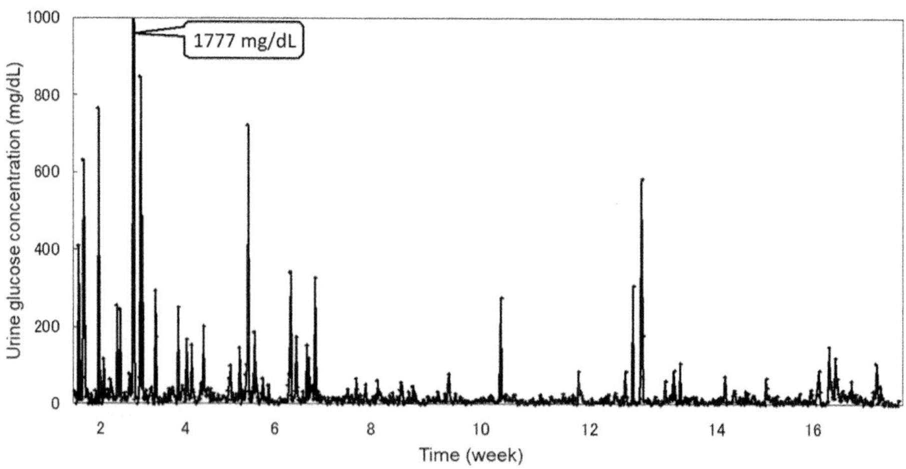

Figure 1.7 Results of SMUG during 4 months.

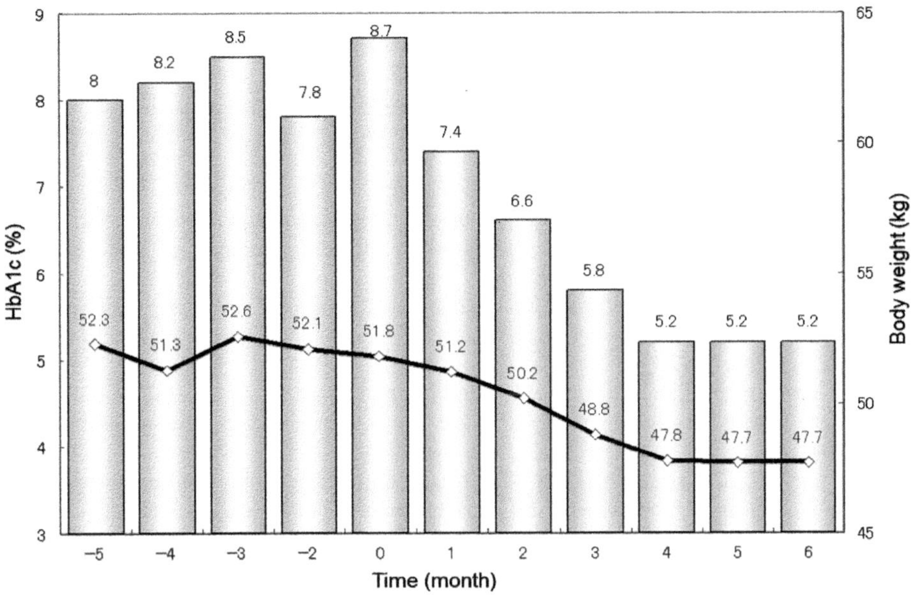

Figure 1.8 Results of HbA1c(JDS) and body weight change.

1.6 Conclusions

In conclusion, (1) urine glucose concentration at 2 hours after a meal is higher in proportion to the amount of ingested carbohydrate; and (2) SMUG is available for control of the meal contents, which suppresses postprandial hyperglycemia, indicating that HbA1c and body weight can be reduced. Recently, clinical trials comparing SMBG and SMUG levels of type 2 diabetes revealed that there is no difference in diet therapy effectiveness [10]. Postprandial hyperglycemia stimulates glucose spikes to vascular endothelial cells. As a result, it has been clarified that not only diabetes but also arteriosclerosis causing stroke and myocardial infarction can be incubated. As well, it is one of risk factors for dementia. Monitoring postprandial hyperglycemia with a urine glucose meter from the earliest stage of diabetes is recommended as a noninvasive healthcare tool that helps modify lifestyle of diet and exercise. Furthermore, a portable urine glucose meter integrating IoT and AI not only supports meals and exercise, but is thought to become an effective diabetes prevention tool customized to characteristics of individuals.

References

[1] H.M. Heise, et al., Noninvasive blood glucose sensors based on near-infrared spectroscopy, Artif Organs 18 (1994) 439.

[2] US Patent 5,553,616, Determination of concentrations of biological substances using raman spectroscopy and artificial neural network discriminator.

[3] US Patent 2005/0215872, Monitoring of physiological analytes.
[4] N. Ito, et al., Development of a transcutaneous blood-constituent monitoring method using a suction effusion fluid collection technique and an ion-sensitive field-effect transistor glucose sensor, Med. Biol. Comput. 32 (1994) 242.
[5] W.F. March, et al., Ocular glucose sensor, Trans. Am. Soc. Artif. Intern. Organs 28 (1982) 232.
[6] WO2014/113174, Encapsulated electronics.
[7] M. Miyashita, et al., Development of urine glucose meter based on micro-planer amperometric biosensor and its clinical application for self-monitoring of urine glucose, Biosensors Bioelectr. 24 (2009) 1336.
[8] I. Yamaguchi, et al., Performance evaluation of urine glucose meter: repeatability, effects of interferential substances, and comparison with clinical glucose analyzer, Rinsyoukensa. 53 (2009) 237.
[9] A. Ohashi, et al., Effect of food intake and its contents on postprandial urine glucose in diabetes candidates by digital urine glucose meter, Japan. Soc. Med. Biol. Eng. 42 (2004) 280.
[10] J. Lu, et al., Comparable efficacy of self-monitoring of quantitative urine glucose with seif-monitoring of blood glucose on glycemic control in non-insulin-treated type 2 diabetes, Diab. Res. Clin. Pract. 93 (2011) 179.

Design, application, and integration of paper-based sensors with the Internet of Things

2

Jen-Hsuan Hsiao[1], *Yu-Ting Tsao*[2], *Chung-Yao Yang*[3] *and Chao-Min Cheng*[4]

[1]Department of Power Mechanical Engineering, National Tsing Hua University, Hsinchu, Taiwan, [2]School of Traditional Chinese Medicine, Chang Gung University College of Medicine, Taoyuan, Taiwan, [3]Hygeia Touch Inc., Taipei, Taiwan, [4]Institute of Biomedical Engineering, National Tsing Hua University, Hsinchu, Taiwan

2.1 Introduction

Paper-based analytical devices (PADs) have attracted significant and growing attention since their early development by the Whitesides research group [1]. Since that time, a number of fabrication methods, applications, and technological integrations, including integration with smartphones, have been investigated, demonstrated, and applied. Common fabrication methods for PADs now include photolithography [1], wax printing [2], wax dipping [3], screen printing [4–6], inkjet printing [7–10], plasma treatment [11–13], laser-based fabrication procedure [14], and the utilization of craft-cutting tools [15]. In addition to the significantly impactful development of two-dimensional PADs, the development of three-dimensional PADs has produced a crop of new tool platforms for the collection of rich data and valuable, health-altering diagnostics [16–19]. Paper-based sensors have proven themselves to be useful in a broad spectrum of fields. Multiple applications have been developed for both human diagnostics and environmental analysis. Among the diagnostic developments, a variety of tests examining urine, blood, and even very small amounts of aqueous humor have been successfully demonstrated. They have targeted a variety of diseases and disease states including paraquat poisoning, Alzheimer's disease, age-related macular degeneration, glaucoma, corneal dystrophy, and chronic wound care to name but a few. Among the environmental analytical devices developed, these have successfully demonstrated the capacity to detect a host of target items including contaminants such as metals, nonmetals, organic

Chemical, Gas, and Biosensors for the Internet of Things and Related Applications.
DOI: https://doi.org/10.1016/B978-0-12-815409-0.00002-4

molecules, pesticides, and microorganisms. All of these tools have the potential for integration with additional tools as well as, and perhaps most significantly, the Internet of Things.

2.2 Bioapplications of paper-based analytical devices

PAD bioapplications are continuing to be developed with a focus on health diagnostics for hastening and improving care via early diagnosis and disease state monitoring. Among these tools, an enzymatic assay for determining urinary creatinine was developed by Talalak et al. [20] that proved useful for investigating kidney function. When creatinine was present in the urine sample, the assays that had immobilized creatinine reagents would react and form pink-red quinoneimine dye. Color intensity was proportional to the creatinine concentration in the sample, and a linear range of 2.5–25, 2.0 mg/dL of detection limit was achieved (illustrated in Fig. 2.1A).

Accurate blood typing is essential for blood transfusion, tissue, and organ transplantation. In the research conducted by Al-Tamimi et al. [24], Kleenex paper was selected from a list of different paper substrates for its porous characteristic. Using this substrate, nonagglutinated red blood cells (RBCs) were eluted following application of 0.9% NaCl buffer. This paper-based assay was able to accurately detect the blood types of 100 samples within 1 minute, including 4 weak AB and 4 weak RhD samples. Another paper-based assay that could perform Rh and forward and reverse ABO blood typing was developed by Noiphung et al. [21]. They used a combination of wax printing and wax dipping methods to fabricate the PADs depicted in Fig. 2.1B. The advantage of this assay was that results could be maintained at room temperature for at least 7 days, a timespan that conventional slide or tube methods cannot achieve. Songjaroen and Laiwattanapaisal [25] described a PAD for simultaneous forward and reverse ABO blood group typing. The results were designed to be a barcode-like chart that could be read via comparison to a Tween-20 buffer wicking path.

Apart from blood typing applications, whole blood or serum samples are often used for various clinical diagnostics. Vella et al. [22] described a vertical flow paper-based assay that could detect alkaline phosphatase and aspartate aminotransferase, the two biomarkers of liver function, and total serum protein. This PAD comprised of a complete assay that could perform the functions of sample separation, distribution, and detection (illustrated in Fig. 2.1C). Test results could be captured using a smartphone and data could be sent to trained professionals for further analysis. Zhang et al. [26] developed a double-channel PAD that was coated with an enzyme–starch solution containing glucose oxidase, lactate oxidase, and horseradish peroxidase. To demonstrate the efficiency of detecting glucose and lactate, individual, mixed, and samples with RBCs were tested. Noiphung et al. [27] used

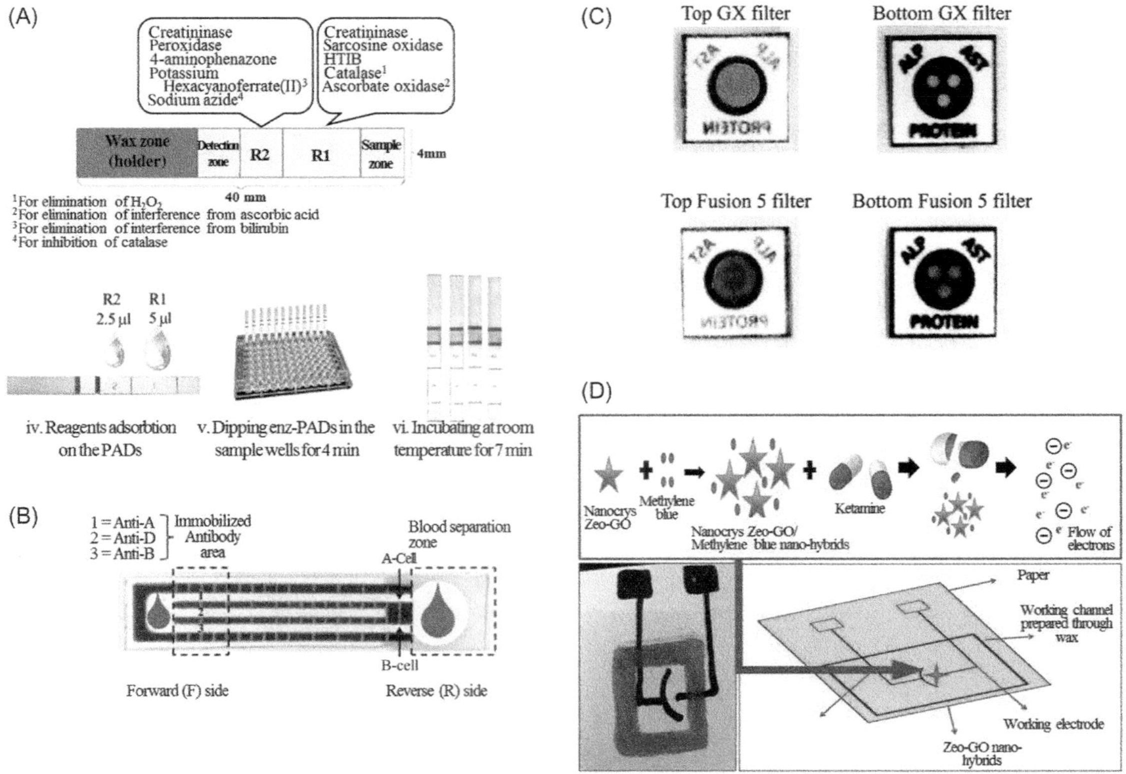

Figure 2.1 The design of PAD approaches for human usage. (A) The design and working principle of an ePAD for the detection of urine creatinine [20]. (B) The schematic of a distance-based PAD for ABO and Rh blood groups detection [21]. (C) The schematic illustration of the fabrication and sensing mechanism of an electrochemical-based ketamine detection PAD [22]. (D) Image of micropatterened liver function test (LFT) PADs assembled with a GX PSM and a Fusion 5 filter [23].

an electrochemical detection method in their enzymatic PADs (ePADs) to measure the current of H_2O_2 at the optimum potential of -0.1 V (vs Ag/AgCl), which was proportional to whole blood glucose concentrations. This ePAD could also be used to measure other biochemical markers if H_2O_2 was a product of its enzymatic reaction.

In addition to general blood and serum test, PADs have also been developed to assist diagnosis in many medical fields such as infection, infertility, and rapid drug test. Tsai et al. [28] used unmodified gold nanoparticles and single-stranded detection oligonucleotides on a paper assay platform to detect target TB DNA sequences. These PADs could complete the diagnostic assay within 1 hour and the detection limit was 2.6 nM *mycobacterium tuberculosis* complex (MTBC) target sequences. They also introduced a model using a smartphone camera to take the diagnostic results and then transmit them for cloud computing. Matsuura et al. [29] used thiazine dye to stain sperm and evaluate sperm concentration in semen. They also used a tetrazolium-based colorimetric assay (MTT (3-(4,5-dimethyl thiazol-2-yl)-2,5-diphenyl tetrazolium bromide) assay) to estimate human sperm motility in semen. The calculation of sperm concentration and motility by PADs helps men to evaluate their sperm quality in an easy way with increased privacy. Wang et al. [30] demonstrated a lateral flow immunoassay to detect dengue virus serotype-2 nonstructural protein-1 antigens in a buffer system. This lateral flow immunoassay could provide real-time diagnosis of dengue fever and create an opportunity for early remedial action. PADs have also been developed for isolating extracellular vesicles, which contain considerable data regarding intercellular communications. This approach may be further developed for nucleic acids analysis [31].

Regarding the development of PAD-based rapid drug tests, Narang et al. [23] investigated the potential for using zeolite nanoflakes and graphene-oxide nanocrystals (Zeo-GO) to amplify the electrochemical sensing signal of ketamine (Fig. 2.1D). This PAD demonstrated a low detection limit of 0.001 nM/mL, and showed a 99% correlation when tested in alcoholic and nonalcoholic drinks. Another example of a PAD that targeted a sedative drug, diazepam, was introduced by Narang et al. [32]. The electrochemical behavior of silica gold-coated nanorods (Si@GNRs) was studied and used to amplify the sensing signal. This PAD showed a low detection limit of 1.5×10^{-9} M, and a linear range of 3.5 nM to 3.5 mM. Apart from therapeutic drug analyses, PADs have also been developed for paraquat poisoning diagnosis. Kuan et al. [33] performed a colorimetric sodium dithionite assay and an ascorbic acid assay on 96-well-patterned papers to detect serum paraquat levels. These PADs could diagnose paraquat intoxication within 10 minutes. In addition to the aforementioned poisoning tests, Yen et al. [34] explored the development of a PAD for rapid detection of organophosphate both for poisoning diagnosis and pesticide residue detection. By estimating the activity of acetylcholinesterase, which is the biomarker of organophosphate, this PAD provided results comparable to the standard testing method.

2.3 Environmental analysis of paper-based analytical devices

A number of PADs have been successfully developed in the field of environmental analysis. Hossain and Brennan [35] proposed an ePAD that could detect heavy metals including Hg(II), Ag(I), Cu(II), Cd(II), Pb(II), Cr(VI), and Ni(II) through colorimetric visualization. When heavy metals were not present in the sample, β-galactosidase (B-GAL) would react with the chromogenic substrate to form red-magenta product. Because heavy metals are strong inhibitors of B-GAL, color intensity is inversely proportional to the concentration of heavy metals in a sample. Cate et al. [36] used distance-based detection in their multiplexed PADs to detect Ni, Cu, and Fe. The blue dye used in these PADs acted as a passive timer, when presented, meaning that the assay was completed and we could estimate the level of Ni, Cu, and Fe by measuring the distance of the color band (Fig. 2.2A). Selectivity was demonstrated by a tolerance study over 11 other metal species, and utility was also demonstrated through measurement of a certified welding fume sample. Another distance-based assay for the detection of copper in drinking water was introduced by Quinn et al. [40]. A total of 34 samples were tested using this device. Results compared favorably with results from inductively coupled plasma mass spectrometry, showing particularly close alignment within the 0.375 ppm concentration.

In regards to the development of nonmetal paper-based sensors, Hao et al. [37] synthesized a type of naphthalimide-based azo dye (probe 1) with a two-step procedure for the colorimetric detection of cyanide anions (CN^-). The activated aldehyde group of the probe 1 would attract the cyanide anions, and an apparent color change from yellow (left side of Fig. 2.2B) to red (right side of Fig. 2.2B) could be completed in a few seconds (Fig. 2.2B). This paper-based strip demonstrated a detection limit of 20 μM for the direct detection of CN^-. Feasibility and reliability in mixed sample were also tested. Weng et al. [41] reported a two-dimensional and a three-dimensional assay for the simultaneous detection of nitrite and oxalate. A total of 12 combinations of wax color and concentration gradients were studied to understand the principle of time-delay valves. Cinti et al. [42] fabricated a reagentless electrochemical PAD to detect phosphate by measuring the electroactive phosphomolybdic complex. This PAD had a detection limit of 4 μM and a linear response of up to 300 μM. Performance was evaluated in distilled water and untreated river water samples. Another PAD to detect organophosphate pesticides was described by Hossain et al., who used indophenyl acetate as the substrate to be hydrolyzed by acetylcholinesterase (AChE), which broke down into blue-purple indophenoxide anion (IDO^-). The blue-purple IDO^- was then trapped by the cationic polymer, polyvinyl amine (PVAm) placed over the detection zone. The mechanism of the PADs is demonstrated in Fig. 2.2C [38]. For the detection of volatile organic compounds (VOCs), Yoon et al. [43] evaluated both

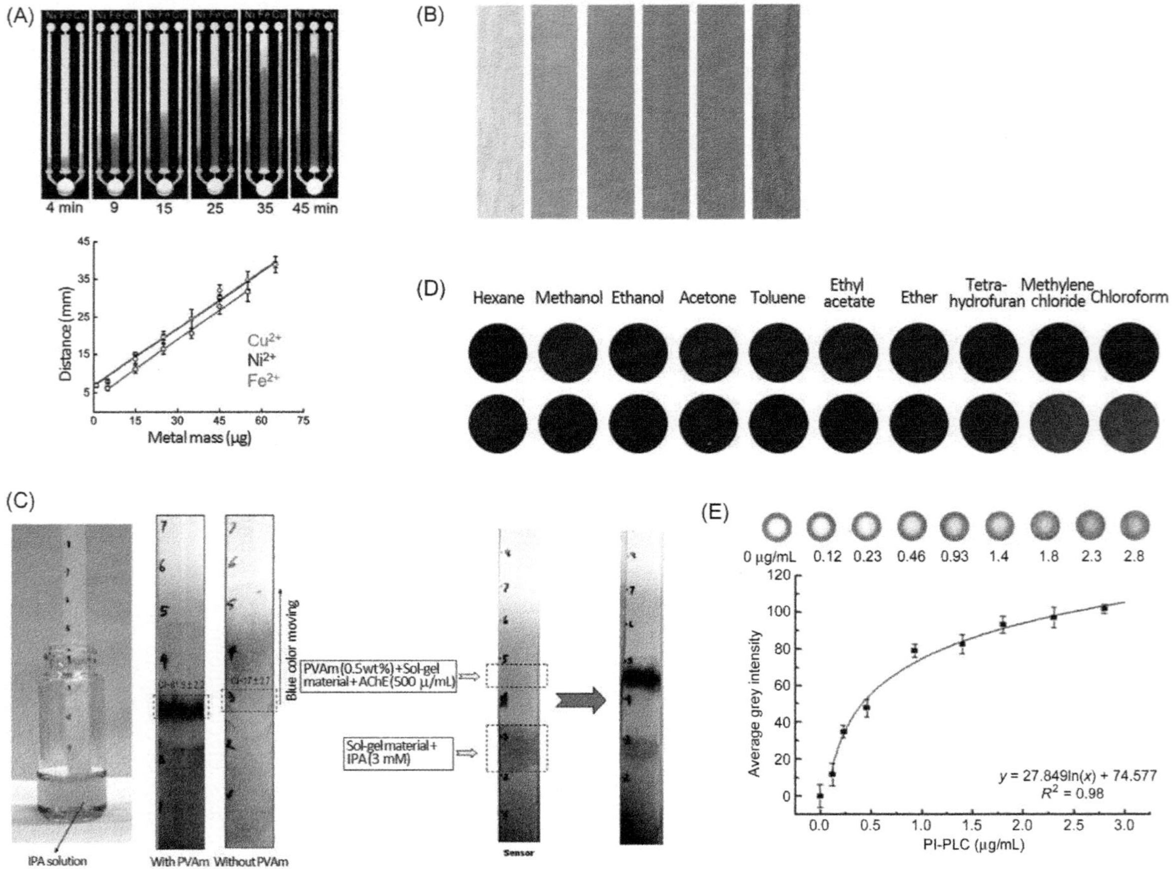

Figure 2.2 Readings and results of different PADs for environmental analysis. (A) The time lapse image and linear dynamic ranges of Fe, Ni, and Cu in a multichannel paper analytical device [36]. (B) Colorimetric results for the detection of blank, 20, 40, 60, 80, and 100 μm CN^- using naphthalimide-based azo dye (probe 1) [37]. (C) The use of PVAm could concentrate the yellow-to-blue (yellow in the lower part and blue in the upper part) color change of IDO − product in a bounded region, and enhances the signal intensity, which is inversely proportional to pesticide concentration [38]. (D) Fluorescence images of the reaction of the PAD and vaporized organic solvents under 546 (first row) and 488 (second row) nm UV irradiation [43]. (E) The response curve of the limit of detection of PI-PLC enzymes, an enzyme−substrate pair with *Listeria monocytogenes* [39].

the colorimetric (from blue-to-red-to-yellow) and fluorometric results of the interaction of polydiacetylenes and VOCs on paper substrates. The red (first row of the test results) and green (second row of the test results) fluorescence intensity distributions of different VOCs are illustrated in Fig. 2.2D.

PADs have also been used to detect pathogenic bacteria in food. Jokerst et al. [39] created a paper-based microspot assay to detect the colorimetric results of targeted pathogen and enzyme–substrate pairs. When used to detect phospholipase C (PI-PLC) with 5-bromo-4-chloro-3-indolyl-myo-inositol phosphate (X-InP), this PAD could detect *Listeria monocytogenes* with a detection limit of 0.12 ± 0.08 μg/mL. The calibration curve for this is illustrated in Fig. 2.2E.

2.4 Integration with smartphone devices

Martinez et al. [44] were the earliest to combine a paper-based sensor with a camera phone, which provided a framework for the development of detecting and diagnosing diseases in remote locations. They carried out a colorimetric assay spotted with glucose and protein reagents, and used a camera phone to capture images and transmit them to trained professionals. In their study, the glucose image was converted to 8-bit grayscale in AdobePhotoshop, and the protein image was converted to CMYK color with the use of the cyan channel. The color intensity was quantified using calibration curves and testing required less than 5 μL.

Colorimetric measurements are one of the most common detection methods chosen when integrating PADs with smartphones. Lopez-Ruiz et al. [45] coupled smartphones and a paper-based sensor for simultaneous pH and nitrite detection (Fig. 2.3A). After samples reacted with the immobilized reagents, images were taken using the flash of the smartphone and no other external equipment. Because some of the chemicals used were photosensitive, the detection process had to be done in a dark environment. The designed Android application employed a system of displayed marks to facilitate alignment, which simplified the process and ensured image processing accuracy. The results showed a resolution of 0.04 units, a pH accuracy of 0.09, and a detection limit of 0.52 mg/L for nitrite. Another large-scale water quality monitoring tool was reported by Sicard et al. [46], who created a device comprising two paper-based sensors (a test strip and a control strip), a smartphone and processing app, and a central website for uploading the measured data. In this study, 4 μL of sample was added prior to a 15-minute incubation. The assay platform was then dipped in ddH_2O to move the substrate to the detection area. If organophosphate pesticides were present in the sample, the colorless indoxyl would not be able to convert into a blue colored product. The image of both the test and control strips were recorded in a vertical position with a white background to minimize the effect of ambient light, and processed with a pixel-counting algorithm after selecting the area of interest (Fig. 2.3B). Finally, the analyzed results were labeled with GPS features and uploaded to the developed website for large-scale environmental contaminants monitoring. Apart from two-dimensional colorimetric

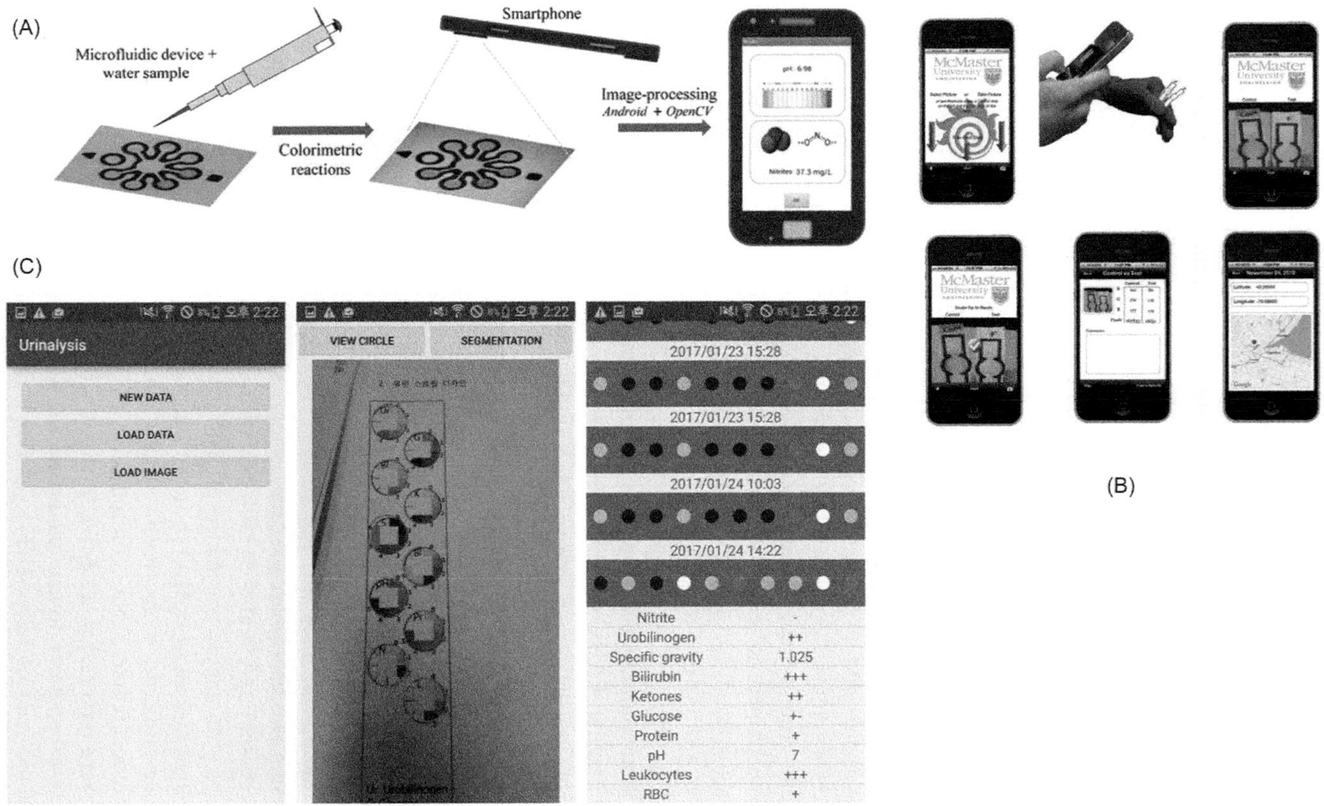

Figure 2.3 Colorimetric measurements of paper-based sensors on smartphone devices. (A) The process flow of a simultaneous pH and nitrite detection noting the marks that facilitate alignment and accuracy [45]. (B) When uploading the detected organophosphate pesticides concentration to the server, the analyzed results could be labeled with GPS features for large-scale water contaminants monitoring [46]. (C) The demonstration of the Android software from loading, detecting, and displaying the results of the 10 different urine tests [47].

assays that were coupled with smartphones, Wang et al. [48] fabricated a three-dimensional PAD that could detect four kinds of heavy metals (Cu(II), Ni(II), Cd (II), and Cr(VI)) in mixed samples, and showed no interference between concomitant ions, chosen masking reagents, and metal selective chromogenic reagents. When analyzing the chromogenic results, the images were first captured by a smartphone and transmitted to a personal computer for further analysis. The quantitative method used showed good alignment when compared with the conventional atomic absorption spectrometer method. Additionally, Ra et al. [47] introduced a smartphone-based point-of-care urinalysis method that could deliver the diagnostic results of a urine test via smartphone. Their manipulating methods are shown in Fig. 2.3C.

The detection of fluorescence-based paper arrays by the smartphone was also recently developed. Thom et al. [49] designed a 3D paper-based microfluidic device that consisted of an internal fluidic battery, a surface-mounted LED and a built-in cuvette for the assay solution. They also designed the corresponding smartphone equipment, which consisted of a poly(ethylene) tube and an optical filter to promote detection quality (Fig. 2.4A) They successfully created a β-D-galactosidase enzyme calibration curve with a 0.7-nM limit of detection. The result is illustrated in Fig. 2.4B. Additionally, Noor et al. [50] employed immobilized quantum dots as donors in fluorescence resonance energy transfer on dry paper to analyze nucleic acid. The results were recorded via a digital camera with R–G–B color selectivity (Fig. 2.4C).

In closing this review, we would like to talk about a particularly well-meaning, health-focused integration of PADs with cellphone, but one not directed at humans. This is the development of the app Petgeia. For the last two decades, the growth of "pet parenting," the tendency of some couples to treat their pets as surrogate children, has continued to drive the pet economy. Health and wellness is paramount to success in pet innovation, and pet trends are closely following human ones. Hygeia Touch Inc., is one of the companies combining colorimetrically based urine strips and smartphone analysis for pet owners to regular check the health of their pets at home. Many different smartphone cameras and infinite lighting conditions make accurate strip reading highly challenging. However, Hygeia Touch has created computer vision algorithms and unique calibration methods to make accurate testing as easy as taking a selfie. Their urine strip assays 10 parameters: leukocytes, nitrites, urobilinogen, protein, pH, blood, specific gravity, ketones, bilirubin, and glucose. All of the above analytes are early disease indicators and for metabolic or systemic diseases that affect kidney function, endocrine, and urinary tract diseases or disorders. After applying urine to the strip or immersing the test areas of the strip in fresh urine, pet owners can place the strip onto the color card. When analyzing the chromogenic results, the images are first captured by a smartphone and transmitted to a cloud server for further analysis. In a few minutes, the analyzed results will be shown on the smartphone and can be saved for further discussion with the vets online or in the clinics (Fig. 2.5). Test results from this process showed good alignment with standard lab-based urinalysis, and the approach allows pet health information to be recorded and saved for further and ongoing analysis.

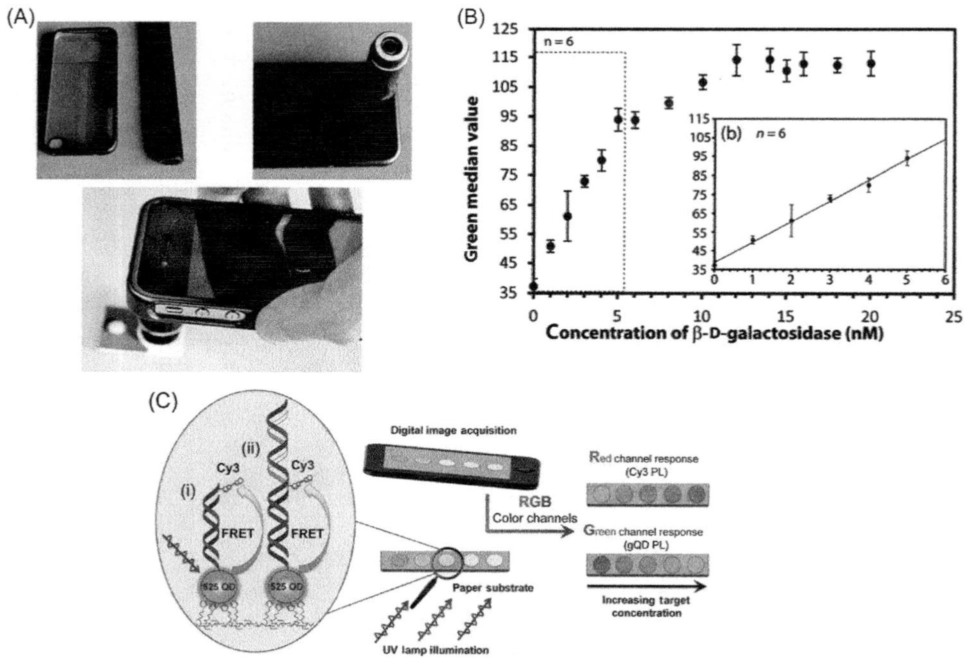

Figure 2.4 The integration of smartphone devices with fluorescence-based paper arrays. (A) A 510 nm cutoff length filter was attached to the cellphone to minimize LED interference (the excitation source) with the *fluorescence* assay color [49]. (B) The calibration curve for the intensity of the *green* color versus the concentration of the β-D-galactosidase enzyme demonstrates the combined use of a fluorescence assay with a smartphone device [49]. (C) The schematic of using a cellphone camera to quantify the ratiometric transduction of nucleic acid hybridization on a paper array with quantum dots [50].

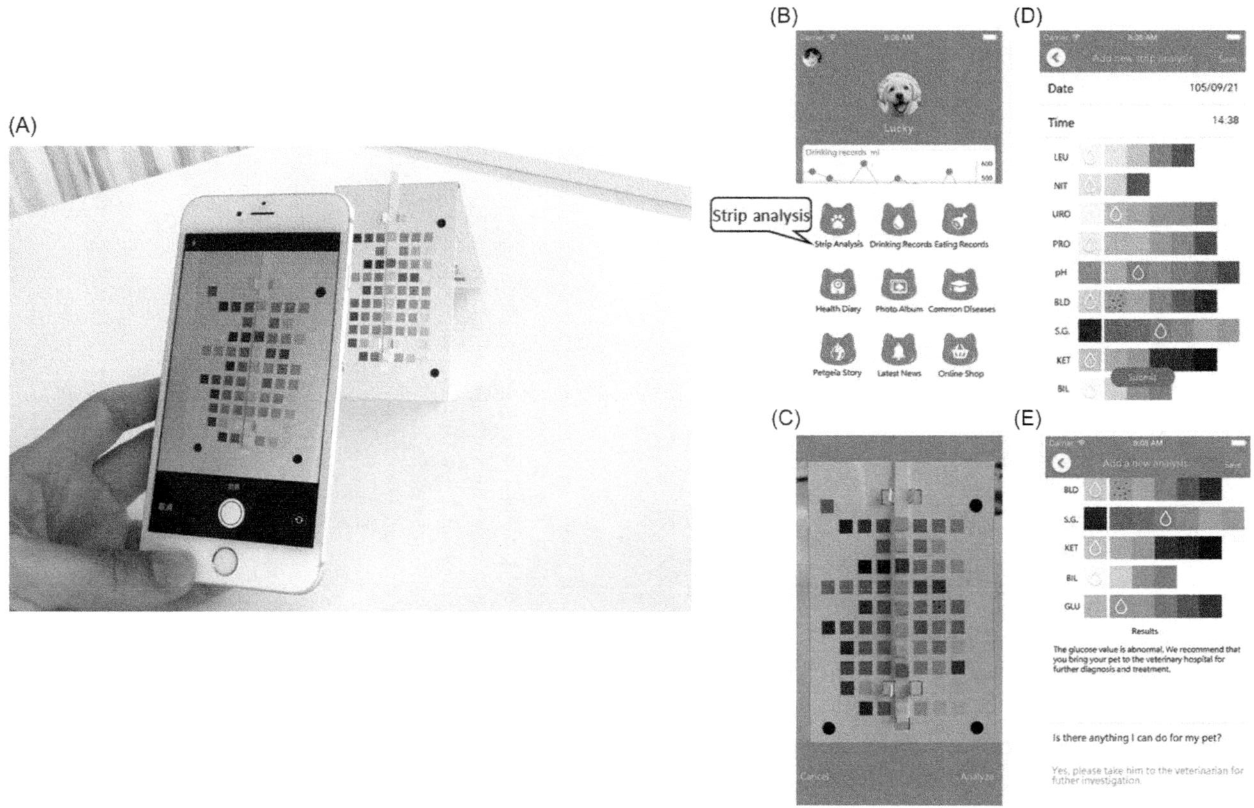

Figure 2.5 (A) Illustration of urine strip colorimetric analysis via smartphone app and color card provided by Hygeia Touch Inc. (B) The various functions of the app including strip analysis and healthcare management. (C) Test strip analysis function combined with a designed color card for use with a smartphone camera. (D) The analyzed results of 10 urine strip parameters after auto analysis via the cloud server. (E) The app is combined with online QA with a qualified veterinarian.

2.5 Conclusion

Petgeia is a good example of PAD integration with the Internet of Things as facilitated by camera phones. Colorimetric assays of PADs have great potential for recording and analysis by image processing devices and applications. Results from such tests can be further transferred to a remote laboratory for analysis or, with the development of artificial intelligence, analyzed by a built-in app. The issue of reducing measurement error in out-laboratory environments is challenging, but many studies overcame such issues with calibration curves or compensating equations [44,51]. We believe that exploring combinations of PADs with the Internet of Things for humans, animals, and the environment will be a continuing and highly impactful trend that will affect a spectrum of fields including health, biochemistry, and environmental analysis.

Author disclosure statement

No competing financial interests exist.

References

[1] A.W. Martinez, et al., Patterned paper as a platform for inexpensive, low-volume, portable bioassays, Angewan. Chem. Int. Ed. 46 (2007) 1318.

[2] E. Carrilho, A.W. Martinez, G.M. Whitesides, Understanding wax printing: a simple micropatterning process for paper-based microfluidics, Anal. Chem. 81 (2009) 7091.

[3] T. Songjaroen, et al., Novel, simple and low-cost alternative method for fabrication of paper-based microfluidics by wax dipping, Talanta 85 (2011) 2587.

[4] S. Chaiyo, et al., Highly selective and sensitive paper-based colorimetric sensor using thiosulfate catalytic etching of silver nanoplates for trace determination of copper ions, Anal. Chim. Acta 866 (2015) 75.

[5] S. Cinti, et al., Fully integrated ready-to-use paper-based electrochemical biosensor to detect nerve agents, Biosens. Bioelectron. 93 (2017) 46.

[6] W. Dungchai, O. Chailapakul, C.S. Henry, A low-cost, simple, and rapid fabrication method for paper-based microfluidics using wax screen-printing, Analyst 136 (2011) 77.

[7] K. Abe, K. Suzuki, D. Citterio, Inkjet-printed microfluidic multianalyte chemical sensing paper, Anal. Chem. 80 (2008) 6928.

[8] V. Rajendra, et al., Printing silicone-based hydrophobic barriers on paper for microfluidic assays using low-cost ink jet printers, Analyst 139 (2014) 6361.

[9] T. Soga, et al., Inkjet-printed paper-based colorimetric sensor array for the discrimination of volatile primary amines, Anal. Chem. 85 (2013) 8973.

[10] J. Wang, et al., Hydrophobic sol–gel channel patterning strategies for paper-based microfluidics, Lab. Chip. 14 (2014) 691.

[11] P.K. Kao, C.C. Hsu, One-step rapid fabrication of paper-based microfluidic devices using fluorocarbon plasma polymerization, Microfluid. Nanofluidics 16 (2014) 811.

[12] X. Li, et al., Paper-based microfluidic devices by plasma treatment, Anal. Chem. 80 (2008) 9131.
[13] J. Tian, et al., Thread as a versatile material for low-cost microfluidic diagnostics, ACS Appl. Mater. Interfaces 4 (2012) 6573.
[14] C.L. Sones, et al., Engineering fluidic delays in paper-based devices using laser direct-writing, Lab. Chip. 14 (2014) 4567.
[15] A.C. Glavan, et al., Rapid fabrication of pressure-driven open-channel microfluidic devices in omniphobic RF paper, Lab. Chip. 13 (2013) 2922.
[16] Y. He, W.B. Wu, J.Z. Fu, Fabrication of paper-based microfluidic analysis devices: a review, RSC Adv. 5 (2015) 2694.
[17] G.G. Lewis, et al., High throughput method for prototyping three-dimensional, paper-based microfluidic devices, Lab. Chip. 12 (2012) 2630.
[18] K.M. Schilling, D. Jauregui, A.W. Martinez, Paper and toner three-dimensional fluidic devices: programming fluid flow to improve point-of-care diagnostics, Lab. Chip. 13 (2013) 628.
[19] D. Sechi, et al., Three-dimensional paper-based microfluidic device for assays of protein and glucose in urine, Anal. Chem. 85 (2013) 10733.
[20] K. Talalak, et al., A facile low-cost enzymatic paper-based assay for the determination of urine creatinine, Talanta 144 (2015) 915.
[21] J. Noiphung, et al., A novel paper-based assay for the simultaneous determination of Rh typing and forward and reverse ABO blood groups, Biosens. Bioelectron. 67 (2015) 485.
[22] S.J. Vella, et al., Measuring markers of liver function using a micropatterned paper device designed for blood from a fingerstick, Anal. Chem. 84 (2012) 2883.
[23] J. Narang, et al., Electrochemcial sensing of anti-diabetic drug using hierarchically deposited PB nanocubes on carbon nanospheres layered on transparent glass electrode, Biosens. Bioelectron. 88 (2017) 249.
[24] M. Al-Tamimi, et al., Validation of paper-based assay for rapid blood typing, Anal. Chem. 84 (2012) 1661.
[25] T. Songjaroen, W. Laiwattanapaisal, Simultaneous forward and reverse ABO blood group typing using a paper-based device and barcode-like interpretation, Anal. Chim. Acta 921 (2016) 67.
[26] L. Zhang, et al., Printed microwells with highly stable thin-film enzyme coatings for point-of-care multiplex bioassay of blood samples, Analyst 140 (2015) 4105.
[27] J. Noiphung, et al., Electrochemical detection of glucose from whole blood using paper-based microfluidic devices, Anal. Chim. Acta 788 (2013) 39.
[28] T.T. Tsai, et al., Paper-based tuberculosis diagnostic devices with colorimetric gold nanoparticles, Sci. Technol. Adv. Mater. 14 (2013) 044404.
[29] K. Matsuura, et al., Paper-based diagnostic devices for evaluating the quality of human sperm, Microfluid. Nanofluidics 16 (2014) 857.
[30] H.K. Wang, et al., Cellulose-based diagnostic devices for diagnosing serotype-2 dengue fever in human serum, Adv. Healthcare Mater. 3 (2013) 187.
[31] C. Chen, et al., Paper-based devices for isolation and characterization of extracellular vesicles, J. Visualized Exp. 98 (2015) 52722.
[32] J. Narang, et al., Lab on paper chip integrated with Si@GNRs for electroanalysis of diazepam, Anal. Chim. Acta 980 (2017) 50.
[33] C.M. Kuan, et al., Paper-based diagnostic devices for clinical paraquat poisoning diagnosis, Biomicrofluidics 10 (2016) 034118.
[34] T.H. Yen, et al., Optical paper-based sensor for ascorbic acid quantification using silver nanoparticles, Talanta 144 (2015) 189.

[35] S.M.Z. Hossain, J.D. Brennan, β-Galactosidase-based colorimetric paper sensor for determination of heavy metals, Anal. Chem. 83 (2011) 8772.

[36] D.M. Cate, et al., Multiplexed paper analytical device for quantification of metals using distance-based detection, Lab. Chip. 15 (2015) 2808.

[37] Y. Hao, et al., A naphthalimide-based azo colorimetric and ratiometric probe: synthesis and its application in rapid detection of cyanide anions, Analyt. Methods 6 (2014) 2478.

[38] S.M.Z. Hossain, et al., Development of a bioactive paper sensor for detection of neurotoxins using piezoelectric inkjet printing of sol–gel-derived bioinks, Anal. Chem. 81 (2009) 9055.

[39] J.C. Jokerst, et al., Development of a paper-based analytical device for colorimetric detection of select foodborne pathogens, Anal. Chem. 84 (2012) 2900.

[40] C.W. Quinn, et al., Contemporary mobilization of legacy Pb stores by DOM in a Boreal Peatland, Environ. Sci. Technol. 52 (2018) 3567.

[41] C.H. Weng, et al., Colored wax-printed timers for two-dimensional and three--dimensional assays on paper-based devices, Biomicrofluidics 8 (2014) 066502.

[42] S. Cinti, et al., Novel reagentless paper-based screen-printed electrochemical sensor to detect phosphate, Anal. Chim. Acta 919 (2016) 78.

[43] B. Yoon, et al., A Litmus-type colorimetric and fluorometric volatile organic compound sensor based on inkjet-printed polydiacetylenes on paper substrates, Macromol. Rapid Commun. 34 (2013) 731.

[44] A.W. Martinez, et al., Simple telemedicine for developing regions: camera phones and paper-based microfluidic devices for real-time, off-site diagnosis, Anal. Chem. 80 (2008) 3699.

[45] N. Lopez-Ruiz, et al., Smartphone-based simultaneous pH and nitrite colorimetric determination for paper microfluidic devices, Anal. Chem. 86 (2014) 9554.

[46] C. Sicard, et al., Tools for water quality monitoring and mapping using paper-based sensors and cell phones, Water Res. 70 (2015) 360.

[47] M. Ra, et al., Smartphone-based point-of-care urinalysis under variable illumination, IEEE J. Transl. Eng. Health Med. 6 (2018) 1.

[48] H. Wang, et al., Paper-based three-dimensional microfluidic device for monitoring of heavy metals with a camera cell phone, Anal. Bioanal. Chem. 406 (2014) 2799.

[49] N.K. Thom, et al., Quantitative fluorescence assays using a self-powered paper-based microfluidic device and a camera-equipped cellular phone, RSC Adv. 4 (2014) 1334.

[50] M.O. Noor, et al., Camera-based ratiometric fluorescence transduction of nucleic acid hybridization with reagentless signal amplification on a paper-based platform using immobilized quantum dots as donors, Analyt. Chem. 86 (2014) 10331.

[51] L. Shen, J.A. Hagen, I. Papautsky, Camera-based ratiometric fluorescence transduction of nucleic acid hybridization with reagentless signal amplification on a paper-based platform using immobilized quantum dots as donors, Lab. Chip. 12 (2012) 4240.

Membrane-type Surface stress Sensor (MSS) for artificial olfactory system

3

Huynh Thien Ngo[1,2], *Kosuke Minami*[1,2], *Gaku Imamura*[1,2], *Kota Shiba*[1,2] *and Genki Yoshikawa*[1,2,3]
[1]Center for Functional Sensor & Actuator (CFSN), National Institute for Materials Science (NIMS), Tsukuba, Japan, [2]International Research Center for Materials Nanoarchitectonics (WPI-MANA), National Institute for Materials Science (NIMS), Tsukuba, Japan, [3]Materials Science and Engineering, Graduate School of Pure and Applied Science, University of Tsukuba, Tsukuba, Japan

3.1 Introduction

Over 400,000 kinds of molecules have been known to contribute to the smells/odors that we experience everyday [1]. In most cases, tens to thousands of different kinds of molecules simultaneously interact with our olfactory receptors to produce the sensation of a single smell/odor. This complexity of smells/odors prevents the efficient development of artificial olfactory sensors. While other sensors that can perceive physical stimuli [e.g., light (eye), sound (ear), and pressure (skin)] have been extensively studied and developed, olfactory sensors have not reached their full commercial potential because of the lack of comprehensive knowledge on the chemical/physical interactions between the receptors and the analytes.

One of the standard techniques to analyze smells/odors is gas chromatography combined with mass spectrometry (GC/MS), which allows the isolation and quantification of most compounds consisting of smells/odors. Although this approach can determine the molecular information, the obtained data requires specific handling and the expensive/bulky instrumentation, preventing it from being implemented in daily use. Furthermore, GC/MS could miss some odorous molecules below ppm level without careful concentration/extraction processes. Thus, it is still difficult to achieve sensitivity, selectivity, cost efficiency, and compactness at the same time in a single module for the simultaneous measurements of a smell/odor composed of multiple kinds of molecules.

An attractive approach to tackling this problem could be found in the *olfactory sensors*, which measure each smell as single samples without the separation of its components. In particular, the sensor arrays composed of multiple sensing elements with diverse chemical selectivity have gained significant attention. As such arrays mimic the mammalian smelling process, this is called "electronic nose" or

Chemical, Gas, and Biosensors for the Internet of Things and Related Applications.
DOI: https://doi.org/10.1016/B978-0-12-815409-0.00003-6

"artificial olfaction." Recent achievements in miniaturization of sensor elements as well as peripheral electronic components have boosted the olfactory sensor technology to be suitable for mobile and Internet of Things (IoT) type applications.

One of the promising technologies for olfactory sensors is a nanomechanical sensor. This type of sensors detects volume- and/or mass-induced mechanical changes of a sensing element. These mechanical changes caused by sorption of target analytes are transduced into electrical signals. The nanomechanical sensor can be, therefore, regarded as a mechanical transducer with nanometer precision. Gimzewski and his coworkers reported the first chemical sensing application using a nanomechanical cantilever sensor in 1994 [2]. They utilized the static bending of a cantilever for the measurement of the catalytic reaction taking place on the surface of a cantilever. Large numbers of research groups have demonstrated that the nanomechanical sensors can be employed to detect a variety of targets and phenomena; for example, mercury vapor [3], formation of self-assembled monolayers [4], hybridization of DNA [5]. In this chapter, we focus on a specific nanomechanical sensor with superior performance: Membrane-type Surface stress Sensor (MSS).

3.2 Membrane-type Surface stress Sensor (MSS)

The MSS is an optimized nanomechanical sensor platform operated in a so-called static mode [6–9]. MSS is composed of a thin silicon membrane suspended by four sensing bridges, in which piezoresistors are embedded. The receptor layer deposited on the membrane interacts with analytes, deforming the receptor layer. This deformation induces surface stress on the membrane. The piezoresistors efficiently transduce such stresses into electrical signals (Fig. 3.1).

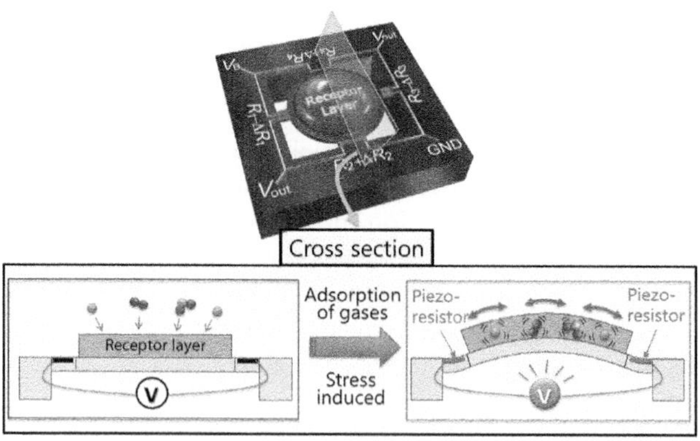

Figure 3.1 Schematic illustrations of the MSS (top) and the working principle of MSS (bottom). Surface stress is induced by sorption of gaseous molecules on a receptor layer coated on the center membrane of MSS. The surface stress is electrically detected with the piezoresistors (resistances change by mechanical stress) embedded on the narrow supporting beams.

Initially, the piezoresistor-based electric readout could not gain high sensitivity, while the small size of the device is an attractive advantage over the more sensitive conventional optical readout cantilever sensors, which require large instrumentation. Yoshikawa made comprehensive optimization and developed the MSS in collaboration with Dr. Heinrich Rohrer and with Dr. Terunobu Akiyama [10] and his colleagues in the MEMS team at École Polytechnique Fédérale de Lausanne (EPFL), achieving both high sensitivity and miniaturization [6]. This structure allows highly sensitive electric readouts of stress/strain induced by the sorption of gaseous molecules on the receptor layer. Besides the high sensitivity (with the limit of detection ~0.1 mN/m), MSS also has high mechanical stability (without free-ends), electric/thermal stability (with self-compensation of full Wheatstone bridge), an easy coating of the various receptor materials, compact system, quick response, and the capability of mass production.

3.3 Receptor materials

Gas sensors can have a wide range of potential applications in different industries: food, cosmetic, medical, etc. Each application requires different selectivity and sensitivity to different gases. This requires a large versatility of the sensors. Since it has been confirmed that almost all solid materials exhibit mechanical deformation by gas sorption, MSS can detect diverse analytes by utilizing various kinds of materials as a receptor layer: polymers, organic and inorganic materials, or hybrids of them. The selection of the receptor materials is determined by the selectivity and sensitivity of the materials for the target analytes. The sensitivity of nanomechanical sensors depends on the receptor material properties, especially its Young's modulus and thickness [11]. The analytical model, which describes the relationship between deflection of a cantilever and various physical parameters of a cantilever itself and receptor layer on it, provides a practical guideline to optimize the thickness and Young's modulus of a receptor layer. Using the Timoshenko's beam theory, which was originally developed to analyze a bimetal strip, an analytical model for the static deflection of a cantilever sensor coated with a solid layer was formulated. For a simple cantilever coated with a solid receptor layer, the deflection of the cantilever (Δz) is described as:

$$\Delta z = \frac{3l^2(t_f + t_c)}{(A+4)t_f^2 + (A^{-1}+4)t_c^2 + 6t_f t_c}\varepsilon_f$$

$$A = \frac{E_f w_f t_f(1-\nu_c)}{E_c w_c t_c(1-\nu_f)}$$

where E_f, ν_f, and t_f are the Young's modulus, the Poisson's ratio, and the thickness of a receptor layer; and E_c, ν_c, and t_c are the Young's modulus, the Poisson's ratio, and the thickness of the cantilever; and w_c and w_f are the width of a cantilever and a receptor layer, respectively. Based on this model, we can derive the following guideline: the optimum thickness of a receptor layer depends on its Young's modulus. This optimum coating thickness can be attributed to a specific point where the additional stiffness due to the additional thickness of coating film becomes dominant over the effective force induced in the entire coating film.

Fig. 3.2 shows the overall trend of higher sensitivity with stiffer materials having proper thickness. One of the most promising stiff materials is inorganic nanoparticles. They are stable and easily functionalized. To confirm their superior performance as nanomechanical sensing materials, nanoparticles with different surface functionalities were synthesized (Fig. 3.3A) and employed for MSS. The discrimination of various vapors from 23 different chemicals (Fig. 3.3B) was demonstrated. Principal component analysis (PCA) showed well-separated clusters of different species according to their functional groups: alcohols, ketones, alkanes, aromatic compounds, and so on.

In addition to the inherent stiffness of inorganic nanoparticles, another important feature that affects sensitivity is the nanostructure of a receptor material. Large surface area and high porosity are effective for enhanced adsorption/desorption of gas molecules. Shiba et al. developed a synthetic process to produce monodispersed metal oxide–based porous nanoparticles [12]. Organic materials lack the chemical stability but offer a rich chemistry to target specific analyte and an easy modification of the functionality. To combine the advantages of both systems, porous inorganic nanoparticles with organic functionalities on the surface were synthesized as a useful receptor materials platform. One example is the silica flake shell with metalloporphyrin on its surface (Fig. 3.4) [13]. This receptor material was used to detect acetone at low concentration of 50 ppm with a high signal-to-noise ratio. The

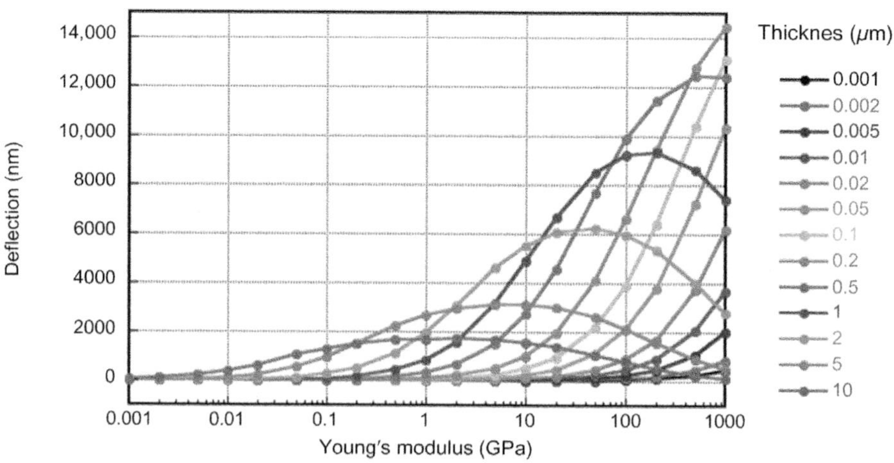

Figure 3.2 Deflection of a cantilever operated in static mode as a function of Young's modulus and thickness of a receptor layer.

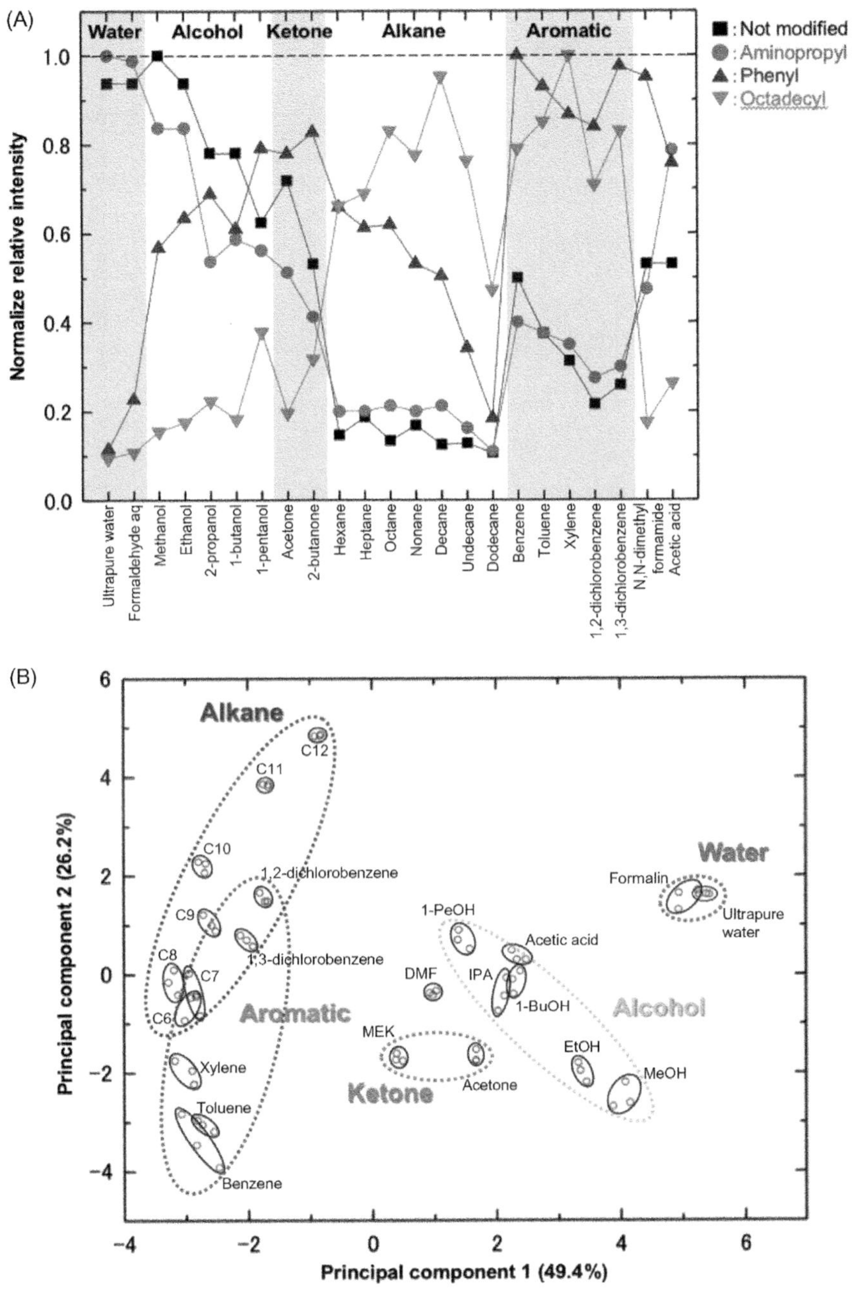

Figure 3.3 (A) Sensitivity of the nanoparticles containing different surface modifications to different gases. (B) PCA of the 23 different gases categorized with their functional groups.

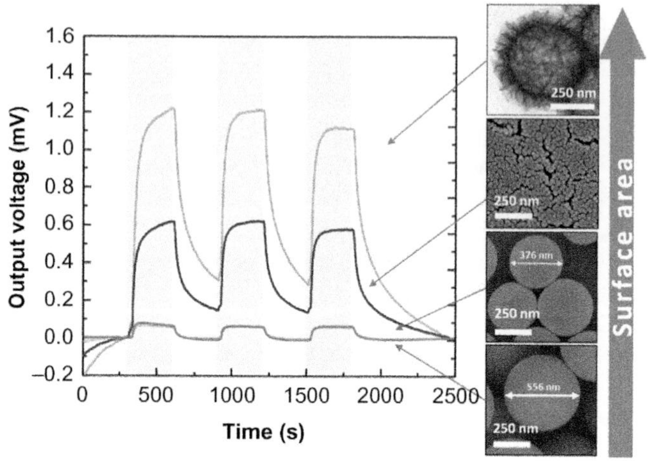

Figure 3.4 Sensitivity of the nanoparticles functionalized with metalloporphyrin. Sensitivity was enhanced with the porosity and surface area.

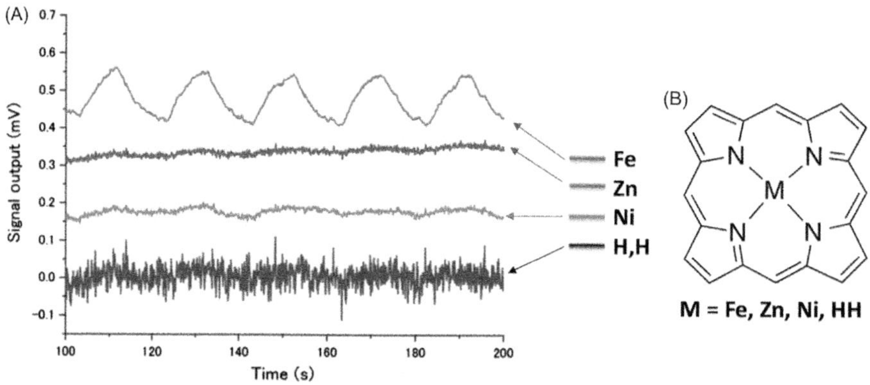

Figure 3.5 (A) Sensing signals of metalloporphine derivatives to propionic acid vapor at 70% relative humidity. (B) Structure of metalloporphine derivatives.
Note that "H,H" and "HH" correspond to the free-base porphine.

sensitivity to acetone was enhanced owing to the large surface area of the receptor materials.

Recently, Ngo and Imamura et al. focused on single complex molecular systems such as porphyrinoids. They reported the selectivity and high robustness to humidity of metalloporphine as receptor materials for MSS (Fig. 3.5) [14]. Porphine is the central structure of porphyrins—a porphyrin without substituents. The use of a porphine and metalloporphines can exclude the effects of peripheral substituents, focusing on the sole effect of the center metal ions and the central polypyrrole macrocycle on gas uptake. The sensitivity could be fine-tuned at atomic scale by alteration of one single central metal atom. The metal insertion leads to significant changes in sensitivity, reflecting the different coordination preference of center metal ions to axial guest molecules in the porphine complex. Gas sensing properties of four porphine derivatives were measured: non-metallated free-base (FB) porphine, nickel porphine, zinc porphine, and iron porphine. The results of gas sensing measurements showed that nickel and zinc insertions into porphines gave rise to equivalent or less sensitivity than FB porphine. In contrast, iron porphine exhibited enhanced sensitivity owing to its high capability of ligand binding and the modified electronic structure according to the iron center. Moreover, this study demonstrated the high humidity-resistant sensing capability of iron porphine. Even in a humidified condition (70% RH), propionic acid at the concentration as low as 19 ppm was clearly detected with iron porphine. The study can be used as a reference for future investigations where the influence of metals in porphyrinoids plays an important role, leading to advanced sensing applications including receptor materials for an artificial olfactory sensor system in ambient conditions.

3.4 Machine learning

As mentioned before, smells/odors are known to be composed of tens to thousands of chemicals with various concentrations, and thus, the extraction of specific information from such a complex system is challenging. However, machine learning can be useful to unravel hidden patterns from the complexity of the obtained sensing data. Shiba, Tamura, and their coworkers combined the MSS platform with machine learning to predict alcohol content of liquors from their smells [15]. They used an array of MSS with four channels, each of which was coated with silica/titania hybrid nanoparticles with a different surface functionality or hydrophobic polymer. Each sensor element in the sensor array responds differently when it is exposed to a smell because of their diverse chemical properties. In the study, it was demonstrated that such a sensor array system combined with a machine learning techniques can be utilized to derive quantitative information, for example, alcohol content as an example, from smell. For this purpose, they employed the kernel ridge regression, which can be applied to a nonlinear problem with avoiding overfitting. The combination of the advanced sensor platform with the machine learning technique led to the successful extraction of specific information. As shown in Fig. 3.6, the model predicted alcohol content of unknown samples from their smells with high accuracy. With increasing training data, the accuracy can be further improved. This approach can be extended to more detailed investigation of complex liquid samples such as blood, coffee, and environmental water. Recently, quantitative prediction of concentrations of each component in a ternary mixture was also achieved through systematic material design–based nanomechanical sensing combined with machine learning. Ternary mixtures consisting of water, ethanol, and methanol with various concentrations were selected as a model system where a target molecule coexists with structurally similar species in a humidified condition. The fine tuning of

Figure 3.6 Alcohol content prediction of liquors based on their smells: *blue* (gray in print version) and *red* (black in print version) dots correspond to training and test data, respectively.

receptor materials based on the feedback obtained by the machine learning achieved high prediction accuracy for each species [16].

3.5 Applications

The MSS system with variety of receptor materials provides a sensing platform that can be applied in many fields such as food, agriculture, environment, cosmetics, healthcare, and medicine. The MSS has been employed in various applications to determine the qualitative differences of the related products.

In the food industry, for example, it is possible for MSS to determine various spices based on their volatile organic compounds (VOCs) having different functional groups. The smells of seven herbs and spices, namely cinnamon, parsley, nutmeg, oregano, garlic, rosemary, and yuzukosho (Japanese spice made from local citrus fruit), were separated and categorized by PCA on the basis of their major components (Fig. 3.7) [17].

One of the most advanced applications of artificial olfactory sensors is a measurement of health conditions for medical diagnosis. Healthcare strategies are currently oriented towards noninvasive techniques for an early diagnosis. For example, cancer is one of the most severe diseases in the world today with the highest mortality. Early diagnosis of cancer increases the chances of successful treatment, improving the survival rate of the patients. Most of the current diagnosis is relying

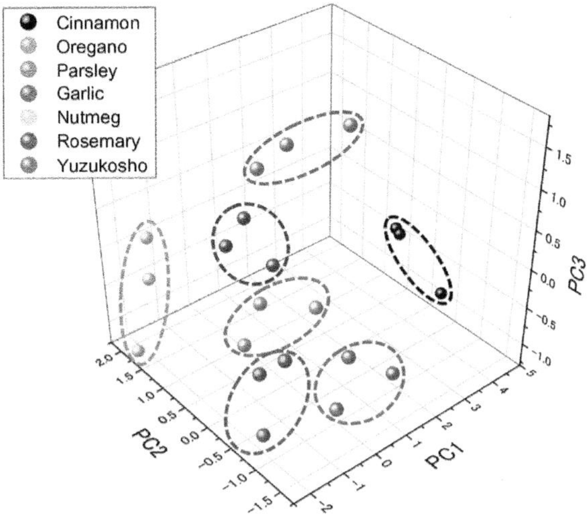

Figure 3.7 Applications of MSS in the food industry for the differentiation of spices.

on the invasive approaches for the detection of specific biomarkers in blood or endoscopy. In contrast to the conventional approaches, it is known that some of the smells including breath, sweat, urine, and blood change depending on patients' health conditions as well as their medical diseases. Thus, there is a chance for artificial olfaction to contribute to the detection of certain diseases for the early diagnosis as a noninvasive method. Disease-related VOCs in patients' breath would exhibit only a slight difference, especially at the early stages of diseases. With the high sensitivity, the MSS is a powerful platform to detect such traces. As a demonstration, Loizeau et al. utilized an array of 16 MSS channels coated with 16 different polymers, which have various chemical/physical properties [18]. Each MSS responds differently to the same analyte, creating a unique "fingerprint." Breath samples of both healthy people and cancer patients (head and neck cancer) were analyzed in a double-blind trial. The breath samples were transported and stored in Tedlar bags at a temperature of 4 °C for 5 hours before analysis. Each sample was pumped into a gas chamber containing the functionalized MSS array. The sensors' responses were recorded six times and analyzed by PCA to reveal the differences between samples. The results show a good discrimination between the breath samples of healthy persons and the ones of patients suffering from head and neck cancer (Fig. 3.8). Moreover, when the cancer patients underwent surgery to remove the tumors, their breath after surgery was categorized as those of healthy persons. Although a limited number of samples were examined in this study, the discrimination accuracy is expected to be improved by increasing the number of samples in combination with machine learning technique, as described above. Furthermore, MSS was demonstrated to be utilized for the mobile olfaction system. Thus, it would be applicable to the daily healthcare and monitoring.

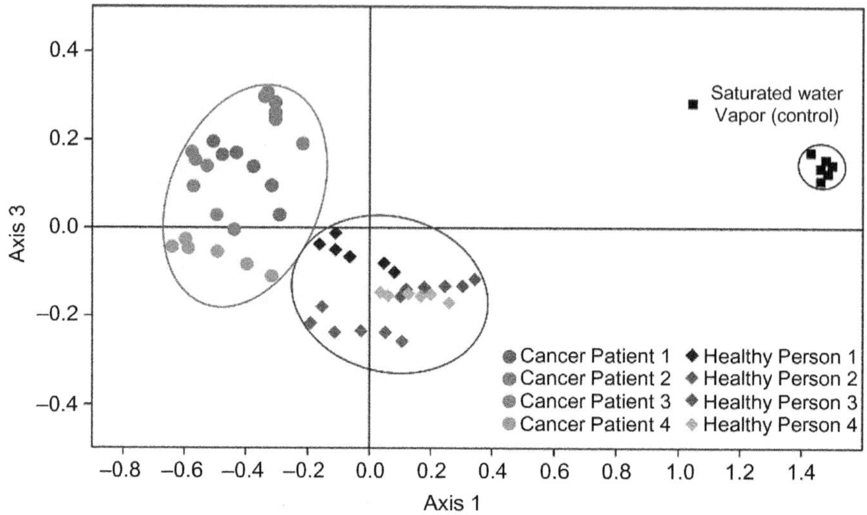

Figure 3.8 Use of MSS in the biomedical application: discrimination of cancer through breath analysis.

3.6 Internet of Things and MSS Alliance/Forum

It is no surprise that sensors in general are considered as one of the core devices in the IoT concept. Artificial olfactory sensors can act as an interface between physical and cyber spaces. The MSS sensing technology combined with information and communication technology would make a significant contribution to implementing this concept into our daily life. However, there are still several challenging issues that need to be overcome to establish real practical devices: optimization of receptor layer coating, precise calibration with standard gases, mass production of sensor chips and devices, and efficient connections with edge computing and cloud systems, etc. These challenges require the integration of science and technologies in collaboration with academia and industry. To integrate all the required cutting-edge technologies, the MSS Alliance was launched by NIMS, Kyocera, Osaka University, NEC, Sumitomo Seika, Asahi Kasei (joined from April, 2017), and NanoWorld in 2015 [19]. Based on their own expertise, each member of the MSS Alliance has been contributing to the development of various technologies required for practical artificial olfaction. To encourage interested companies and institutes to perform demonstration experiments, the MSS Forum was launched in November 2017 [20,21]. The MSS Alliance and Forum aim at developing a practical olfactory IoT sensor toward a de facto standard of artificial olfaction.

3.7 Conclusion

The MSS has achieved high sensitivity compared with the conventional cantilever-type nanomechanical sensors with piezoresistive readout in addition to various practical advantages. The versatility with diverse types of receptor materials for gas sensing makes MSS an effective platform for artificial olfactory sensing. The demonstrations in food, cosmetic, and biomedical applications proved the high potential of MSS as a practical technology in various applications. Along with the MSS Alliance/Forum, the MSS integrated with related cutting-edge hardware/software technologies will soon become a standard platform for olfactory sensing systems.

References

[1] J.A. Gottfried, Function follows form: ecological constraints on odor codes and olfactory percepts, Curr. Opin. Neurobiol. 19 (2009) 422–429.
[2] J.K. Gimzewski, C. Gerber, E. Meyer, R.R. Schlittler, Observation of a chemical-reaction using a micromechanical sensor, Chem. Phys. Lett. 217 (1994) 589–594.
[3] T. Thundat, R.J. Warmack, G.Y. Chen, D.P. Allison, Thermal and ambient-induced deflections of scanning force microscope cantilevers, Appl. Phys. Lett. 64 (1994) 2894–2896.
[4] R. Berger, E. Delamarche, H.P. Lang, C. Gerber, J.K. Gimzewski, E. Meyer, et al., Surface stress in the self-assembly of alkanethiols on gold, Science 276 (1997) 2021–2024.
[5] J. Fritz, M.K. Baller, H.P. Lang, H. Rothuizen, P. Vettiger, E. Meyer, et al., Translating biomolecular recognition into nanomechanics, Science 288 (2000) 316–318.
[6] G. Yoshikawa, T. Akiyama, S. Gautsch, P. Vettiger, H. Rohrer, Nanomechanical membrane-type surface stress sensor, Nano Lett. 11 (2011) 1044–1048.
[7] G. Yoshikawa, T. Akiyama, F. Loizeau, K. Shiba, S. Gautsch, T. Nakayama, et al., Two dimensional array of piezoresistive nanomechanical membrane-type surface stress sensor (MSS) with improved sensitivity, Sensors 12 (2012) 15873–15887.
[8] G. Yoshikawa, F. Loizeau, C.J.Y. Lee, T. Akiyama, K. Shiba, S. Gautsch, et al., Double-side-coated nanomechanical membrane-type surface stress sensor (MSS) for one-chip–one-channel setup, Langmuir 29 (2013) 7551–7556.
[9] R.J.S. Guerrero, F. Nguyen, G. Yoshikawa, Real-time gas identification on mobile platforms using a nanomechanical membrane-type surface stress sensor, EPJ Tech. Instrum 1 (2014) 9.
[10] T. Akiyama, NanoWorld AG.
[11] G. Yoshikawa, Mechanical analysis and optimization of a microcantilever sensor coated with a solid receptor film, Appl. Phys. Lett. 98 (2011) 173502.
[12] K. Shiba, M. Ogawa, Microfluidic syntheses of well-defined sub-micron nanoporous titania spherical particles., Chem. Commun. (2009) 6851–6853.

[13] I. Osica, G. Imamura, K. Shiba, Q. Ji, L.K. Shrestha, J.P. Hill, et al., Highly networked capsular silica–porphyrin hybrid nanostructures as efficient materials for acetone vapor sensing., ACS Appl. Mater. Interfaces 9 (2017) 9945–9954.
[14] H.T. Ngo, K. Minami, G. Imamura, K. Shiba, G. Yoshikawa, Effects of center metals in porphines on nanomechanical gas sensing, Sensors 18 (2018) 1640.
[15] K. Shiba, R. Tamura, G. Imamura, G. Yoshikawa, Data-driven nanomechanical sensing: specific information extraction from a complex system., Sci. Rep. 7 (2017) 3661.
[16] K. Shiba, R. Tamura, T. Sugiyama, Y. Kameyama, K. Koda, E. Sakon, et al., Functional nanoparticles-coated nanomechanical sensor arrays for machine learning-based quantitative odor analysis, ACS Sens. 3 (2018) 1592–1600.
[17] G. Imamura, K. Shiba, G. Yoshikawa, Smell identification of spices using nanomechanical membrane-type surface stress sensors, Japan. J. Appl. Phys. 55 (2016) 1102B3–1102B5.
[18] Loizeau, F., Lang, H.P., Akiyama, T., Gautsch, S., Vettiger, P., Tonin A., et al. 2013 IEEE 26th International Conference on Micro Electro Mechanical Systems (MEMS), 2013, pp. 621–624.
[19] MSS alliance (Press release: http://www.nims.go.jp/eng/news/press/2015/10/201510130.html).
[20] MSS forum (Press release: http://www.nims.go.jp/eng/news/press/2017/201710160.html).
[21] MSS forum (Web page: http://mss-forum.com/en/).

Sensing technology based on olfactory receptors

4

Hidefumi Mitsuno[1], Takeshi Sakurai[2] and Ryohei Kanzaki[1]
[1]Research Center for Advanced Science and Technology, The University of Tokyo, Tokyo, Japan, [2]Department of Agricultural Innovation for Sustainability, Faculty of Agriculture, Tokyo University of Agriculture, Kanagawa, Japan

4.1 Olfactory mechanisms in biological systems

Olfactory information is vital for survival and successful reproduction in both vertebrates and insects. Accordingly, animals have evolved highly sophisticated olfactory systems that can detect and discriminate a vast array of odorants around their environments. In this section, we introduce the basic principles of odorant detection and olfactory transduction in vertebrates and invertebrates (insects) with an emphasis on differences of odorant receptors (ORs) and transduction mechanisms between these animals.

4.1.1 Olfactory mechanisms in vertebrates

4.1.1.1 Anatomy of olfactory organs in mammals

Terrestrial vertebrates have two olfactory organs, the main olfactory epithelium (MOE) inside the nasal cavity for general odorants and the vomeronasal organ close to the vomer for pheromones (Fig. 4.1). In this chapter, we will focus on olfactory systems in the MOE, because utilization for odorant sensors has been limited to this system to date. The MOE consists of olfactory receptor neurons (ORNs), basal cells, and support cells. ORNs constitute bipolar neurons that extend axons to the first olfactory center, the olfactory bulb (OB). Numerous tiny cilia protrude from the dendrites of ORNs into the olfactory mucus, which covers the surface of the olfactory epithelium. Odorant molecules that enter the nasal cavity are dissolved in the olfactory mucus and detected by ORs expressed on the cilia membrane of ORNs.

4.1.1.2 Odorant detection and signal transduction

In vertebrates, ORs belong to the seven transmembrane G-protein-coupled receptor (GPCR) superfamily [2]; thus odorant signals are transduced into electrical signals through the heterotrimeric G-protein-mediated second messenger pathway (Fig. 4.2). Olfactory transduction is initiated by the binding of odorants with ORs

Chemical, Gas, and Biosensors for the Internet of Things and Related Applications.
DOI: https://doi.org/10.1016/B978-0-12-815409-0.00004-8

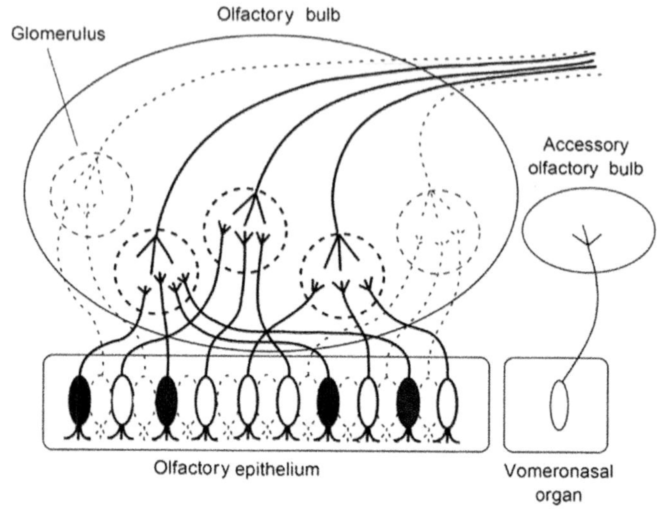

Figure 4.1 Olfactory organs in vertebrates. Each ORN in the main olfactory epithelium expresses only a single OR gene. ORNs expressing the same OR converge onto the same glomerulus in the olfactory bulb. Vomeronasal sensory neurons in vomeronasal organ project their axon into the accessory olfactory bulb.
Source: After R. Kanzaki, K. Nakatani, T. Sakurai, N. Misawa, H. Mitsuno, Physiology of chemical sense and its biosensor application, in: T. Nakamoto (Ed.), Essentials of Machine Olfaction and Taste, New York, Wiley, 2016 [1].

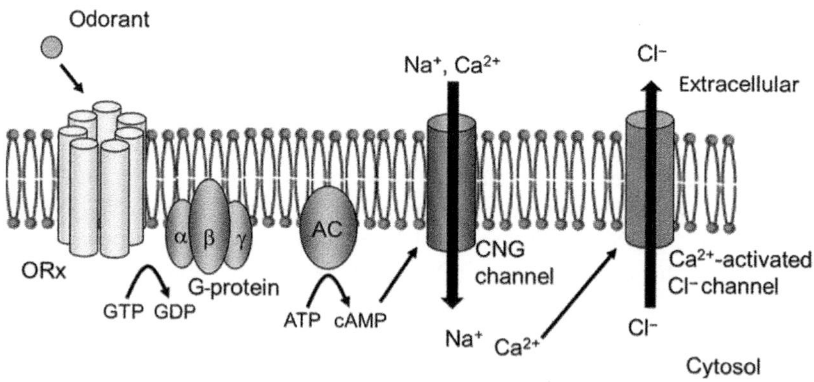

Figure 4.2 Olfactory signal transduction in vertebrates. Odorant signals are transduced into electrical signals via the heteromeric G-protein-mediated metabotropic system. *AC*, Adenylyl cyclase, *CNG channel*, cyclic nucleotide-gated channel.

expressed on the membrane of olfactory cilia, which triggers conformational changes of the OR to activate Gαolf, the olfactory-specific G-protein α (Gα) subunit [3]. Consequently, activated Gαolf hydrolyzes the GTP that is bound to the nonactive form of Gαolf into GDP, which in turn triggers dissociation of Gαolf from the Gβ and Gγ subunits. Dissociated Gαolf activates adenylyl cyclase, which then synthesizes cAMP from ATP [4]. cAMP plays a central role in olfactory transduction as a second messenger [5,6]. Binding of cAMP to cyclic nucleotide-gated ion channels on the cilia membrane opens the channel and induces an influx of cations, which causes depolarization of the olfactory cilia membrane [7,8]. Specifically, Ca^{2+} flow into the cell results in the opening of Ca^{2+}-activated Cl^- channels to induce the efflux of Cl^-, which induces membrane depolarization [9]. After a series of metabotropic signal transduction cascades, action potentials occur in ORNs, which are then transmitted to and processed by the first olfactory center, the OB.

4.1.1.3 Odorant receptors and odor coding in mammals

OR genes were first discovered in rat by Buck and Axel [2]. ORs form a multigene family that consists of approximately 1000 genes in rat and around 400 genes in human [10,11]. Olfactory epithelium consists of four zones, with expression of each OR gene restricted a single zone [12–14]. Each individual ORN expresses only a single OR gene from among the large repertory of OR genes, which is known as the "one ORN-one OR" rule [15]. ORNs expressing the same OR convergently project to the same glomerulus in the OB [16,17]. Generally, ORs can recognize multiple odorants [18–20]. A single odorant can also activate more than one OR. Therefore each odorant or odorant mixture is encoded by a combination of activation patterns of multiple ORs (Fig. 4.3).

The combinatorial nature of olfactory coding makes it possible for animals to discriminate a huge number of odorants that is much larger than the number of ORs. For example, it has been suggested that humans can discriminate over 400,000 odors, despite expressing only 400 ORs [21]. Moreover, it has been more recently estimated that humans can discriminate at least 1 trillion olfactory stimuli [22], although this estimation remains controversial [23,24]. However, combinatorial olfactory coding by ORs constitutes an important characteristic for the development of OR-based odorant sensors with high discriminatory power.

In addition to ORs, vertebrates utilize other receptor families to detect olfactory signals as well, such as vomeronasal receptors (reviewed in Ref. [25]) and trace amine associated receptors (reviewed in Ref. [26]).

4.1.2 Olfactory mechanisms in insects

4.1.2.1 Anatomy of olfactory organs in insects

Insects detect odorants via a pair of antennae on the head. In some insects, maxillary palps on maxillae also function as olfactory organs. The surface of these organs is covered by a large number of cuticular specializations, termed sensilla (Fig. 4.4), which comprise sensory neurons for various stimuli such as odorants, tastes, humidity, or mechanosensation. Among these, olfactory sensilla are characterized by the presence of numerous small pores on the cuticle termed olfactory pores (Fig. 4.4C)

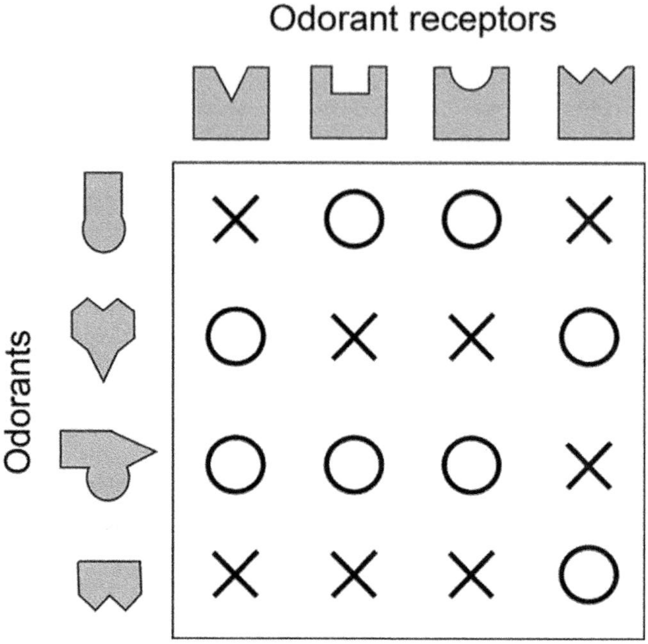

Figure 4.3 Schematics of combinatorial odor coding. ORs can recognize multiple odorants and a single odorant can also activate multiple ORs.
Source: After R. Kanzaki, K. Nakatani, T. Sakurai, N. Misawa, H. Mitsuno, Physiology of chemical sense and its biosensor application, in: T. Nakamoto (Ed.), Essentials of Machine Olfaction and Taste, New York, Wiley, 2016.

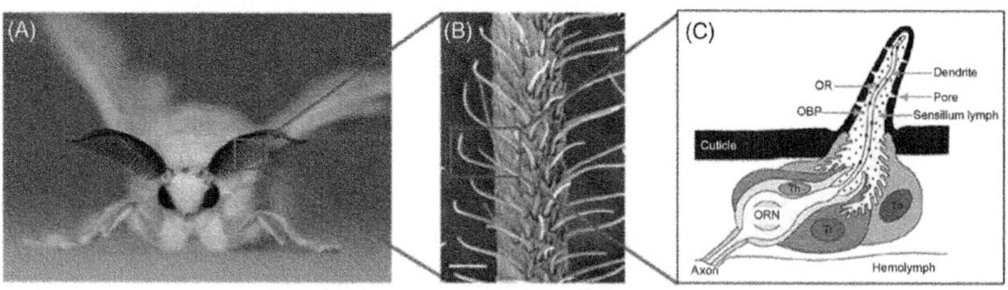

Figure 4.4 Olfactory organ of insects. Antennal structure of male silkmoth *B. mori* is shown as an example. (A) A male silkmoth with its antennae for odorant detection. (B) Scanning electron micrograph of an antenna of male silkmoth. Scale bar: 25 μm. (C) Schematic diagram of an olfactory sensillum. ORNs are wrapped by three types of accessory cell: tormogen (To), trichogen (Tr), and thecogen cells (Th).
(C) Source: Modified from E. Jacquin-Joly, C. Merlin, J. Chem. Ecol. 30 (2004) 2359 [27].

[28,29]. Most olfactory sensilla contain several ORNs. Cell bodies of ORNs reside at the base of the sensillum and extend their outer dendrite, the site of odorant detection, into the sensillum lumen and their axons to the antennal lobe (AL), the first olfactory center in insects and the analog of the OB in mammals [30]. ORN cell bodies are wrapped by three types of accessory cells: thecogen, trichogen, and tormogen cells (Fig. 4.4C). The sensillum lumen is filled with sensillum lymph, which is secreted by trichogen and tormogen cells. Accessory cells isolate sensillum lymph space around the outer dendrites of the ORNs from the hemolymph. Different chemical compositions between the sensillum lymph and the hemolymph generate a standing electrical potential difference, termed the transepithelial potential (see Refs. [31,32]). Odorant stimulation generates a receptor potential in the outer dendritic membrane, which can induce the generation of action potentials in a more proximally located spike-generating zone.

There are several types of olfactory sensilla, which are categorized by their outer structure (e.g., s. trichodea, s. basiconica, s. coeloconica, and s. placodea). In some cases, sensillum type is correlated with its function. For example, in male moths, s. trichodea contain ORNs specific to conspecific sex pheromones [33], whereas other sensillum types contain ORNs for the so-called general odorants such as from foods or plants [34].

4.1.2.2 Odorant detection by olfactory sensilla

Odorant molecules in the air are first absorbed and diffuse on the cuticular surface of sensillum, and then enter the sensillum interior though olfactory pores [35,36]. Because odorants are generally hydrophobic, they have difficulty in passing through the sensillum lymph to reach the dendritic membrane. Insects solve this problem by secreting large amounts of small soluble proteins, termed odorant binding proteins (OBPs) in the sensillum lymph [37]. It is believed that hydrophobic odorant molecules are dissolved into the sensillum lymph by binding with OBPs and then delivered to the dendritic membrane. However, a recent study that examined knockout flies lacking a specific type of OBP demonstrated that such OBPs are not important for the solubilization of odorants but rather for broadening the dynamic range to various odorant concentrations by buffering odorants within sensillum lymph [38]. Thus a comprehensive understanding of the function of OBPs requires further investigation.

ORs are expressed, together with their odorant receptor coreceptor (Orco), on the dendritic membrane of ORNs [39]. OR and Orco form a heteromeric OR complex that functions as an odorant-gated nonselective cation channel [40]. Binding of odorants with OR triggers conformational changes in this complex and induces an influx of cations, which evoke depolarization of the receptor membrane potential (Fig. 4.5). This electrical response propagates to the base of the axon and finally elicits action potentials in ORNs (see Section 4.1.2.1).

4.1.2.3 Odorant receptors and signal transduction

In contrast to vertebrate ORs that belong to GPCRs, insect ORs form odorant-gated nonselective cation channels with Orco. Each insect species possesses multiple ORs

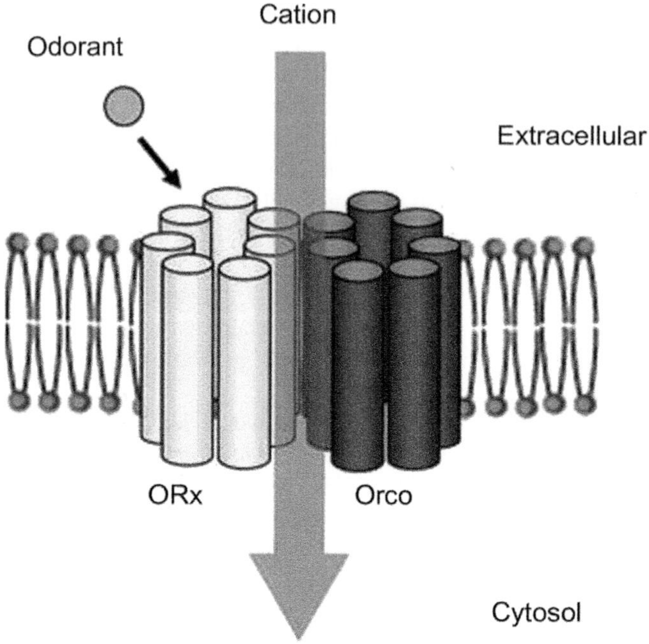

Figure 4.5 Olfactory signal transduction in insects. Insect odorant receptor (OR) forms a heteromeric odorant-gated nonselective cation channel with an Orco family protein. Although a single OR and Orco are depicted in this figure for simplicity, actual stoichiometry of an OR–Orco heteromeric complex has not been revealed yet.
Source: Modified from T. Sakurai, S. Namiki, R. Kanzaki, Front. Physiol. 5, (2014) 125 [41].

with different odorant selectivity and a single Orco. In the OR–Orco complex, the OR determines odorant selectivity, whereas both OR and Orco contribute to formation of the ion channel pore and thus determination of channel properties [42]. Orco is not only required for the formation of a functional OR complex but also for the transportation of ORs to the dendritic membrane [39]. In a heterologous expression system such as cultured cells and *Xenopus* oocytes, coexpression of Orco enhances membrane localization of ORs and is required for the formation of functional heteromeric OR complexes [40,43]. Thus Orco is considered to be important for the successful functional reconstruction of insect ORs. Because insect ORs function as odorant-gated ion channels, they can transduce odorant signals into electrical signals by themselves and do not require additional signal transduction machineries such as the G-proteins, enzymes, and ion channels required in vertebrate OR transduction systems. This simple machinery of olfactory transduction renders insect ORs as good candidates for constructing sensor elements in cultured cells or artificial membranes (see Section 4.2).

4.1.2.4 Odor coding by olfactory receptor neurons

The total number of ORs differs considerably among species, ranging from 10 in the body louse [44] to over 300 in ants [45]. Insect ORs are roughly categorized into two types, one is specialist that responds to a specific odorant such as sex pheromones, and the other is generalist that responds to multiple odorants. In principal, similar to vertebrate ORs, each generalist OR can bind different odorants and each odorant can be recognized by multiple generalist ORs and thus odor information is coded by combinations of activated ORs as in vertebrates (Fig. 4.3). Comprehensive functional analyses of the ORs from the fruit fly (*Drosophila melanogaster*) and the African malaria mosquito (*Anopheles gambiae*) revealed that the response spectrum of generalist ORs was distributed continuously from narrowly to broadly tuned spectra [46–48]. As specialist ORs, highly specific ORs including sex pheromone receptors in moths have been reported. In many cases, the activation of such receptors is closely linked to the initiation of a specific behavior such as courtship or avoidance behavior. For example, it has been shown that activation of the sex pheromone receptor BmOR1 and corresponding ORNs by the sex pheromone bombykol in the silkmoth (*Bombyx mori*) is necessary and sufficient to drive full pheromone source orientation behavior in male silkmoths [49–51].

Comprehensive analysis using the fruit fly has uncovered the relationship between ORs and ORNs as well as ORNs and the glomerulus in the AL [52,53]. Similar to vertebrate olfactory systems, most ORNs selectively express one of many ORs, with ORNs expressing the same OR projecting into a single defined glomerulus in the AL (Fig. 4.6). As each generalist OR normally responds to various odorants and each odorant is detected by various generalist ORs, odorant information is represented as a combination of activated glomeruli.

4.2 Biosensing technologies based on odorant receptors

Odorant sensing technologies using olfactory signals derived from ORs have been broadly researched and various types of technologies involving the direct use of living bodies, as well as the use of neuronal cells and tissues isolated from living organisms, or of biological materials to interact with odorant molecules have been proposed [1]. With the advance of genome analysis and genetic engineering, considering that biological molecules are derived from genetic information, numerous studies have artificially produced ORs for use in odorant detection (reviewed in Ref. [55]). To use the OR proteins as sensing elements, it is necessary to reconstruct, produce, and utilize functional receptor proteins. As described in Section 4.1, mammalian ORs comprise GPCRs, whereas insect ORs are ligand-gated ion channels. Because of differences in signal transduction mechanisms of these ORs, each study needs to be optimally designed in terms of particular cell

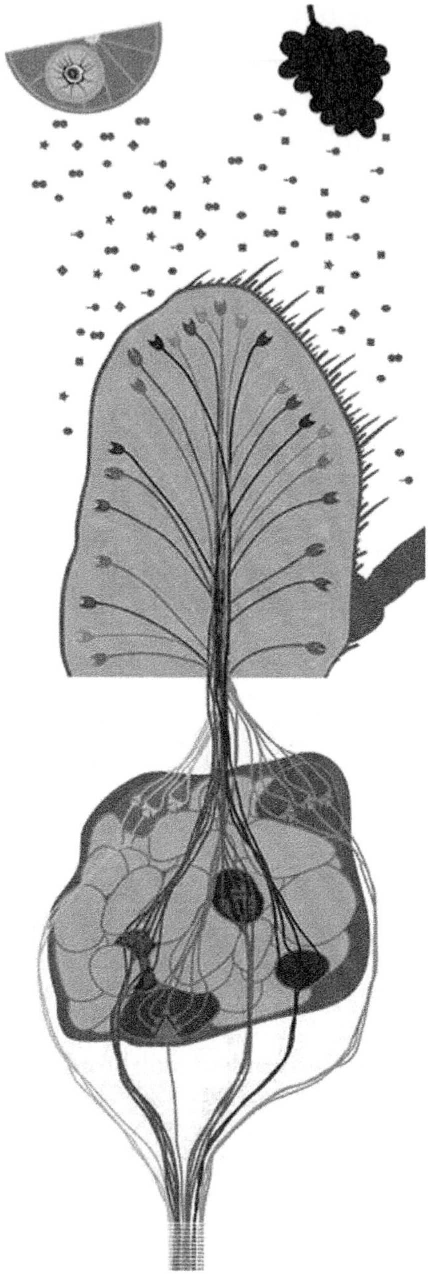

Figure 4.6 Schematic representation of projection pattern of ORNs in *D. melanogaster*. Odorants in the air are detected by ORNs in antennae (top). Colors of ORNs indicate ORs they express. ORNs expressing the same OR (indicated by the same color) convergently project their axons to the same glomerulus in the antennal lobe (bottom).
Source: Reproduced from V. Grabe and S. Sachse, Biosystems 164 (2018) 94–101 [54].

types for producing proteins and measurement systems for ascertaining response when ORs are utilized as sensing elements. In this section, within a specific focus on the artificial reconstruction of ORs, we will introduce sensing technologies that incorporate each of these two different types of ORs: mammalian and insect ORs.

4.2.1 Mammalian odorant receptors

4.2.1.1 Cell-based expression systems

During investigation of the interaction of ORs with odorant molecules, cell-based expression systems have been primarily used for OR reconstruction. Accordingly, based on these systems, sensing elements have been developed by using cells expressing ORs and their derivatives. To facilitate the expression of mammalian ORs, various cell types have been used to date: *Escherichia coli* (bacteria) in prokaryotes, *Saccharomyces cerevisiae* (yeast), *Xenopus laevis* oocytes (frog), human embryonic kidney 293 (HEK293T) cells, and Sf9 cells (insect) in eukaryotes (reviewed in Ref. [56]). Details of studies using each of these cell types are provided in the following sections.

4.2.1.1.1 Bacterial cells

A bacterium, *E. coli*, is commonly used for overexpression of various heterologous proteins in prokaryotic cells owing to the associated high productivity of a large amount of target proteins. In contrast, it is often problematic that the majority of expressed proteins are localized into inclusion bodies in a nonfunctional state, according to the characteristics of the protein. In particular, the challenges associated with the functional expression and purification of mammalian ORs in *E. coli.* have been addressed by the laboratory of Prof. Tai Hyun Park. For example, Kim et al. expressed the human OR, hOR2AG1, as a fusion protein with a glutathione-*S*-transferase (GST) tag at the N-terminus to allow the collection of a membrane fraction with embedded ORs as odorant sensing elements [57]. Moreover, they immobilized the membrane fractions onto single-wall carbon nanotube (swCNT)-field effect transistor (FET) and measured the source–drain current. The swCNT-FET sensor was reported to be able to selectively detect amyl butyrate, a target odorant of hOR2AG1, with detection limit of 100 fM (Fig. 4.7). Similarly, Yoon et al. conjugated hOR2AG1, which was expressed in *E. coli*, with conducting polymer nanotubes, placed these onto FETs, and measured the responses to amyl butyrate [58]. Notably, they were able to detect the odorants at the lower concentration of 40 fM. In addition, Lee et al. reported that by using chemically modified swCNT to effect reliable attachment of hOR2AG1, the sensitivity of the sensor system could be improved to concentrations as low as 1 fM [59,60].

There is also evidence that an OR from *Caenorhabditis elegans*, which functions as a GPCR similar to mammalian ORs, could be expressed in *E. coli* for use in sensor elements. Sung et al. expressed a fusion protein of ODR-10 of the nematode with GST and a 6x His tag to immobilize the crude membrane fraction onto quartz crystal microbalances (QCM). The ODR-10-conjugated QCM sensor could

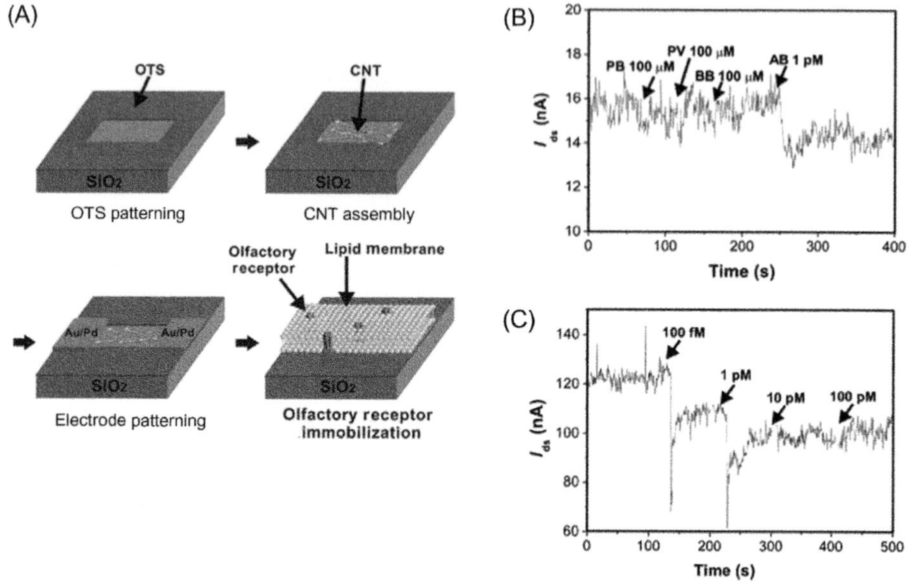

Figure 4.7 Human olfactory receptor (hOR2AG1)-functionalized swCNT-FET sensors. (A) Immobilization of membrane fraction derived from *E. coli* expressing hOR2AG1 onto swCNT-FET. (B) Odorant selectivity. *AB*, Amyl Butyrate; *BB*, butyl butyrate; *PB*, propyl butyrate; *PV*, pentyl valerate. (C) Conductance changes to various concentrations of amyl butyrate.
Source: Reproduced from T.H. Kim, S.H. Lee, J. Lee, H.S. Song, E.H. Oh, T.H. Park, et al., Adv. Mater. (Weinheim, Germany) 21 (2009) 91.

detect a natural ligand, diacetyl (2,3-butanedione), which was detectable as low as 1 pM [61]. Together, these studies demonstrated that various ORs from human and nematode could be reconstituted by expressing the functional proteins in *E. coli* and purifying the membrane fractions, rendering them available as odorant sensing elements.

4.2.1.1.2 Yeast cells

Yeast, *S. cerevisiae*, has been frequently utilized owing to its advantageous characteristic of being suitable for the overexpression of heterologous proteins including eukaryote proteins. In addition, yeast cells are used as a functional expression system for membrane proteins such as ORs and ion channels. For example, Pajot-Augy et al. proposed a method for acquiring the odorant responses of expressed mammalian ORs, focusing on the signal transduction of mating pheromones in yeast [62]. The budding yeast *S. cerevisiae* receives the mating pheromone peptides α-factor or a-factor, which are emitted by mating partners, through GPCRs expressed on the cell membrane of each mating type to recognize opposite mating partners via the resulting signal transduction cascade. As the GPCRs are similar to those in mammalian cells, the signal transduction pathway should therefore be activated by expressing mammalian ORs along with mammalian or yeast/mammalian chimeric G-protein α-subunits. Based on this hypothesis, Pajot-Augy et al. coexpressed rat OR, I7, and Gα proteins in the yeast cells and induced cell growth according to the expression of a resistance gene induced by the interaction between I7 and heptanal under a selective medium lacking histidine [62].

In comparison, Vidic et al. expressed ORI7 or OR17−40 along with the Gα protein in yeast cells to isolate nanosomes and immobilized them onto the gold electrode of the BIAcore3000 surface plasmon resonance (SPR) analyzer [63,64]. They demonstrated that the SPR sensor chip with immobilized nanosomes was able to detect ligand odorants as a shift of SPR response level in the presence of GTP-γS and discriminate between ligand and nonligand in a similar manner as the whole cell. This indicated that the responses of the receptors from the isolated nanosomes could potentially be acquired without the cellular signaling cascade of the yeast cells. Subsequently, Benilova et al. prepared yeast nanosomes from cells coexpressing OR17−40 with the Gαolf protein and immobilized these as olfactory biofilms for detection using the NanoSPR-6 dual channel electrochemical SPR spectrometer, which succeeded in measuring the same response pattern of helional during storage periods of 2 days [65,66]. Furthermore, as another measurement principle, Marrakchi et al. immobilized *S. cerevisiae* yeast cells expressing OR17−40 into a working electrode on interdigitated microconductometric electrodes to detect the target ligand, helional, with high sensitivity (lower threshold; 10^{-14} M), but not a nonligand, heptanal, with high selectivity [67].

As an application toward the detection of warfare agents, Radhika et al. reported an engineered yeast strain that detected an explosive compound, 2,4-dinitrotoluene [68]. They constructed "olfactory yeast" that contained Gαolf, Gβ, Gγ, and ACIII, components of the mammalian olfactory signal pathway, as well as cAMP-response element (CRE) binding protein and CRE-driven green fluorescence protein (GFP) to screen ORs using the fluorescence of GFP expression. Using this method, they identified a rat OR, Olfr226, as a 2,4-dinitrotoluene-responsive receptor (Fig. 4.8). Fukutani et al. also reported a bioluminescence-based yeast sensing system expressing the OR226 chimeric receptor protein with an N-terminus of I7 based on a firefly luciferase (*luc*) reporter assay [69]. They also demonstrated that accessory proteins, consisting of receptor transporting protein 1 short (RTP1s) from the mammalian olfactory epithelium and insect OBP, improved odorant sensitivity of the engineered yeast [70]. Thus these studies demonstrate that yeast cells possess numerous merits for the expression of various types of proteins utilizing previously developed genetic engineering methods.

4.2.1.1.3 Mammalian cultured cells

Mammalian cultured cells have been generally used for the expression of mammalian ORs owing to their capability to facilitate the glycosylation and proper folding of eukaryotic proteins. In particular, HEK293 and HeLa cells are frequently used as expression systems for sensor elements. Wetzel et al. first reported that a human OR, OR17−40, was able to be functionally expressed in HEK293 cells and *X. laevis* oocytes, indicating that these expression methods could be used for the

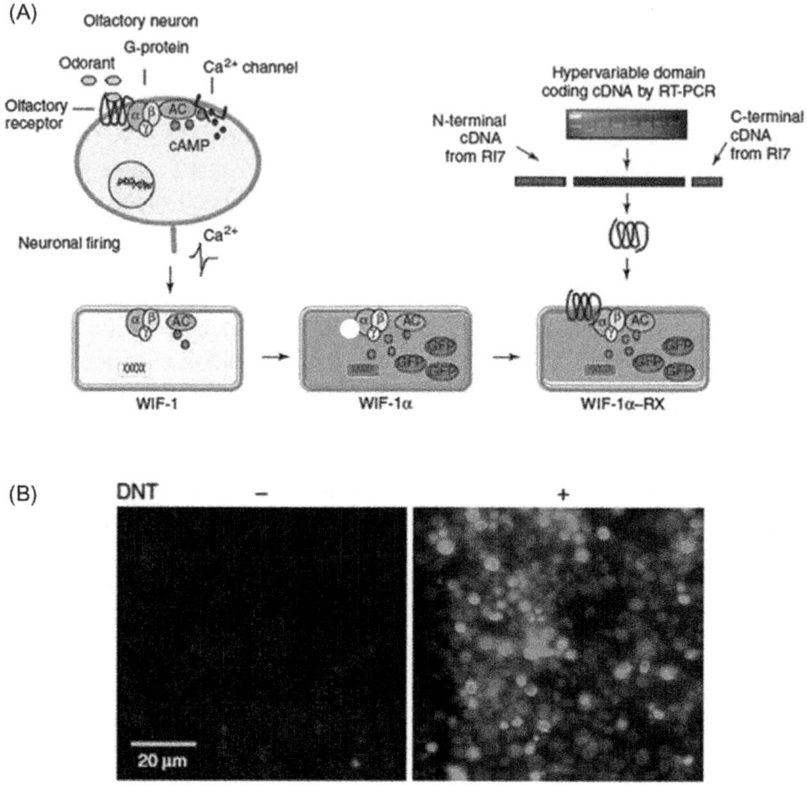

Figure 4.8 Engineering of "olfactory yeast." (A) Schematic illustration showing the construction of olfactory yeast. (B) Response of olfactory yeast expressing the OR to DNT (25 µM).
Source: Reproduced from V. Radhika, T. Proikas-Cezanne, M. Jayaraman, D. Onesime, J.H. Ha, D.N. Dhanasekaran, Nat. Chem. Biol. 3 (2007) 325.

characterization of mammalian ORs through the measurement of calcium imaging or electrophysiological recording [71]. Subsequently, the expression of mammalian ORs has been conducted using additional mammalian cultured cells including Chinese hamster ovary (CHO) and African green monkey kidney fibroblast (COS-7) cells (reviewed in Ref. [56]).

For the functional analysis of mammalian ORs, researchers mainly use optical methods for measurements of OR activity, such as calcium imaging and reporter assays. In contrast, for applications as sensing elements, the materials derived from the cells expressing mammalian ORs are conjugated with various measurement systems, such as QCM, SPR, or CNT-FET. For example, Ko and Park reported that the rat OR, I7, could be expressed in HEK293 cells as a fusion protein with a Rho-tag import sequence at the N-terminus [72]. They cultured the cell line onto the gold electrode of QCM and measured the resonant frequency changes to octyl aldehyde (octanal). In addition, Lee et al. applied the same cell line toward SPR (SPRi; K-MAC, Korea), which indicated that the selective responses of I7 could be monitored as a reflectance shift produced by the intracellular calcium ion concentration changes [73]. Lee et al. also used microfabricated planar electrodes in conjunction with HEK293 cells coexpressing a gustatory cyclic nucleotide-gated channel as well as I7 to develop a cell-based odorant sensor system that enabled them to measure the odorant responses as electrical signals, that is, extracellular field potentials [74]. Notably, Lee et al. further demonstrated that HEK293 cells expressing mammalian ORs were also available suitable for the detection of gaseous odorants [75]. They fabricated a microfluidic system that was divided into two compartments, one side being a gaseous phase and the other a liquid phase, via a porous polycarbonate membrane with 2 μm pore size, and cultured hOR-expressing HEK293 cells with a calcium indicator, Fluo-4, on the liquid phase side of the membrane. By monitoring the fluorescence intensity changes, they demonstrated that the cell exhibited fluorescence responses to gaseous odorants when they were supplied into the gaseous phase side of the membrane (Fig. 4.9).

Additionally, nanovesicles are able to be isolated from HEK293 cells and used as a sensing layer. Jin et al. isolated nanovesicles by centrifugation from HEK293 cells expressing hOR2AG1, which were cultured in medium including the actin filament polymerization inhibitor, cytochalasin B, and immobilized the nanovesicles onto swCNT-FET [76]. The nanovesicle-coated swCNT-FET was reported to detect amyl butyrate with a detection limit of 1 fM. Similarly, Ahn et al. reported that swCNT-FET, which had been coated with nanovesicles expressing OR8H2, allowed the detection of 1-octen-3-ol derived from fungal contamination in grain [77]. Thus various mammalian ORs can be functionally expressed in several types of mammalian cultured cells and partially used for detection of odorants in actual samples.

4.2.1.2 Other (noncell-based expression system) applications

Cell-free protein synthesis can also be utilized for the production of proteins. In this method, target proteins are synthesized in vitro by mixing extracts derived from *E. coli*, rabbit reticulocytes, human cells, wheat germ, and insect cells with, for example, amino acids, an energy source, and DNA. Kaiser et al. reported that correctly folded human OR 17-4 could be generated using a wheat germ cell-free method [78]. They selected proper detergent and its concentration to purify hOR17-4 protein in a soluble and nonaggregated state and immobilized it onto an SPR sensor chip (Biacore T100) to successfully measure the responses to a known ligand, undecanal.

Artificially synthesized peptides of OR binding sites could also be potentially used as sensing elements. For example, Wu and Lo predicted the most plausible binding sites for a target odorant [trimethylamine (TMA)] in a dog OR, *olfd canfa* OR, via computer docking simulation using the GRAMM program [79]. Subsequently, they synthesized two short oligopeptides and immobilized them onto the surface of piezoelectric gold electrodes, demonstrating that the polypeptides possessed a high binding capability for TMA. Sankaran et al. also designed

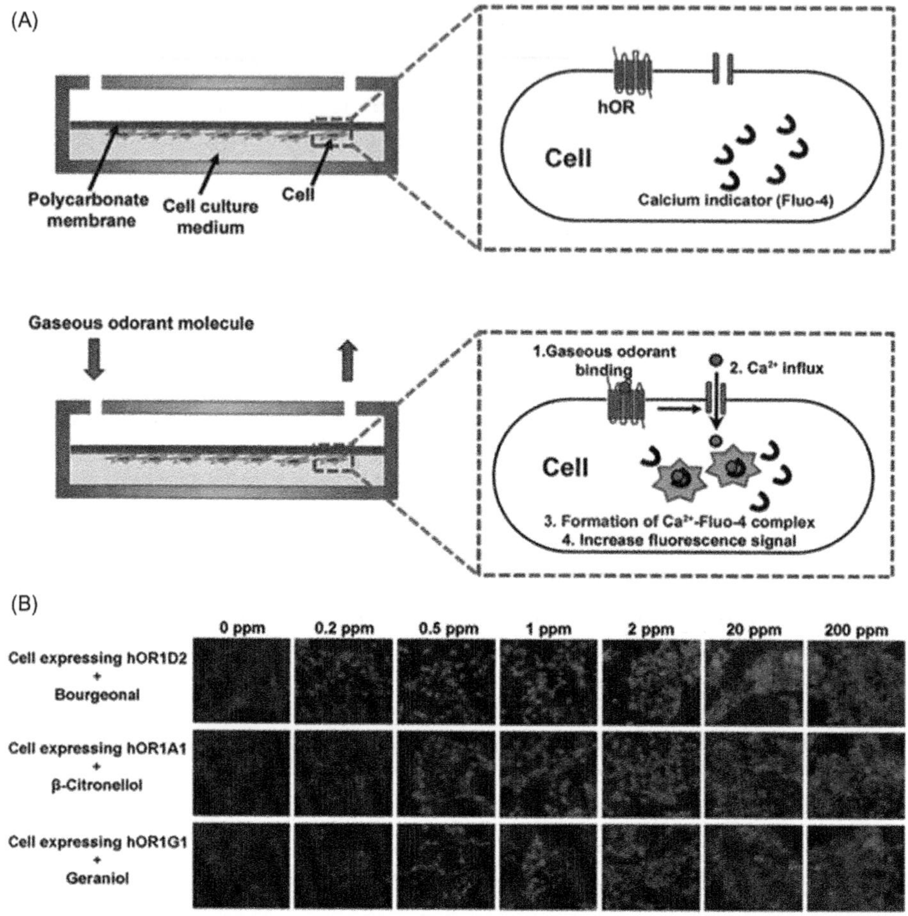

Figure 4.9 Microfluidic device for the detection of gaseous odorants. (A) Schematic illustration of the fabrication of the microfluidic device. Upper figure indicates the sectional view of the device and lower figure indicates the gaseous odorants detecting mechanism. (B) Dose-dependent responses of HEK293 cells expressing human ORs to the target gaseous odorants.
Source: Reproduced from S.H. Lee, E.H. Oh, T.H. Park, Biosensor Bioelectron. 74 (2015) 554.

a polypeptide as a sensing material by predicting the binding sites of an olfactory receptor and determining their binding affinity through computational simulation using the Tripos software Sybyl 8.0 [80]. They reported that a QCM sensor coated with the synthesized polypeptide was able to detect 1-hexanol and 1-pentanol with a detection limit of 2–3 and 3–5 ppm, respectively. More recently, Lim et al. reported that the real-time and high sensitivity (detection limit of 10 fM) sensing of TMA is possible by conjugating a synthesized polypeptide with SWNT-FET [81] (Fig. 4.10). Furthermore, the peptide-based sensor allowed the

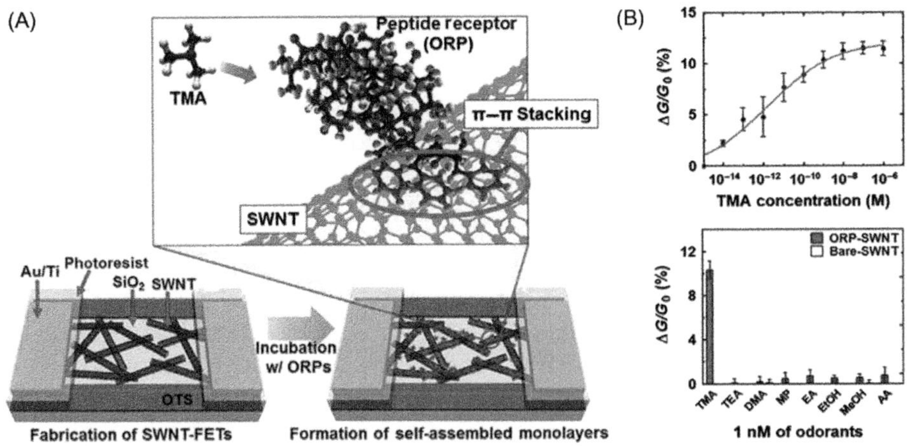

Figure 4.10 Peptide receptor-based bioelectronic nose. (A) Schematic illustration of swCNT-FET immobilized peptide receptor. (B) Conductance changes to various concentration of TMA (upper) and various kinds of odorants (lower). *AA*, Acetic acid; *DMA*, dimethylamine; *EA*, ethylacetate; *EtOH*, ethanol; *MeOH*, methanol; *MP*, 2-methyl-1-propanol; *TEA*, trimethylamine; *TMA*, trimethylamine.
Source: Reproduced from J.H. Lim, J. Park, J.H. Ahn, H.J. Jin, S. Hong, T.H. Park, Biosensor Bioelectron. 39 (2013) 244.

selective detection of TMA and the distinguishing of specific spoiled seafood from among other types of spoiled seafood without any pretreatment. These studies indicate that OR proteins themselves along with OR-derived polypeptides also exhibit potential as sensing elements, suggesting that these may represent more readily available and useful sensing materials than those derived from cell-expression systems owing to the simple production methods and lack of requirement of a lipid bilayer.

4.2.2 Insect odorant receptors

Insects can sensitively and selectively detect various types of odorants present in their environment. Therefore sensing technologies focusing on insect olfaction have also recently attracted attention. Both the living insect body and its tissues have been utilized for odor-sensing technologies. Park et al. used electroantennogram (EAG) measurement technology to construct a portable Quadro-probe EAG recording system by conjugating four different antennae with Quadro-probes [82]. They demonstrated that this system discriminated 20 different volatile compounds from sex pheromones to general odorants based on the EAG amplitudes of each antenna in real time. Moreover, Myrick et al. reported that using the insect antenna–based chemosensor system could discriminate two different odorants delivered from closely located odorant sources, suggesting that this system might be used for detecting more complex odor plumes, as performed by various insects [83].

Honeybees, *Apis mellifera*, are also utilized as a biological system for odor sensing. Through typical association learning of an odorant with a sugar solution, the bee can detect the odorant in conjunction with the proboscis extension response, which is utilized as an index of the odorant detection. Based on this principle, trained bees appear to serve as a highly sensitive detector for explosive chemicals [84]. Thus the results from studies incorporating the living bodies and tissues of insects have demonstrated the effectiveness of insect olfaction as a higher performance sensing technology.

Concurrent with the advances of genome sequencing, numerous OR genes have been identified from various insect species and their functions characterized through assorted genetic engineering methods (see Section 4.2.1). In parallel, associated techniques have been developed as odor-sensing technologies incorporating the functions of insect ORs.

4.2.2.1 Cell-based expression systems

Insect ORs have been functionally analyzed by using cell-based expression system such as *X. laevis* oocytes and cultured cells [56]. In particular, *Xenopus* oocytes comprise a single cell, with a size over approximately 1 mm in diameter. By microinjection of the mRNA of target OR genes, functional proteins could be readily expressed in the oocytes. In addition, the weak inward currents that are induced by interaction between ORs and odorants could to be measured using glass electrodes based on the electrophysiological method termed two electrode-voltage clamping. Wetzel et al. demonstrated that *D. melanogaster* OR43a selectively responded to cyclohecanol and cyclohexanone by coexpressing it along with the human Gα subunit, Gα15, in the oocyte [85]. In addition, Sakurai et al. also reported that in oocytes expressing *B. mori* BmOR1 with BmGαq, inward currents could be induced by the silkmoth sex pheromone component, bombykol [49].

Subsequently, Nakagawa et al. demonstrated that coexpression of an Orco with a highly conserved sequence among various insect species promoted the functional expression of ORs in heterologous cells [43]. This study led to the important finding that insect ORs with Orco function as nonselective cation channels [40]. Moreover, these studies indicated that the odorant detection function resulting from both insect ORs and Orco could be reconstructed by using *Xenopus* oocytes. In follow-up studies, Misawa et al. proposed a biohybrid odorant sensor wherein *Xenopus* oocytes functionally expressing insect ORs and Orco are integrated as odorant sensor elements [86]. They prepared two oocytes, each of which expressed BmOR1 or BmOR3 along with BmOrco, to develop an odorant sensor chip by arraying the oocytes onto an acrylic flow channel. The sensor chip could distinguish two pheromone components, bombykol and bombykal, which exhibit slightly different chemical structures, at only a few parts per billion. In addition, they demonstrated that the sensor chip could be integrated into a robotic system owing to its portability, enabling signal measurement without disturbance by electric noise (Fig. 4.11).

Cultured cells derived from insects and mammals have also been used for the functional analysis of insect ORs. For example, Neuhaus et al. used mammalian

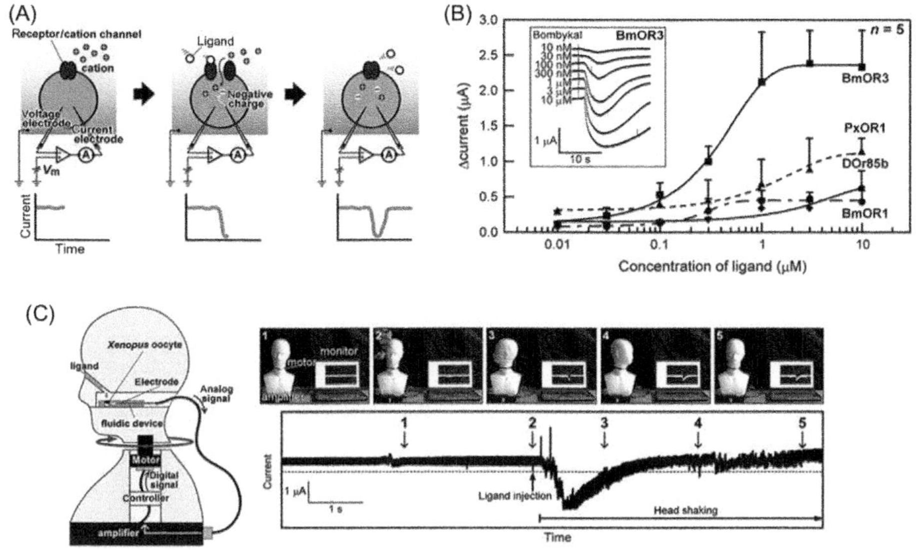

Figure 4.11 An odorant sensor with *Xenopus* oocytes expressing insect odorant receptors. (A) Principle image of two electrode-voltage clamp. (B) Dose-dependent responses of the chemical sensors to the target odorants. (C) Head-shaking robot system integrating the chemical sensor.
Source: Reproduced from N. Misawa, H. Mitsuno, R. Kanzaki, S. Takeuchi, Proc. Natl. Acad. Sci. U.S.A. 107 (2010) 15340.

HEK293 cells for the functional expression of insect ORs and Orco [87]. Through calcium imaging measurements using Fura-2, they indicated that HEK293 cells expressing OR43a alone could be activated by a target odorant, cyclohecanone, whereas coexpression of OR43a and Orco exhibited more sensitive fluorescence responses to the odorants without differences in OR expression level. Among cultured insect cells, Sf9 and Sf21 cell lines, which are derived from the ovary of the armyworm moth, *Spodoptera frugiperda*, have been commonly used for functional OR expression. Kiely et al. reported a functional assay method of insect ORs based on calcium imaging with Fluo-4, which indicated that Sf9 cells expressing *D. melanogaster* OR22a exhibited selective fluorescence responses to a range of odorants, with response profiles similar to those obtained from in vivo assays [88].

Toward the development of sensing technologies, Mitsuno et al. constructed Sf21 cell lines that coexpressed a silkmoth sex pheromone receptor, BmOR1 or BmOR3, along with Orco and the calcium sensitive fluorescence protein, GCaMP3, which exhibited increased fluorescence intensity when exposed to corresponding sex pheromone components [89]. Furthermore, they also indicated that the cell lines were able to detect an odorant without variable responsiveness for at least 2 months and could be integrated into a microfluidic glass chip to create an odorant sensor chip. Similarly, this methodology is applicable to various types of insect ORs for the detection of general odorants. For example, Termtanasombat et al. used

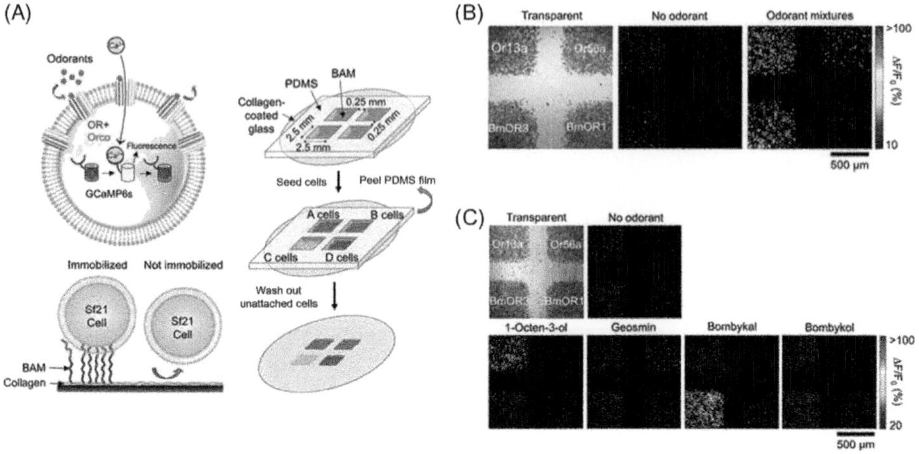

Figure 4.12 Sf21 cell-based odorant sensor array. (A) Schematic illustration of the development of the Sf21 cell-based odorant sensor array. (B) Response pattern of the sensor array to odorant mixture. BmOR1, BmOR3, OR13a, and OR56a respond to bombykol, bombykal, 1-octen-3-ol, and geosmin, respectively. The odorant mixture consists of these four odorants. (C) Response patterns of the sensor array to a single odorant.
Source: Reproduced from M. Termtanasombat, H. Mitsuno, N. Misawa, S. Yamahira, T. Sakurai, S. Yamaguchi, et al., J. Chem. Ecol. 42 (2016) 716.

silkmoth sex pheromone receptors or fruit fly ORs as well as Orco and GCaMP6s to construct four cell lines, each of which detected bombykol (pheromone), bombykal (related-pheromone), geosmin (moldy odor), and 1-octen-3-ol (moldy odor) by increasing their fluorescence intensities [90]. Moreover, they immobilized and arrayed the four cell lines onto a plane glass chip using a biocompatible anchor for membrane (BAM) to successfully discriminate four odorants according to fluorescence intensity pattern (Fig. 4.12).

In addition, cultured cells expressing insect OR complex have also been reported as an applicable technique for detecting chemical vapor. Sato and Takeuchi constructed a hydrogel microchamber to load the spheroid formation of HEK293T cells expressing the *A. gambiae* OR, GPROR2, or *D. melanogaster* OR, OR47a, which respond to 2-methylphenol or pentyl acetate, respectively [91]. The responses of ORs to vapor stimulation derived from the headspace of an odorant solution were able to be acquired from the spheroids by using extracellular field potential recording (Fig. 4.13). Thus cells expressing insect as well as mammalian ORs are also being applied toward the development of biosensing technologies.

4.2.2.2 Other (noncell expression system) applications

Cell-free protein expression systems have also been adopted for the functional analyses of insect membrane proteins including ORs, with the purified in vitro–synthesized OR proteins and their peptides being available as odorant sensing elements. Recently, it

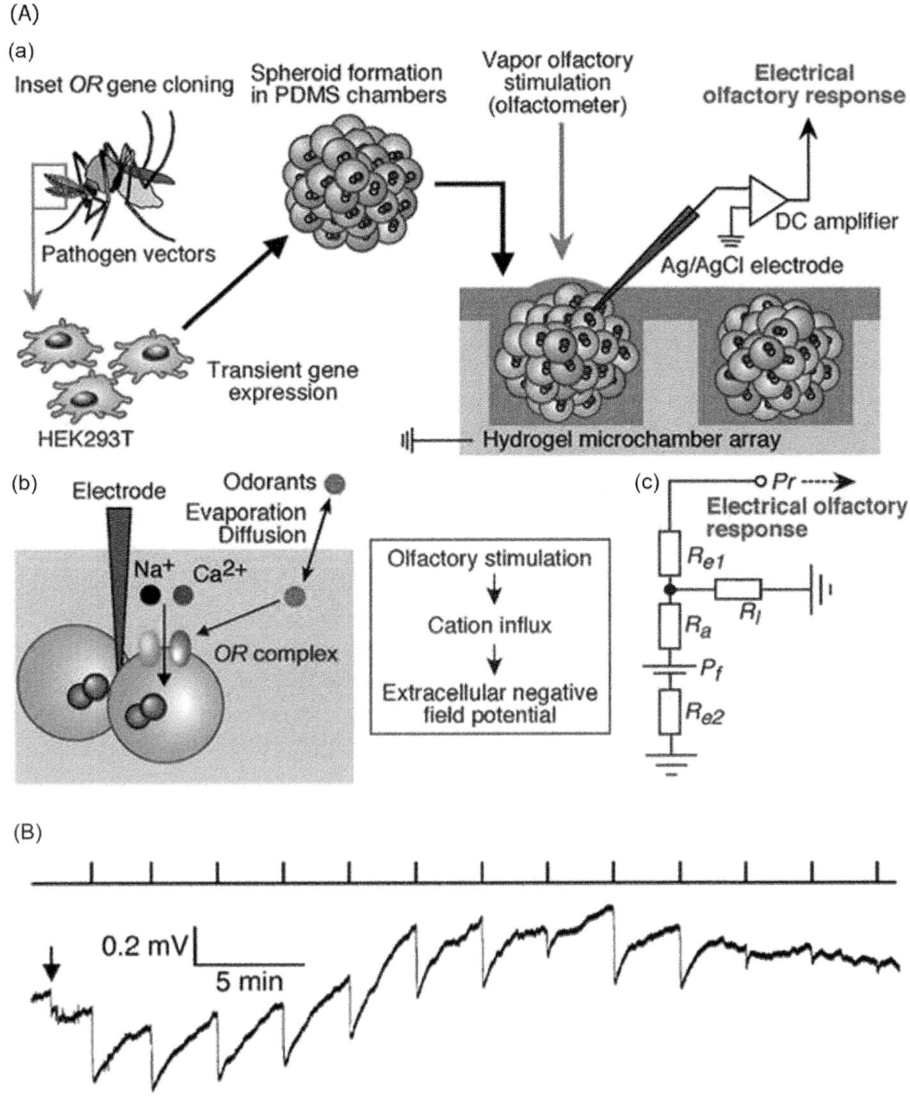

Figure 4.13 Vapor chemical detection by OR-expressing spheroid. (A) Schematic illustration of the construction and measurement of OR-expressing spheroid. (B) Responses of the spheroid expressing OR to the headspace of odorant solution.
Source: Reproduced from K. Sato, S. Takeuchi, Angew. Chem. Int. Ed. 53 (2014) 11798.

was reported that the function of an insect OR complex could be reconstructed in an artificial membrane by integrating the genes and cell extracts into a lipid bilayer capsule. Hamada et al. encapsulated cell-free protein synthesis reagents along with plasmid vectors for the expression BmOR1 and BmOrco inside cell-sized lipid bilayer capsules termed giant vesicles, and demonstrated that BmOR1 and BmOrco were embedded in the artificial membrane [92]. They succeeded in measuring the inward current of the BmOR1–BmOrco complex induced by the cognate ligand, bombykol, from the giant vesicle by using a patch-clamp technique. These results suggest that artificial membranes with embedded insect ORs and Orco may be available as a potential technology for receptor-based odorant sensing.

In insects, current genetic modification techniques have made it possible to create transgenic insects by recombining various desired genes into the genomic DNA. Especially in the silkmoth, use of this technique is reported to be leading toward the application of transgenic silkmoths expressing target ORs in the antenna as a type of odorant sensor. In *B. mori*, a series of mating behaviors is triggered in the male upon reception of a sex pheromone, bombykol, delivered from the female, resulting in localization behavior to the odor source [93]. This behavior is triggered by the activation of ORNs expressing BmOR1, which responds to bombykol [49,50]. Therefore the ectopic expression of target ORs in the ORNs would lead to the production of silkmoths that orient and localize to the target odor source. Accordingly, Sakurai et al. produced a transgenic silkmoth that expressed a sex pheromone receptor PxOR1 from the diamondback moth, *Plutella xylostella*, which is sensitive to Z11–16:Ald, in bombykol receptor neurons [50]. They revealed that the transgenic silkmoth exhibited full mating behavior upon the activation of bombykol receptor neurons with Z11–16:Ald. Furthermore, the transgenic silkmoth localized to female diamondback moths in a wind tunnel. These results suggest that the production of transgenic moths exhibiting ectopic expression of target ORs in the ORNs via genetic modification techniques could enable the generation of "sensor insects" that detect target odorants and localized to the odor source.

4.3 Summary

In this chapter, we summarized the mechanisms of olfaction, especially signal transduction by ORs, in living organisms and introduced proposed sensing technologies based on their ORs. Since the mid-1990s, numerous sensing technologies using ORs have been actively researched and developed, although the studies that were introduced in this chapter represent only a subset of these technologies. Application of ORs may enable the development of innovative odorant sensor platforms that can detect target odorants with high selectivity and selectivity beyond those of conventional odor-sensing systems. Although a number of effective technologies are considered to reach sufficiently high level to allow practical use, few odorant sensors based on ORs are currently available owing to the limitations of biological materials such as problems of, for example, stability or long-term usage. Therefore improvement of such factors would be very important for the realization of practical odor sensors using ORs in the future.

We have introduced sensing technologies based on "mammalian ORs" and "insect ORs" separately. Since mammalian ORs are GPCRs and insect ORs are ligand-gated ion channels, there are the advantages and disadvantages in sensing technologies using each type of ORs in accordance with their original characteristics. For example, one of the advantages of GPCRs is to be capable of signal amplification by the metabotropic signaling pathway. Therefore it might be possible to develop more sensitive odor-sensing systems based on cells expressing mammalian ORs. Nevertheless, it would be difficult to completely reconstruct the complicated signaling pathway, which starts from the GPCRs, in heterologous cells. On the other hand, one of the advantages of ligand-gated ion channels is to be capable of quickly inducing ion influx after the interaction with ligands. Therefore use of insect ORs might lead to development of odor-sensing systems with faster response time, that is, real-time sensing systems. Thus utilization of the advantages of mammalian ORs or insect ORs would enable us to develop various types of artificial sensing systems suitable for our intended purposes.

In addition, the sensing technologies described in this chapter are applicable to numerous ORs from living organisms inhabiting various environments worldwide. The use of optimized ORs for target odorants would lead to the development of specific odorant sensor elements that detect various kinds of desired target odorants for designated applications. However, as the performance of the odorant sensing elements (e.g., sensitivity, selectivity, and specificity) is dependent on the intrinsic characteristics of the expressed ORs, it is crucial to clarify the relationships between odorants and ORs to facilitate the selection of optimized ORs. In insects, over 100 ORs have been characterized from various species. Among these, the fruit fly, *D. melanogaster*, has been utilized in the scientific field of olfaction as a model organism and its ORs have been comprehensively investigated by numerous researchers using various functional analysis methods, such as the fruit fly in vivo expression system (termed the "empty neuron system") or heterologous cell-expression systems. Focusing on these data, the team led by Prof. Giovanni Galizia has developed a database of odor responses (DoOR; http://neuro.uni-konstanz.de/DoOR/default.html), wherein information regarding specific odorant responses for different receptors can be obtained [94,95]. As this database expands, it will facilitate the selection of sets of ORs that are specific to target odorants and are optimized for the discrimination of defined odors. Through conjugating such information with the sensing technologies described in this chapter, it is expected that a new era will soon be attained wherein odorant sensors can be provided for the detection of desired target odorants in various situations.

References

[1] R. Kanzaki, K. Nakatani, T. Sakurai, N. Misawa, H. Mitsuno, Physiology of chemical sense and its biosensor application, in: T. Nakamoto (Ed.), Essentials of Machine Olfaction and Taste, Wiley, New York, 2016.

[2] L. Buck, R. Axel, A novel multigene family may encode odorant receptors: a molecular basis for odor recognition, Cell 65 (1991) 175.

[3] D.T. Jones, R.R. Reed, Golf: an olfactory neuron specific-G protein involved in odorant signal transduction, Science 244 (1989) 790.
[4] H.A. Bakalyar, R.R. Reed, Identification of a specialized adenylyl cyclase that may mediate odorant detection, Science 250 (1990) 1403.
[5] L.J. Brunet, G.H. Gold, J. Ngai, General anosmia caused by a targeted disruption of the mouse olfactory cyclic nucleotide-gated cation channel, Neuron 17 (1996) 81.
[6] H. Takeuchi, T. Kurahashi, Identification of second messenger mediating signal transduction in the olfactory receptor cell, J. Gen. Physiol. 122 (2003) 557.
[7] T. Nakamura, G.H. Gold, A cyclic nucleotide-gated conductance in olfactory receptor cilia, Nature 325 (1987) 442.
[8] S.J. Kleene, The electrochemical basis of odor transduction in vertebrate olfactory cilia, Chem. Senses 33 (2008) 839.
[9] M. Dibattista, S. Pifferi, A. Boccaccio, A. Menini, J. Reisert, The long tale of the calcium activated Cl^- channels in olfactory transduction, Channels (Austin) 11 (2017) 399.
[10] L.R. Saraiva, X. Ibarra-Soria, M. Khan, M. Omura, A. Scialdone, P. Mombaerts, et al., Hierarchical deconstruction of mouse olfactory sensory neurons: from whole mucosa to single-cell RNA-seq, Sci. Rep. 5 (2015) 18178.
[11] Y. Niimura, A. Matsui, K. Touhara, Acceleration of olfactory receptor gene loss in primate evolution: possible link to anatomical change in sensory systems and dietary transition, Mol. Biol. Evol. 35 (2018) 1437.
[12] K.J. Ressler, S.L. Sullivan, L.B. Buck, A zonal organization of odorant receptor gene expression in the olfactory epithelium, Cell 73 (1993) 597.
[13] R. Vassar, J. Ngai, R. Axel, Spatial segregation of odorant receptor expression in the mammalian olfactory epithelium, Cell 74 (1993) 309.
[14] K. Miyamichi, S. Serizawa, H.M. Kimura, H. Sakano, Continuous and overlapping expression domains of odorant receptor genes in the olfactory epithelium determine the dorsal/ventral positioning of glomeruli in the olfactory bulb, J. Neurosci. 25 (2005) 3586.
[15] S. Serizawa, K. Miyamichi, H. Sakano, One neuron-one receptor rule in the mouse olfactory system, Trends Genet. 20 (2004) 648.
[16] F. Wang, A. Nemes, M. Mendelsohn, R. Axel, Odorant receptors govern the formation of a precise topographic map, Cell 93 (1998) 47.
[17] P. Feinstein, T. Bozza, I. Rodriguez, A. Vassalli, P. Mombaerts, Axon guidance of mouse olfactory sensory neurons by odorant receptors and the beta2 adrenergic receptor, Cell 117 (2004) 833.
[18] K. Touhara, S. Sengoku, K. Inaki, A. Tsuboi, J. Hirono, T. Sato, et al., Functional identification and reconstitution of an odorant receptor in single olfactory neurons, Proc. Natl. Acad. Sci. USA 96 (1999) 4040.
[19] K. Kajiya, K. Inaki, M. Tanaka, T. Haga, H. Kataoka, K. Touhara, Molecular bases of odor discrimination: reconstitution of olfactory receptors that recognize overlapping sets of odorants, J. Neurosci. 21 (2001) 6018.
[20] B. Malnic, J. Hirono, T. Sato, L.B. Buck, Combinatorial receptor codes for odors, Cell 96 (1999) 713.
[21] K. Mori, Y. Yoshihara, Molecular recognition and olfactory processing in the mammalian olfactory system, Prog. Neurobiol. 45 (1995) 585.
[22] C. Bushdid, M.O. Magnasco, L.B. Vosshall, A. Keller, Humans can discriminate more than 1 trillion olfactory stimuli, Science 343 (2014) 1370.
[23] R.C. Gerkin, J.B. Castro, The number of olfactory stimuli that humans can discriminate is still unknown, elife 4 (2015) e08127.
[24] M. Meister, On the dimensionality of odor space, elife 4 (2015) e07865.

[25] L. Silva, A. Antunes, Vomeronasal receptors in vertebrates and the evolution of pheromone detection, Annu. Rev. Anim. Biosci. 5 (2017) 353.
[26] S.D. Liberles, Trace amine-associated receptors: ligands, neural circuits, and behaviors, Curr. Opin. Neurobiol. 34 (2015) 1.
[27] E. Jacquin-Joly, C. Merlin, Insect olfactory receptors: contributions of molecular biology to chemical ecology, J. Chem. Ecol. 30 (2004) 2359.
[28] T.A. Keil, Morphology and development of the peripheral olfactory organs, in: B.S. Hansson (Ed.), Insect Olfaction, Springer, 1999, pp. 5–48.
[29] R.A. Steinbrecht, Olfactory receptors, in: E. Eguchi, Y. Tominaga (Eds.), Atlas of Arthropod Sensory Receptors-Dynamic Morphology in Relation to Function, Springer, Tokyo, 1999, pp. 155–176.
[30] J.G. Hildebrand, G.M. Shepherd, Mechanisms of olfactory discrimination: converging evidence for common principles across phyla, Annu. Rev. Neurosci. 20 (1997) 595.
[31] H. Morita, A. Shiraishi, Chemoreception physiology, in: G.A. Kerkut, L.I. Gilbert (Eds.), Comprehensive Insect Physiology, Biochemistry and Pharmacology. Vol. 6: Nervous System, Pergamon Press, Oxford, New York, 1985.
[32] H. Kijima, Y. Okada, S. Oiki, S. Goshima, K. Nagata, T. Kazawa, Free ion concentrations in receptor lymph and role of transepithelial voltage in the fly labellar taste receptor, J. Comp. Physiol. A 177 (1995) 123.
[33] K.E. Kaissling, G. Kasang, H.J. Bestmann, W. Stransky, O. Vostrowsky, A new pheromone of the silkworm moth *Bombyx mori*, Naturwissenschaften 65 (1978) 382.
[34] B. Pophof, Olfactory responses recorded from sensilla coeloconica of the silkmoth *Bombyx mori*, Physiol. Entomol. 22 (1997) 239.
[35] G. Kasang, K.E. Kaissling, Specificity of primary and secondary olfactory processes in Bombyx antennae, in: D. Schneider (Ed.), IVth International Symposium on Olfaction and Taste, Wissenschaftl. Verlagsgesellsch, Stuttgart, 1972, pp. 200–206.
[36] S. Kanaujia, K.E. Kaissling, Interactions of pheromone with moth antennae: adsorption, desorption and transport, J. Insect Physiol. 31 (1985) 71.
[37] R.G. Vogt, L.M. Riddiford, Pheromone binding and inactivation by moth antennae, Nature 293 (1981) 161.
[38] N.K. Larter, J.S. Sun, J.R. Carlson, Organization and function of *Drosophila* odorant binding proteins, eLife 5 (2016) e20242.
[39] R. Benton, S. Sachse, S.W. Michnick, L.B. Vosshall, Atypical membrane topology and heteromeric function of *Drosophila* odorant receptors in vivo, PLoS Biol. 4 (2006) e20.
[40] K. Sato, M. Pellegrino, T. Nakagawa, T. Nakagawa, L.B. Vosshall, K. Touhara, Insect olfactory receptors are heteromeric ligand-gated ion channels, Nature 452 (2008) 1002.
[41] T. Sakurai, S. Namiki, R. Kanzaki, Molecular and neural mechanisms of sex pheromone reception and processing in the silkmoth *Bombyx mori*, Front. Physiol. 5 (2014) 125.
[42] T. Nakagawa, M. Pellegrino, K. Sato, L.B. Vosshall, K. Touhara, Amino acid residues contributing to function of the heteromeric insect olfactory receptor complex, PLoS One 7 (2012) e32372.
[43] T. Nakagawa, T. Sakurai, T. Nishioka, K. Touhara, Insect sex-pheromone signals mediated by specific combinations of olfactory receptors, Science 307 (2005) 1638.
[44] E.F. Kirkness, et al., Genome sequences of the human body louse and its primary endosymbiont provide insights into the permanent parasitic lifestyle, Proc. Natl. Acad. Sci. USA 107 (2010) 12168.
[45] X. Zhou, A. Rokas, S.L. Berger, J. Liebig, A. Ray, L.J. Zwiebel, Chemoreceptor evolution in hymenoptera and its implications for the evolution of eusociality, Genome Biol. Evol. 7 (2015) 2407.

[46] E.A. Hallem, J.R. Carlson, Coding of odors by a receptor repertoire, Cell 125 (2006) 143.
[47] A.F. Carey, G. Wang, C.Y. Su, L.J. Zwiebel, J.R. Carlson, Odorant reception in the malaria mosquito *Anopheles gambiae*, Nature 464 (2010) 66.
[48] G. Wang, A.F. Carey, J.R. Carlson, L.J. Zwiebel, Molecular basis of odor coding in the malaria vector mosquito *Anopheles gambiae*, Proc. Natl. Acad. Sci. USA 107 (2010) 4418.
[49] T. Sakurai, T. Nakagawa, H. Mitsuno, H. Mori, Y. Endo, S. Tanoue, et al., Identification and functional characterization of a sex pheromone receptor in the silkmoth *Bombyx mori*, Proc. Natl. Acad. Sci. USA 101 (2004) 16653.
[50] T. Sakurai, H. Mitsuno, S.S. Haupt, K. Uchino, F. Yokohari, T. Nishioka, et al., A single sex pheromone receptor determines chemical response specificity of sexual behavior in the silkmoth *Bombyx mori*, PLoS Genet. 7 (2011) e1002115.
[51] T. Sakurai, H. Mitsuno, A. Mikami, K. Uchino, M. Tabuchi, F. Zhang, et al., Targeted disruption of a single sex pheromone receptor gene completely abolishes *in vivo* pheromone response in the silkmoth, Sci. Rep. 5 (2015) 11001.
[52] A. Couto, M. Alenius, B.J. Dickson, Molecular, anatomical, and functional organization of the Drosophila olfactory system, Curr. Biol. 15 (2005) 1535.
[53] E. Fishilevich, L.B. Vosshall, Genetic and functional subdivision of the *Drosophila* antennal lobe, Curr. Biol. 15 (2005) 1548.
[54] V. Grabe, S. Sachse, Fundamental principles of the olfactory code, Biosystems 164 (2018) 94–101.
[55] L. Du, C. Wu, Q. Liu, L. Huang, P. Wang, Recent advances in olfactory receptor-based biosensors, Biosensors Bioelectron. 42 (2013) 570.
[56] R. Glatz, K. Bailey-Hill, Mimicking nature's noses: from receptor deorphaning to olfactory biosensing, Prog. Neurobiol. 93 (2011) 270.
[57] T.H. Kim, S.H. Lee, J. Lee, H.S. Song, E.H. Oh, T.H. Park, et al., Single-carbon-atomic-resolution detection of odorant molecules using a human olfactory receptor-based bioelectronic nose, Adv. Mater. (Weinheim, Germany) 21 (2009) 91.
[58] H. Yoon, S.H. Lee, O.S. Kwon, H.S. Song, E.H. Oh, T.H. Park, et al., Polypyrrole nanotubes conjugated with human olfactory receptors: high-performance transducers for FET-type bioelectronic noses, Angew. Chem. Int. Ed. 48 (2009) 2755.
[59] S.H. Lee, J.J. Hye, H.S. Song, S. Hong, T.H. Park, Bioelectronic nose with high sensitivity and selectivity using chemically functionalized carbon nanotube combined with human olfactory receptor, J. Biotechnol. 157 (2012) 467.
[60] S.H. Lee, O.S. Kwon, H.S. Song, S.J. Park, J.H. Sung, J. Jang, et al., Mimicking the human smell sensing mechanism with an artificial nose platform, Biomaterials 33 (2012) 172.
[61] J.H. Sung, H.J. Ko, T.H. Park, Piezoelectric biosensor using olfactory receptor protein expressed in Escherichia coli, Biosensors Bioelectron. 21 (2006) 1981.
[62] E. Pajot-Augy, M. Crowe, G. Levasseur, R. Salesse, I. Connerton, Engineered yeasts as reporter systems for odorant detection, J. Receptors Signal Transduction 23 (2003) 155.
[63] J.M. Vidic, J. Grosclaude, M.A. Persuy, J. Aioun, R. Salesse, E. Pajot-Augy, Quantitative assessment of olfactory receptors activity in immobilized nanosomes: a novel concept for bioelectronic nose, Lab Chip 6 (2006) 1026.
[64] J.M. Vidic, M. Pla-Roca, J. Grosclaude, M.A. Persuy, R. Monnerie, D. Caballero, et al., Gold surface functionalization and patterning for specific immobilization of olfactory receptors carried by nanosomes, Anal. Chem. 79 (2007) 3280.
[65] I. Benilova, V.I. Chegel, Y.V. Ushenin, J. Vidic, A.P. Soldatkin, C. Martelet, et al., Stimulation of human olfactory receptor 17-40 with odorants probed by surface plasmon resonance, Europ. Biophys. J. 37 (2008) 807.

[66] I. Benilova, J. Vidic, E. Pajot-Augy, A.P. Soldatkin, C. Martelet, N. Jafftezic-Renault, Electrochemical study of human olfactory receptor OR 17−40 stimulation by odorants in solution, Mater. Sci. Eng. C. 28 (2008) 633.
[67] M. Marrakchi, J. Vidic, N. Jaffrezic-Renault, C. Martelet, E. Pajot-Augy, A new concept of olfactory biosensor based on interdigitated microelectrodes and immobilized yeasts expressing the human receptor OR17-40, Europ. Biophys. J. 36 (2007) 1015.
[68] V. Radhika, T. Proikas-Cezanne, M. Jayaraman, D. Onesime, J.H. Ha, D.N. Dhanasekaran, Chemical sensing of DNT by engineered olfactory yeast strain, Nat. Chem. Biol. 3 (2007) 325.
[69] Y. Fukutani, J. Ishii, K. Noguchi, A. Kondo, M. Yohda, An improved bioluminescence-based signaling assay for odor sensing with a yeast expressing a chimeric olfactory receptor, Biotechnol. Bioeng. 109 (2012) 3143.
[70] Y. Fukutani, A. Hori, S. Tsukada, R. Sato, J. Ishii, A. Kondo, et al., Improving the odorant sensitivity of olfactory receptor-expressing yeast with accessory proteins, Anal. Biochem. 471 (2015) 1.
[71] C.H. Wetzel, M. Oles, C. Wellerdieck, M. Kuczkowiak, G. Gisselmann, H. Hatt, Specificity and sensitivity of a human olfactory receptor functionally expressed in human embryonic kidney 293 cells and *Xenopus Laevis* oocytes, J. Neurosci. 19 (1999) 7426.
[72] H.J. Ko, T.H. Park, Piezoelectric olfactory biosensor: ligand specificity and dose-dependence of an olfactory receptor expressed in a heterologous cell system, Biosensors Bioelectron. 20 (2005) 1327.
[73] S.H. Lee, H.J. Ko, T.H. Park, Real-time monitoring of odorant-induced cellular reactions using surface plasmon resonance, Biosensors Bioelectron. 25 (2009) 55.
[74] S.H. Lee, S.B. Jun, H.J. Ko, S.J. Kim, T.H. Park, Cell-based olfactory biosensor using microfabricated planar electrode, Biosensors Bioelectron. 24 (2009) 2659.
[75] S.H. Lee, E.H. Oh, T.H. Park, Cell-based microfluidic platform for mimicking human olfactory system, Biosensors Bioelectron. 74 (2015) 554.
[76] H.J. Jin, S.H. Lee, T.H. Kim, J. Park, H.S. Song, T.H. Park, et al., Nanovesicle-based bioelectronic nose platform mimicking human olfacry signal transduction, Biosensors Bioelectron. 35 (2012) 335.
[77] J.H. Ahn, J.H. Lim, J. Park, E.H. Oh, M. Son, S. Hong, et al., Screening of target-specific olfactory receptor and development of olfactory biosensor for the assessment of fungal contamination in grain, Sensors Actuators B: Chem. 210 (2015) 9.
[78] L.G.-B. Kaiser, J. Steuerwald, D. Vanberghem, M. Herlihy, S.G. Zhang, Efficient cell-free production of olfactory receptors: Detergent optimization, structure, and ligand binding analyses, Proc. Natl. Acad. Sci. USA 105 (2008) 15726.
[79] T.Z. Wu, Y.R. Lo, Synthetic peptide mimicking of binding sites on olfactory receptor protein for use in 'electronic nose', J. Biotechnol. 80 (2000) 63.
[80] S. Sankaran, S. Panigrahi, S. Mallik, Olfactory receptor based piezoelectric biosensors for detection of alcohols related to food safety applications, Sensors Actuators B 155 (2011) 8.
[81] J.H. Lim, J. Park, J.H. Ahn, H.J. Jin, S. Hong, T.H. Park, A peptide receptor-based bioelectronic nose for the real-time determination of seafood quality, Biosensors Bioelectron. 39 (2013) 244.
[82] K.C. Park, S.A. Ochieng, J. Zhu, T.C. Baker, Odor discrimination using insect electroantennogram responses from an insect antennal array, Chem. Senses 27 (2002) 343.
[83] A.J. Myrick, K.C. Park, J.R. Hetling, T.C. Baker, Detection and discrimination of mixed odor strands in overlapping plumes using an insect-antenna-based chemosensor system, J. Chem. Ecol. 35 (2009) 118.

[84] J.J. Bromenshenk, C.B. Henderson, G.C. Smith, Biol. System (Paper II) (2003) 273–283.
[85] C.H. Wetzel, H.-J. Behrendt, G. Gisselmann, K.F. Stortkuhl, B. Hovemann, H. Hatt, Functional expression and characterization of a *Drosophila* odorant receptor in a heterologous cell system, Proc. Natl. Acad. Sci. USA 98 (2001) 9377.
[86] N. Misawa, H. Mitsuno, R. Kanzaki, S. Takeuchi, Highly sensitive and selective odorant sensor using living cells expressing insect olfactory receptors, Proc. Natl. Acad. Sci. USA 107 (2010) 15340.
[87] E.M. Neuhaus, G. Gisselmann, W. Zhang, R. Dooley, K. Stortkuhl, H. Hatt, "Odorant receptor heterodimerization in the olfactory system of *Drosophila melanogaster*, Nat. Neurosci. 8 (2005) 15.
[88] A. Kiely, A. Authier, A.V. Kralicek, C.G. Warr, R.D. Newcomb, "Functional analysis of a *Drosophila melanogaster* olfactory receptor expressed in Sf9 cells, J. Neurosci. Methods 159 (2007) 189.
[89] H. Mitsuno, T. Sakurai, S. Namiki, H. Mitsuhashi, R. Kanzaki, Novel cell-based odorant sensor elements based on insect odorant receptors, Biosensors Bioelectron 65 (2015) 287.
[90] M. Termtanasombat, H. Mitsuno, N. Misawa, S. Yamahira, T. Sakurai, S. Yamaguchi, et al., Cell-based odorant sensor array for odor discrimination based on insect odorant receptors, J. Chem. Ecol. 42 (2016) 716.
[91] K. Sato, S. Takeuchi, Chemical vapor detection using a reconstituted insect olfactory receptor complex, Angew. Chem. Int. Ed. 53 (2014) 11798.
[92] S. Hamada, M. Tabuchi, T. Toyota, T. Sakurai, T. Hosoi, T. Nomoto, et al., Giant vesicles functionally expressing membrane receptors for an insect pheromone, Chem. Commun. 50 (2014) 2958.
[93] E. Kramer, Orientation of the male silkmoth to the sex attractant bombykol, in: D.A. Denton, J.P. Coghlan (Eds.), Olfaction and Taste, vol. 5, Academic Press, New York, 1975, pp. 329–335.
[94] C.G. Galizia, D. Munch, M. Strauch, A. Nissler, S. Ma, Integrating heterogeneous odor response data into a common response model: A DoOR to the complete olfactome, Chem. Senses 35 (2010) 551.
[95] D. Munch, C.G. Galizia, DoOR 2.0–Comprehensive mapping of *Drosophila melanogaster* odorant responses, Sci. Rep. 6 (2016) 21841.

Advanced surface modification technologies for biosensors

5

Chun-Jen Huang[1,2,3]
[1]Department of Biomedical Sciences and Engineering, National Central University, Jhong-Li, Taiwan, [2]Department of Chemical and Materials Engineering, National Central University, Jhong-Li, Taiwan, [3]R&D Center for Membrane Technology, Chung Yuan Christian University, Chung-Li City, Taiwan

5.1 Biosensors and biointerfaces

Since the first introduction of a biosensor by Leland C. Clark in 1962, a wide range of biosensor technologies have been developed and applied in medical diagnostics, food control, and environmental monitoring. The biosensor as defined by the International Union of Pure and Applied Chemistry (IUPAC) is an analytical device for the detection of an analyte that specifically associates a biological component (biorecognition element, BRE) immobilized with a physicochemical transducer [1]. As shown in Fig. 5.1, the biosensor comprises:

1. Sample handling system that enables to sample, purify, concentrate, inject, and mix the solutions containing target molecules.
2. BRE that can be isolated from a living system or synthesized in a laboratory. To date, molecular imprints, lectins, enzymes, receptors, antibodies, and nucleic acids have been used as BREs that can specifically interact with the target analytes of interest.
3. Biointerface that is responsible for accommodation of BREs, maintenance of their bioactivity, resistance against nonspecific adsorption, and easy implementation.
4. Transducer that translates the recognition events, for example, analyte capture or enzymatic catalysis, into physical or chemical signals by means of electrical, electrochemical, optical, thermal, acoustic, piezoelectric components. They are designed to react to the presence of analyte sensitively, rapidly, and quantitatively.
5. Amplifier that enhances the responses from transducers and lower the background noise.
6. User interface that displays the results in a digital way.

Among these, the design of interface between BREs and transducer is essential for a stable, sensitive, and accurate biosensor [2]. Biointerface science is defined as the study and control of biomolecular interactions with surfaces and has become a new catalog in an interdisciplinary research area [3]. The biointerfaces are typically highly hydrated to retain the activities of biomolecules. Biointerfaces can be composed of biomaterials, synthesized polymers, or inorganic materials and decorated with BREs via a wide range of chemical and physical approaches, which is the focus of the chapter.

Chemical, Gas, and Biosensors for the Internet of Things and Related Applications.
DOI: https://doi.org/10.1016/B978-0-12-815409-0.00005-X

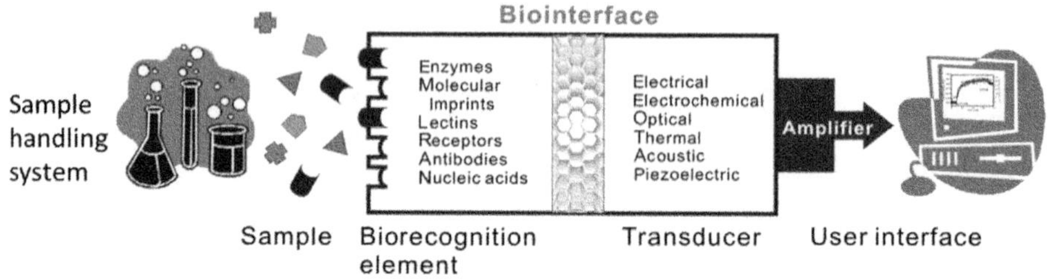

Figure 5.1 Configuration of a biosensor comprising of biorecognition, interface, and transduction elements.

One challenge for the new generation of biointerfaces in applications of biosensing is to bring together the recent advances in material science and molecular biology to pursue ultrasensitive detection in-field. In addition, a careful characterization for biointerfaces can be critical to meet the detection requirement and optimization. In chemical and biological sensors, additionally, stable linkage between BREs and the sensor surface needs to be established. The immobilized biomolecules have to retain their biological activity; the surface architecture should also exhibit nonfouling properties minimizing the nonspecific interactions with analyzed samples. The most common immobilization methods, relying on self-assembled monolayers (SAMs), polymeric hydrogel binding matrix, and bioaffinity binding platforms, are introduced. Therefore the strategies for antifouling properties and coupling chemistries of BREs will be reviewed for reliable and reproducible results.

5.2 Binding platforms based on self-assembled monolayers

SAMs are highly organized layers of amphiphilic molecules and provide a convenient, flexible, and simple platform to modify interfacial properties of metals, metal oxides, and semiconductors [4–6]. SAMs are organic assemblies prepared by the adsorption of molecular constituents from solution or the gas phase onto the surface of solids. SAMs at interfaces have attracted substantial attention due to their unique properties, such as control on the molecular level, ordered structure, high packing, the ease of preparation and functionalization [4]. To date, four types of assemblies have been developed as shown in Fig. 5.2. They are grouped into categories based on the attachment moieties, which are fatty acids [7,8], organosilicon [9,10], organosulfur [4,11,12], and catecholic [13–16] derivatives. These molecules have been employed for modification of substrates, such as metal oxides, novel metals, silicon oxides, and oxidized polymers. Historically, in 1946, Zisman and his coworkers published the preparation of a monomolecular layer by adsorption of a surfactant onto a clean metal surface [17,18]. At that time, the potential of self-assembly was not recognized, and this publication initiated only a limited level of interest. In the late 1970s, Sagiv successfully adsorbed trichlorosilanes onto silicon oxide [19].

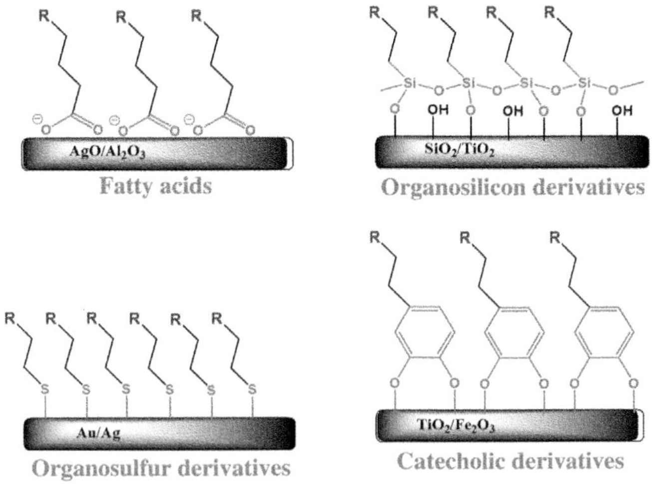

Figure 5.2 Four common types of assemblies and the corresponding materials for deposition of SAMs.

Since then a large variety of self-assembly systems have been studied. Nuzzo and Allara showed in the early 1980s that thiols on gold SAMs can be prepared by the adsorption of di-*n*-alkyl disulfides from dilute solutions [5]. Nowadays, the SAMs with thiols on gold, and alkylsilanes on silicon represent the most popular combinations of self-assembling molecules and substrates. In late 2000, Textor and colleagues developed a series of functional catechols [13,14,20] for modification of metal oxides. They demonstrated that the catechol derivatives as a self-assembling ligands, such as nitrocatechols, are able to covalently link to metal and metal oxide surfaces.

Fig. 5.3 shows the deposition of these polymers and SAMs on solids schematically. A linear polymer is synthesized by polymerization of monomers. Polymer chains are often structurally disordered, indicating that polymer chains are often coiled with a large gyration volume in a solution [21]. Therefore the structure and density of the polymeric coating on surfaces tend to be loose and incomplete. For SAMs, the spontaneous deposition process involves firm anchorage onto surfaces via chemical reactions, adsorption of assemblies to a substrate, and self-organization through noncovalent intermolecular interactions to render an organized and compact thin film on a given surface. In addition, for better reproducibility, durability, and easy implementation, BREs can be attached to SAMs with defined orientation and stable linkage with functional groups. As a result, SAMs are excellent models for the surface chemistry for understanding the physical–chemical phenomena and finely tuning the interfacial functionalities at the molecular level, as well as for developing a biointerface for biosensing applications.

Besides the abovementioned general characteristics of SAMs, they hold important factors to make themselves popular for biosensor development. Firstly, a SAM

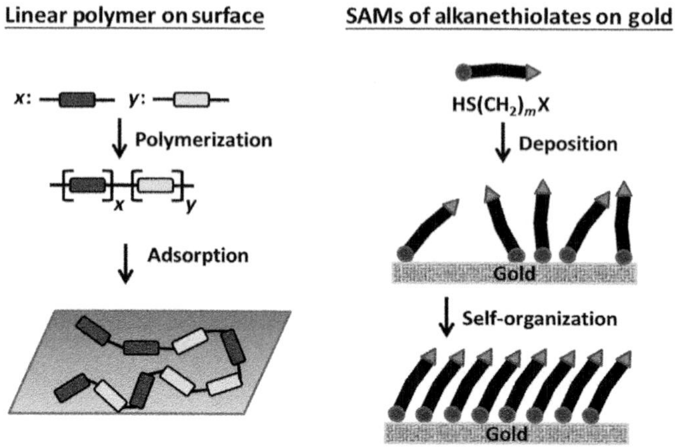

Figure 5.3 Schematic representation of the formation of conventional polymers and of the SAMs.

on a planar substrate can be prepared in the laboratory readily by dipping the substrate in a dilute solution of an organic assembly. Second, the formation of a SAM needs only a very small amount of chemicals. Therefore it is economically viable to use even expensive compounds for the development of a functional biointerface in a sensor.

5.2.1 Organosulfur derivatives

The most studied and best understood system among assembling ligands is the alkanethiol at Au(111) surfaces, which takes advantage of the environmental inertness of gold substrate, the high affinity of the sulfur–gold bonding and the formation of ordered structure by the van der Waals forces between long carbon chains [4–6]. One example in the wide spectrum of its biological applications is a tailor surface with alkanethiols with various functional head groups for manipulating the adhesion of proteins, cells, bacteria, and even large marine organisms [22,23].

In a conventional protocol for the preparation of biointerfaces for sensing purposes, the mixed SAMs, prepared by codeposition of different thiols, provide a practical experimental system in which the surface can be modified by biomolecules of interest and repel the nonspecific adsorption. The versatile mixed SAMs can be achieved by using suitable bifunctional molecules of different terminal functional groups, which are highly desirable for development of biosensors and nanomaterials for medical uses [24,25]. For example, oligo(ethylene glycol) (OEG)-terminated thiolates have been used to form stealth SAMs to improve biocompatibility, resist nonspecific adsorption, and achieve good wetting property. The functional groups on SAMs were endowed by adding carboxylate- [26–28] or amine-terminated [26,27] thiolates. Dostalek and colleagues carried out the sensitive detection of pathogenic bacteria using a long-range surface plasmon resonance

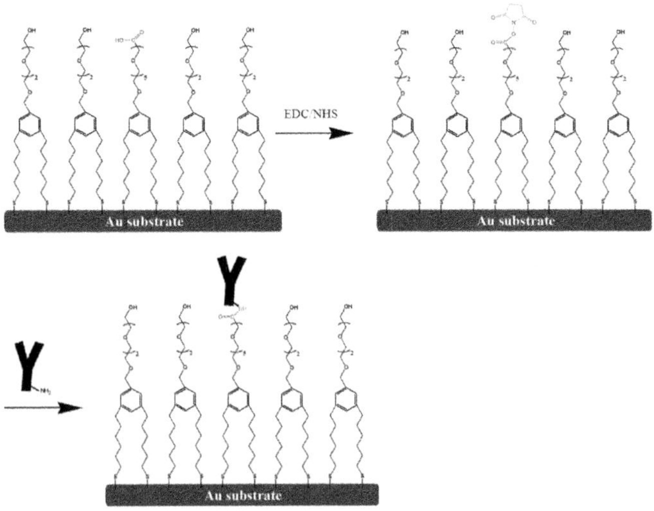

Figure 5.4 BRE immobilization on carboxylic acid–terminated SAMs through EDC/NHS amine coupling chemistry.

(SPR) biosensor through a sandwich immunoassay on a functional SAMs with oligo(ethylene) glycol- and carboxylic acid–terminated thiolates. The antibodies against bacteria were immobilized by facile 1-ethyl-3-(-3-dimethylaminopropyl) carbodiimide hydrochloride (EDC)/*N*-hydroxysuccinimide (NHS) coupling chemistry as shown in Fig. 5.4 [28].

5.2.2 Organosilicon derivatives

Salinization is the spontaneous formation of silanes through alkoxy hydrolysis and condensation to afford siloxane bonds with hydroxyl groups on the surface. Therefore the modification with organosilanes on oxide surfaces provides advantageous features, such as stable covalent conjugation, postmodification without deterioration of the coatings, and high compatibility with silicon-based sensor technologies, such as semiconductor field-effect transistor (FET) [29]. The FET is strongly associated with the Debye length, which is about 1 nm in a 0.1 M aqueous NaCl solution at 25°C [30]. However, good control over the structure and formation of functional organosilane coatings appears to be of fundamental importance and challenge [31,32].

Generally, the surface is activated by strong acids or an oxygen plasma to clean the surface and generate silanol groups at surfaces [33]. The treatment renders the surface hydrophilic and prone to the formation of a thin water layer, which is essential for the formation of well-packed silane film. Many studies indicate that the deposition process highly depends on solvent [34], deposition time [35,36], temperature [37], water content, and solution age [38]. In addition, the storage of organosilanes typically subjects the formation of polysiloxane aggregates in solutions, particularly in the presence of water [39]. The aggregates lose the reactivity to covalently conjugate on surfaces, but are physically deposited on the surfaces, giving rise to an unstable and inhomogeneous film. Very recently, advanced modification approaches have been developed for homogeneous and well-ordered organic adlayers on silicon substrates. Functional 1-alkenes were grafted by thermally induced or visible light–induced hydrosilylation on a Si–H substrate [40–43]. The process relies on the mechanism in which the nucleophilic attack of positively charged surface Si sites by alkenes leads to the formation of a stable Si–C bond. The unique feature of this technology allows temporal and spatial controlled modification on silicon via UV irradiation [44]. Moreover, heterocyclic silanes containing Si–N or Si–S in the ring were developed for the modification of surface silanols via a ring-opening click reaction [45]. The heterocyclic silanes permit rapid modification of silicon porous nanostructures and functionalization. In addition, amino- or thiol-terminated silatranes have been applied for DNA conjugation and nanoparticle immobilization [46–50]. The five-membered tricyclic silatranes were synthesized by treatment of triethanolamine with an alkoxysilanes. In the silatrane structures, the N atom is forced to approach toward the Si, resulting in the formation of strong transannular bonds (N→Si) and high stability toward hydrolysis [51]. Therefore the precision deposition of silatrane was accomplished for uniform and thin adlayers with various functionalities to allow accurate interpretation of the interfacial phenomena for sensing and nanomaterial applications.

5.2.3 Catechol derivatives

Catechols represent an important and versatile building block for the design of mussel-inspired synthetic coatings. The coating concept was originated from biomimetic strategies from mussel adhesive protein sequences mimicking mussel adhesion [52,53]. Catechols enable to establish strong interactions with different substrates and have promoted catechol as an important anchor for surface modifications. Compared with other types of ligands, catecholic ligands are relatively new. Textor and colleagues have demonstrated that the catechol derivatives as a self-assembling ligands are able to covalently link to metal and metal oxide surfaces [13,14,20]. Because of the stability, surface ligands are highly resistant to oxidation at high pH values. Jiang's group developed a biomimetic zwitterionic carboxybetaine polymers growing from catecholic initiators via atom transfer radical polymerization for graft-on surface modification [54–56]. In addition, a zwitterionic catecholic compound was synthesized as a surface ligand for modification of the contrast agent (e.g., superparamagnetic iron oxide nanoparticles) for magnetic resonance imaging [57–59]. It was found that the assembly of catechols on TiO_2 surfaces is a pH-dependent behavior [15]. The zwitterionic catechol adsorbs on surfaces via hydrogen bonding with hydroxyl groups on oxidized TiO_2 surface at low pH in the initial stage. The bonding of adsorbed catechols converts to bidentate bonds by the replacement of surface hydroxyl groups through deprotonating of catecholic hydroxyl groups after transferring the substrate to a high pH solution. However, to date, the potential of the assembling catechols in molecular detection has not been shown, likely due to limited availability.

5.3 Binding matrix based on polymeric hydrogels

Hydrogels are insoluble, cross-linked water-swollen polymer networks of hydrophilic homopolymers or copolymers. Owing to their highly open structure and large inner surface, hydrogels can accommodate large amounts of molecules with specific functions. In addition, they can be tailored in various structures and integrated into (micro-)systems. The chemical nature and physical structures of hydrogels can be sensitive to external stimuli. Therefore hydrogels become irreplaceable materials in numerous important fields including pharmaceutical applications (e.g., drug delivery or tissue engineering) [60,61] and biosensor technologies for detection of chemical or biological analytes [62,63].

In biosensor applications, hydrogel materials are employed at an interface between an analyzed aqueous sample and a transducer. Typically, the hydrogel interface is modified with BREs such as antibodies, enzymes, or with biomimetic molecular imprinted moieties to specifically recognize target analytes. Compared with the other types of biointerfaces (e.g., based on SAMs), hydrogels can accommodate orders of magnitude larger amounts of recognition elements [64,65], provide more natural microenvironment for remaining the bioactivity of biomolecules, and offer routes for implementing of additional functionalities (e.g., separation of target analyte from other molecules in a sample) [66]. In addition, the class of "smart" gels that respond to external stimuli has become of increasing interest in biosensor research. Various organic polymers were synthesized and applied for hydrogel biointerface architectures as reported in literature. These include chitosan [67], dextran [64], poly(vinylalcohol) [68], 2-hydroxyethyl methacrylate [69], poly (2-vinylpyridine) [70], poly(ethylene glycol) (PEG) [71], and poly(*N*-isopropylacrylamide) (NIPAAm) [65]. The sensing mechanisms based on physicochemical behaviors and biochemical responses of hydrogels will be reviewed.

5.3.1 Physicochemical sensing mechanisms

Stimulus-sensitive hydrogels used in sensors and actuators has been well documented [72]. The hydrogels themselves are applied as recognition elements in responses to various stimuli to induce physicochemical changes of gels. For example, temperature sensors are important for all aspects, and sensors are needed that precisely detect the exact temperature under different environmental conditions. The hydrogel-based temperature sensing systems in which hydrogels were used as 3D sensing networks formed by temperature-sensitive polymers and as appropriate matrices for temperature sensing probes were developed. The principles for swelling dependent pH value detection can be various changes of the holographic diffraction wavelength in optical Bragg grating sensors [73], and shifts of the resonance frequency of a quartz crystal microbalance in microgravimetric sensors [74,75].

For hydrogel-based pH sensors, the transduces are able to convert the changes of the physical properties of stimuli responsive hydrogels into electrical/optical signals. An intelligent polymerized crystalline colloidal array was designed by photocross-linking of acrylamide and NIPAAm with bisacrylaminde [76]. The hydrogel was polymerized around a face-centered cubic (fcc) array of monodispersed negatively charged polystyrene colloidal particles. Hence, change in pH-induced swelling/deswelling of the hydrogel due to protonation/deprotonation of the particle surface and the hydrogel can follow. The other study was to combine piezoresistive-responsive elements as mechanoelectrical transduces and the pH-responsive hydrogels for pH detection [77].

For ion-sensing applications, the detection of Pb^{2+} over a broad range of ion strengths was shown by chelating Pb^{2+} with crown ether groups in the hydrogel network. When the chelation of Pb^{2+} occurred, the hydrogel started to swell driven by osmotic pressure, leading the shift in the diffracted light [78]. Zhang et al. used a hydrogel matrix based on acrylamide and (3-acrylamidopropyl) triethyl ammonium chloride for sensitive detection of CrO_4^{2-} with a limited detection of 10^{-11} M [79].

In the field of gas sensing, hydrogels can be modified by gas sensitive molecules, for example, special fluorescent dyes, which make them sensitive for certain gases. One example for the use as protective layer is the internal hydrogel separation layer of the planar microelectrode nitric oxide (NO) sensor to measure local NO surface concentrations [80]. The working electrode and the gas permeable membrane were separated by an internal hydrogel layer. It was proved that the NO-selective microelectrode sensor is an appropriate tool for determining accurate surface NO concentrations.

5.3.2 Biochemical sensing mechanisms

In biochemical sensing mechanisms, more complex detection methods were developed by integrating hydrogels and BREs. The hydrogels were intensively applied into two types of biosensors, catalytic and bioaffinity, for developing new schemes of biosensing technologies.

In the metabolism sensors, enzymes are immobilized for detection of target substrates. Enzymes are highly specific and efficient in their reactions with substrates. Jang and Koh developed a multiplexed enzyme-based assay within a microfluidic device using shape-coded PEG hydrogel microparticles [81]. The microparticles with immobilized enzymes were fabricated in the patterning chamber by photolithography and collected in the detection chamber by pressure-driven flow. Simultaneous detection of glucose and ethanol was performed using glucose oxidase and alcohol oxidase in the hydrogel microparticles.

In the bioaffinity sensing systems, the detection occurs due to the bioaffinity interactions, such as antibody–antigen interaction or DNA hybridization. Therefore BREs were attached in the 3D hydrogel binding matrix. Dostalek and colleagues demonstrated substantial potential of functional NIPAAm-based hydrogels on SPR-originated biosensors [64,65,82–85]. They developed a label-free biosensor based on hydrogel optical waveguide spectroscopy for the sensitive detection of 17 β-estradiol (E2). The approach was implemented by using a thin hydrogel layer of a carboxylated PNIPAAm terpolymer that was attached to a metallic sensor surface to simultaneously serve as a binding matrix and an optical waveguide. Refractive index changes that are accompanied with the specific capture of E2 molecules from a sample on the sensor surface were probed by resonantly excited hydrogel optical waveguide modes [85].

5.4 Coupling chemistries for immobilization of biorecognition elements

Determining the optimal biointerface for immobilization of BREs is critical and influences the reliability of detections. Though many attachment approaches have been proposed, issues, such as denaturation, orientation, binding density, site specificity of BREs, on sensors remain problematic. In this chapter, current frequently used immobilization strategies for BREs are described, which can be classified into three categories, that is, physical, covalent, and bioaffinity immobilization. The question of which strategy can be considered as the best choice is still open, but it is unlikely that one scheme will be the best for all types of BREs. Therefore the extrapolation of all immobilization approaches is difficult and mostly unsuccessful due to the wide subset of characteristics and functional properties of BREs and substrates.

5.4.1 Physical immobilization

The physical immobilization was accomplished by attaching the molecules on surface via intermolecular forces, mainly ionic bonds, hydrophobic, and polar interactions. The physical interaction depends on the particular BRE and surface. The attached BRE adlayer is likely to be heterogeneous and randomly oriented, since each molecule can form many contacts in different orientations for minimizing surface energy. In addition, the adsorption capacity of physical immobilization is limited by the geometric size of the BREs. High-density packing may sterically block active sites of BREs, interfering with recognition interactions.

Although physical immobilization provides a fast and easy approach for sensor development, the shortcomings are random orientation and weak attachment. The adsorbed BREs can be removed by washing with buffers or detergents when performing the assays. Moreover, problems relating to mass transport effects and high background signals emanating from nonspecific interactions can result in false detection.

5.4.2 Amine chemistry

Amine groups on BREs are the most commonly used as anchoring points. Fig. 5.5 illustrates the use of amine chemistry, in which carboxylic acid on surfaces is activated by NHS to react with nucleophilic amine to form a stable amide bond.

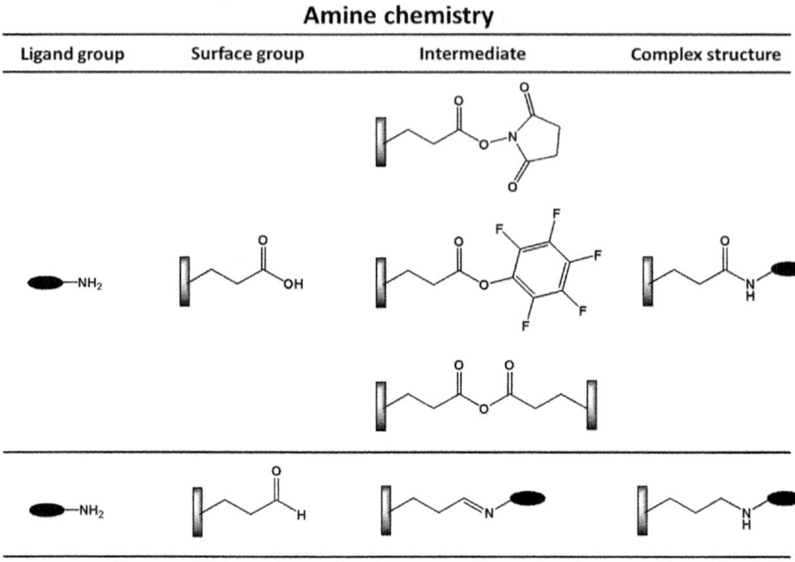

Figure 5.5 Amine chemistry on NHS-, pentafluorophenol-, and anhydride-derived surfaces and andehyde-derivatized surfaces.

The immobilization efficiency strongly depends on experimental conditions, such as pH, concentration, ionic strength, and reaction time. Alternatively, in Fig. 5.5, pentafluorophenol increases the reactivity of the activated ester on the surface by approximately an order of magnitude [86]. Moreover, broadly applicable method is the formation of amide linkages via an interchain anhydride intermediate [87]. In this method, carboxylic acids are dehydrated with trifluoroacetic anhydride to yield an interchain anhydride. Exposure of this activated surface to amines generates amide bonds. The other amine chemistry relies on the interaction between amine and aldehyde groups, leading to the formation of a labile Schiff's base. The linkage can be further stabilized by reduction to form a stable secondary linkage [88]. Aldehyde-amino chemistry has been extensively used for protein immobilization on different surfaces.

5.4.3 Thiol chemistry

Thiol groups are abundant in the protein structure, such as cysteine, which can be used for ligand immobilization. The maleimide group undergoes an addition reaction with thiol groups to form stable thioether bonds in Fig. 5.6 [89]. The reaction is suitable at a pH range of 6.5–7.5, while at higher pH values some cross-reactivity with amine has been observed [90]. Maleimide reacts rapidly, but it undergoes slow hydrolysis at aqueous conditions, which leads to problems in long-term operation.

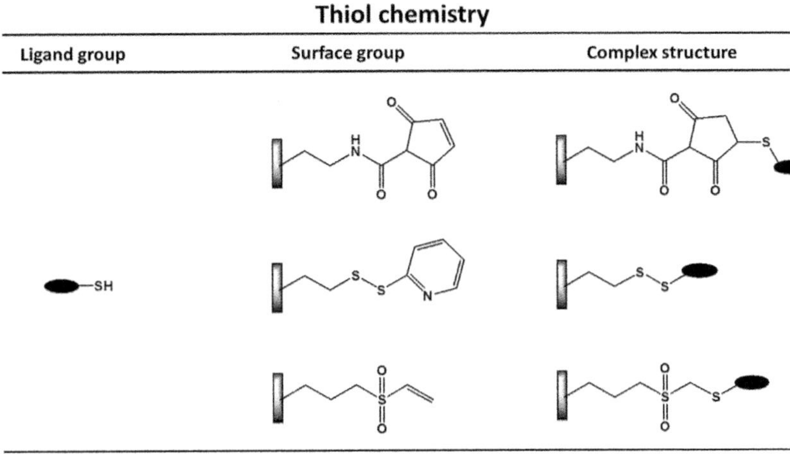

Figure 5.6 Thiol chemistry on maleimide-, disulfide-, and vinyl sulfone-derived surfaces.

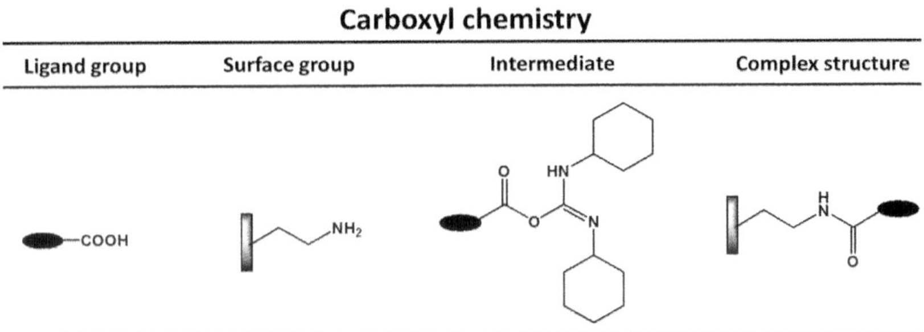

Figure 5.7 Carboxyl chemistry using carbodiimide activation.

Disulfide reagents participate in disulfide exchange reactions with another thiol, resulting in the formation of a new mixed disulfide [91]. Reversibility of the linkage by exposure to reducing reagents can become a problem in attempting stable immobilization. Pyridyil disulfides are commonly used since they contain a leaving group that is easily transformed into a nonreactive compound. Vinyl sulfone reacts with thiol groups by a Michael addition reaction. Thiol selectivity and water stability are the main advantages of this reagent, which has been frequently used for pegylation of proteins using vinyl sulfone PEG derivatives [92].

5.4.4 Carboxyl chemistry

Immobilization via carboxyl groups on ligands provokes a good opportunity for surface modification. An effective strategy to covalently attach ligands via carboxyl groups using carbodiimide for activation of carboxyl group is shown in Fig. 5.7 [93]. The ligands were attracted to the surface electrostatically and then transformed to covalent attachment. The low-pK aminated surfaces allow covalent immobilization using low concentrations of carbodiimide. However, a high concentration of carbodiimide dramatically decreases enzymatic activity [94].

5.4.5 Epoxy chemistry

Epoxy chemistry provides easy protocols for ligand immobilization due to its stability at neutral pH values, wet conditions, and reactivity with several nucleophilic groups, such as amine, hydroxyl, and thiol groups, to form strong bonds (Fig. 5.8). However, the covalent reactions between ligands and epoxy surfaces are extremely slow. Previous works developed a two-step protocol to firstly attract ligands to the surface by hydrophobic interaction to enhance surface concentration of ligands, and then react effectively with epoxy groups [95]. Commercially available epoxy surface reveals several experimental limitations. Epoxy-agarose shows negligible immobilization both at high and at low ionic strength conditions, due to the lack of a hydrophobic core for the adsorption step.

5.4.6 Click chemistry

Click chemistry is the 1,3-dipolar cycloaddition of an azide and alkyne to form 1,2,3-triazole, which has been applied for a wide range of applications due to its simple workup and purification steps, rapidly creating new products (Fig. 5.9) [96,97]. Since the introduction of click chemistry into macromolecules and surface chemistry, it is expected to provide a useful strategy for uniform, high-density surface immobilization of BREs in a covalent selective fashion. Click chemistry holds excellent stability because triazole formation is irreversible and quantitative. In situ preparation of azide-terminated monolayers has been performed for creating an ideal surface for BRE attachment, meeting the ordered structure of SAM with the

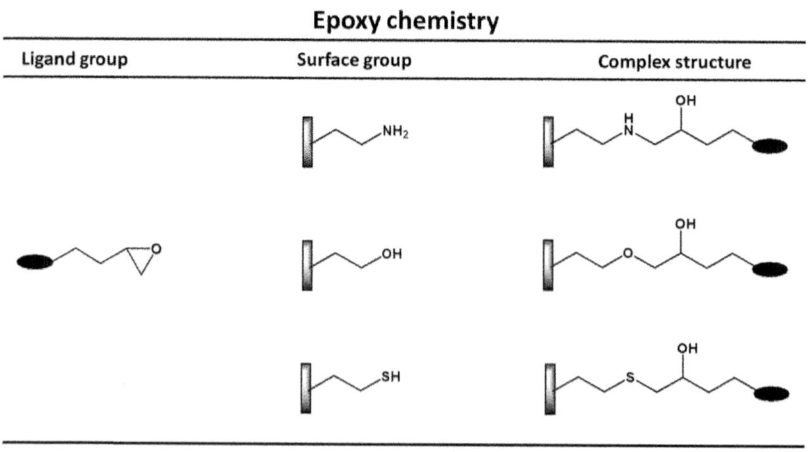

Figure 5.8 Epoxy chemistry on amine-, hydroxyl-, and thiol-derived surfaces.

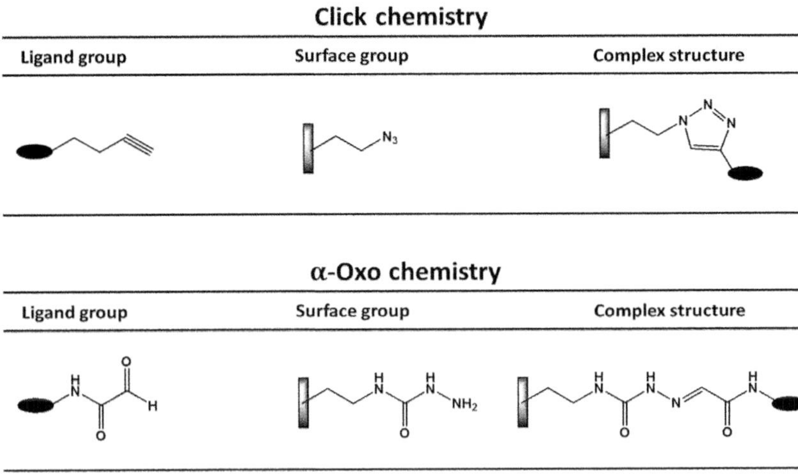

Figure 5.9 Click chemistry with alkyne-terminated molecule onto an azide-functionalized surface. α-Oxo semicarbazone immobilization chemistry.

specificity of click chemistry. Therefore a various series of (bio)molecules, including PEG, lactose, DNA, biotin, and recombinant proteins, have been stably immobilized without formation of side products.

5.4.7 α-Oxo semicarbazone chemistry

α-Oxo semicarbazone chemistry is the condensation reaction between α-oxo aldehyde-functionalized ligands and a semicarbazone-functionalized surface (Fig. 5.9). Because it applies synthetic glyoxylic acid moiety rather than native residue functional groups in the natural products, and the formed linkage is resistant against hydrolysis, this chemistry is extremely versatile and has been used to various bioconjugations [98]. The reaction was performed in acetate buffer at pH 5.5 at room temperature. The method has proven to be applicable to conjugation of antibody [99], peptide [100], and biotin [101].

5.4.8 Bioaffinity conjugation

The mostly used bioaffinity immobilization is based on avidin-biotin, which interact strongly and noncovalently ($Kd = 10^{13}-10^{15}\,M^{-1}$) [102]. This bond is stable enough in harsh conditions and this specific interaction permits well-oriented biomolecule immobilization. Avidin is a tetrameric glycoprotein soluble in aqueous solutions and stable over a wide range of pH values and temperatures. It can bind up to four molecules of biotins. Streptavidin is a closely related tetrameric protein with similar affinity to biotin. Biotin is a natural vitamin found in all living cells. Since biotin is a small molecule, its conjugation to biomolecules does not affect their conformation, size, or functionality.

Other bioaffinity immobilization methods are suitable for antibodies based on specific interaction of protein A/protein G with the Fc region of IgG molecules. The advantage of this method is that the binding sites of antibody, located on the Fab variable region, remain well accessible [103]. However, the disadvantage of the protein A-mediated immobilization is that only certain classes of antibodies are able to bind protein A. The protein G shows a broad binding activity for immunoglobulins of different species and subclasses.

5.5 Antifouling materials

To date, good nonfouling surfaces have been prepared when the following conditions were fulfilled: (1) zero net charge, (2) hydrophilic nature, and (3) presence of groups that are hydrogen acceptors instead of hydrogen bound donors [104–106]. Though the last one remains controversial, the molecular design for antifouling materials has been advancing rapidly to meet more and more complicate applications.

5.5.1 Poly(ethylene glycol) antifouling materials

The alkanethiols terminated with tri- or hexa-(ethylene glycol) groups are the most popular materials used to resist the nonspecific adsorption of other biomolecules for biosensor applications. Poly(ethylene oxide) or PEG has the same repeating unit ($-CH_2CH_2O-$). PEG has a structure with desirable properties, such as hydrophilicity and biocompatibility. It has been intensively used in surface modification for improving nonspecific adsorption [107,108]. Furthermore, PEG is a well-known antifouling material due to its weakly basic ether linkages and its low value of polymer–water interfacial energy (<5 mJ/m^2) [109]. In addition, the terminal ethylene glycol end-group adopts either a helical conformation aligned perpendicular to the surface or an amorphous conformation [110,111]. Therefore PEG possesses a balloon conformation that increases solution viscosity and acts as a water barrier preventing proteins from approaching the surface. However, it was found that PEG-coated nanoparticles tend to aggregate under high salt conditions and high temperature [12,112,113]. In addition, other concerns of OEG materials, such as ease of oxidation, urge for development of alternatives [112–117].

5.5.2 Zwitterionic antifouling materials

The other materials of interest for nonfouling are zwitterionic materials comprising of poly(phosphocholine), poly(sulfobetaine), and poly(carboxybetaine) polymers [118,119]. Dating back to 1980, the phosphocholine headgroup was shown to be nonothromobogenic, and several groups polymerized zwitterionic phosphatidylcholine analogues to create stabilized membranes [120]. Zwitterions are polyelectrolytes that contain both positively and negatively charged groups, and the overall charges are neutral. Zwitterionic materials afford the antifouling properties due to strong ionic structuring of water, creating a highly hydrophilic surface. There is no unusual structuring of water and water itself does not form a physical barrier. There is an energy barrier. Water associated with surface polymers becomes part of the excluded volume portion of the steric mechanism. Water that is not attached is part of the osmotic pressure component of the mechanism [117]. In addition, as a hydrated zwitterionic layer, the steric repulsive forces on compressing a polymer chain such as PEG grafted on a surface, contribute to the resistance of a polymer chain or dense surface layer to invasion by foulants. Therefore the excellent antifouling properties of zwitterionic materials have been demonstrated for various biomedical applications.

5.6 Outlook

In the era of Internet of Things technologies, sensors have been employed for detection and gathering information from users, samples, and the environment. The development of materials for the immobilization of sensing elements onto sensor surfaces has been advanced over the last two decades. Effective, facile, and reliable approaches have been driven by the availability of better surface deposition techniques and material chemistries for the commercialization of sensitive and robust systems and better tools for attachment of BREs. The role of biointerfaces, therefore, can determine the overall performance of a sensor in real-world application. The trend of developing new architectural surface chemistry will continue to fulfill new demands. In the methods for BRE immobilization, orientation and site specificity of attached BREs remain a critical issue for high-throughput and sensitive detection. The optimal accessibility of target molecules to sensors becomes the first obstacle in the detection. Also, in the field of nanomaterials for sensing, good stability and effectiveness of coatings are highly desirable to fully realize their potential.

Studies with zwitterionic materials to protect surfaces are expected to grow for better understanding and wide applications. The structural diversity (the choice and spatial arrangement of charged groups) brings functional versatility to zwitterionic materials beyond fundamental studies and biological activities. This functional aspect will be further emphasized by the ionic nature of electrolyte materials, which enables the adjustment of materials charge density, pH sensitivity, counterion association, etc., thus distinguishing this type of materials from other nonionic ones. The coating technologies with zwitterionic materials will be not only limited to antifouling properties but also intelligent responsive properties to be applied for sensing applications.

References

[1] A.P.F. Turner, Biochemistry—biosensors sense and sensitivity, Science 290 (5495) (2000) 1315–1317.

[2] S.H. North, E.H. Lock, C.R. Taitt, S.G. Walton, Critical aspects of biointerface design and their impact on biosensor development, Anal. Bioanal. Chem. 397 (3) (2010) 925–933.

[3] D. Samanta, A. Sarkar, Immobilization of biomacromolecules on self-assembled monolayers: methods and sensor applications, Chem. Soc. Rev. 40 (5) (2011) 2567–2592.
[4] J.C. Love, L.A. Estroff, J.K. Kriebel, R.G. Nuzzo, G.M. Whitesides, Self-assembled monolayers of thiolates on metals as a form of nanotechnology, Chem. Rev. 105 (4) (2005) 1103–1169.
[5] R.G. Nuzzo, L.H. Dubois, D.L. Allara, Fundamental-studies of microscopic wetting on organic-surfaces. 1. Formation and structural characterization of a self-consistent series of polyfunctional organic monolayers, J. Am. Chem. Soc. 112 (2) (1990) 558–569.
[6] H. Kondoh, C. Kodama, H. Nozoye, Structure-dependent change of desorption species from *n*-alkanethiol monolayers adsorbed on Au(111): desorption of thiolate radicals from low-density structures, J. Phys. Chem. B 102 (13) (1998) 2310–2312.
[7] N. Karsi, P. Lang, M. Chehimi, M. Delamar, G. Horowitz, Modification of indium tin oxide films by alkanethiol and fatty acid self-assembled monolayers: a comparative study, Langmuir 22 (7) (2006) 3118–3124.
[8] S.Y. Lin, T.K. Tsai, C.M. Lin, C.H. Chen, Y.C. Chan, H.W. Chen, Structures of self-assembled monolayers of *n*-alkanoic acids on gold surfaces modified by underpotential deposition of silver and copper: odd-even effect, Langmuir 18 (14) (2002) 5473–5478.
[9] C. Haensch, S. Hoeppener, U.S. Schubert, Chemical modification of self-assembled silane based monolayers by surface reactions, Chem. Soc. Rev. 39 (6) (2010) 2323–2334.
[10] S.B. Yeh, C.S. Chen, W.Y. Chen, C.J. Huang, Modification of silicone elastomer with zwitterionic silane for durable antifouling properties, Langmuir 30 (38) (2014) 11386–11393.
[11] F. Schreiber, Structure and growth of self-assembling monolayers, Prog. Surf. Sci. 65 (5–8) (2000) 151–256.
[12] C.J. Huang, S.H. Chu, L.C. Wang, C.H. Li, T.R. Lee, Bioinspired zwitterionic surface coatings with robust photostability and fouling resistance, ACS Appl. Mater. Interfaces 7 (42) (2015) 23776–23786.
[13] M. Rodenstein, S. Zurcher, S.G.P. Tosatti, N.D. Spencer, Fabricating chemical gradients on oxide surfaces by means of fluorinated, catechol-based, self-assembled monolayers, Langmuir 26 (21) (2010) 16211–16220.
[14] E. Amstad, T. Gillich, I. Bilecka, M. Textor, E. Reimhult, Ultrastable iron oxide nanoparticle colloidal suspensions using dispersants with catechol-derived anchor groups, Nano. Lett. 9 (12) (2009) 4042–4048.
[15] C.J. Huang, L.C. Wang, J.J. Shyue, Y.C. Chang, Developing antifouling biointerfaces based on bioinspired zwitterionic dopamine through pH-modulated assembly, Langmuir 30 (42) (2014) 12638–12646.
[16] C.J. Huang, L.C. Wang, Bio-inspired multifunctional catecholic assembly for photo-programmable biointerface, Colloid Surf. B Biointerfaces 134 (2015) 247–253.
[17] C.O. Timmons, W.A. Zisman, Investigation of fatty acid monolayers on metals by contact potential measurements, J. Phys. Chem. 69 (3) (1965) 984–990.
[18] O. Levine, W.A. Zisman, Physical properties of monolayers adsorbed at the solid-air interface. I. Friction and wettability of aliphatic polar compounds and effect of halogenation, J. Phys. Chem. 61 (8) (1957) 1068–1077.
[19] J. Sagiv, Organized monolayers by adsorption. 1. Formation and structure of oleophobic mixed monolayers on solid surfaces, J. Am. Chem. Soc. 102 (1) (1980) 92–98.
[20] S. Zurcher, D. Wackerlin, Y. Bethuel, B. Malisova, M. Textor, S. Tosatti, et al., Biomimetic surface modifications based on the cyanobacterial iron chelator anachelin, J. Am. Chem. Soc. 128 (4) (2006) 1064–1065.

[21] L. Yan, W.T.S. Huck, G.M. Whitesides, Self-assembled monolayers (SAMS) and synthesis of planar micro- and nanostructures, J. Macromol. Sci. Polym. Rev. C44 (2) (2004) 175–206.
[22] C.S. Chen, M. Mrksich, S. Huang, G.M. Whitesides, D.E. Ingber, Geometric control of cell life and death, Science 276 (5317) (1997) 1425–1428.
[23] I. Banerjee, R.C. Pangule, R.S. Kane, Antifouling coatings: recent developments in the design of surfaces that prevent fouling by proteins, bacteria, and marine organisms, Adv. Mater. 23 (6) (2011) 690–718.
[24] N.K. Chaki, K. Vijayamohanan, Self-assembled monolayers as a tunable platform for biosensor applications, Biosens. Bioelectron. 17 (1–2) (2002) 1–12.
[25] K.P. Garcia, K. Zarschler, L. Barbaro, J.A. Barreto, W. O'Malley, L. Spiccia, et al., Zwitterionic-coated "stealth" nanoparticles for biomedical applications: recent advances in countering biomolecular corona formation and uptake by the mononuclear phagocyte system, Small. 10 (13) (2014) 2516–2529.
[26] N. Islam, P.V. Gurgel, O.J. Rojas, R.G. Carbonell, Effects of composition of oligo(ethylene glycol)-based mixed monolayers on peptide grafting and human immunoglobulin detection, J. Phys. Chem. C 118 (10) (2014) 5361–5373.
[27] H. Wang, S.F. Chen, L.Y. Li, S.Y. Jiang, Improved method for the preparation of carboxylic acid and amine terminated self-assembled monolayers of alkanethiolates, Langmuir 21 (7) (2005) 2633–2636.
[28] C.J. Huang, J. Dostalek, A. Sessitsch, W. Knoll, Long-range surface plasmon-enhanced fluorescence spectroscopy biosensor for ultrasensitive detection of *E. coli* O157:H7, Anal. Chem. 83 (3) (2011) 674–677.
[29] B.J.R. Steffen Onclin, N. David, Reinhoudt engineering silicon oxide surfaces using self-assembled monolayers, Angewan. Chem. Int. Ed. 44 (39) (2005) 6282–6304.
[30] B. Veigas, E. Fortunato, P.V. Baptista, Field effect sensors for nucleic acid detection: recent advances and future perspectives, Sensors 15 (5) (2015) 10380–10398.
[31] W. Peng, S.M. Rupich, N. Shafiq, Y.N. Gartstein, A.V. Malko, Y.J. Chabal, Silicon surface modification and characterization for emergent photovoltaic applications based on energy transfer, Chem. Rev. 115 (23) (2015) 12764–12796.
[32] A.J. Nijdam, M. Ming-Cheng Cheng, D.H. Geho, R. Fedele, P. Herrmann, K. Killian, et al., Physicochemically modified silicon as a substrate for protein microarrays, Biomaterials. 28 (3) (2007) 550–558.
[33] L.T. Zhuravlev, Concentration of hydroxyl groups on the surface of amorphous silicas, Langmuir 3 (3) (1987) 316–318.
[34] M.C.G. Noémi Rozlosnik, B. Niels, Larsen effect of solvents and concentration on the formation of a self-assembled monolayer of octadecylsiloxane on silicon (001), Langmuir 19 (4) (2003) 1182–1188.
[35] Y. Liu, L.K. Wolf, M.C. Messmer, A study of alkyl chain conformational changes in self-assembled *n*-octadecyltrichlorosilane monolayers on fused silica surfaces, Langmuir 17 (14) (2001) 4329–4335.
[36] A.M. Salvi, R. Pucciariello, M.R. Guascito, V. Villani, L. Intermite, Characterization of the interface in rubber/silica composite materials, Surface Interface Anal. 33 (10–11) (2002) 850–861.
[37] C. Carraro, O.W. Yauw, M.M. Sung, R. Maboudian, Observation of three growth mechanisms in self-assembled monolayers, J. Phys. Chem. B 102 (23) (1998) 4441–4445.
[38] T. Vallant, H. Brunner, U. Mayer, H. Hoffmann, Formation of self-assembled octadecylsiloxane monolayers on mica and silicon surfaces studied by atomic force microscopy and infrared spectroscopy, J. Phys. Chem. B 102 (37) (1998) 7190–7197.

[39] A. Glaser, J. Foisner, H. Hoffmann, G. Friedbacher, Investigation of the role of the interplay between water and temperature on the growth of alkylsiloxane submonolayers on silicon, Langmuir 20 (13) (2004) 2299–5604.

[40] M.R. Linford, P. Fenter, P.M. Eisenberger, C.E.D. Chidsey, Alkyl monolayers on silicon prepared from 1-alkenes and hydrogen-terminated silicon, J. Am. Chem. Soc. 117 (11) (1995) 3145–3155.

[41] B.J. Eves, Q.-Y. Sun, G.P. Lopinski, H. Zuilhof, Photochemical attachment of organic monolayers onto H-terminated Si(111), J. Am. Chem. Soc. 126 (44) (2004) 14318–14319.

[42] L. Scheres, M. Giesbers, H. Zuilhof, Self-assembly of organic monolayers onto hydrogen-terminated silicon: 1-alkynes are better than 1-alkenes, Langmuir 26 (13) (2010) 10924–10929.

[43] Y. Li, S. Calder, O. Yaffe, D. Cahen, H. Haick, L. Kronik, et al., Hybrids of organic molecules and flat, oxide-free silicon: high-density monolayers, electronic properties, and functionalization, Langmuir 28 (26) (2012) 9920–9929.

[44] A.A. Luc Scheres, H. Zuilhof, Self-assembly of high-quality covalently bound organic monolayers onto silicon, Langmuir 23 (16) (2007) 8343–8346.

[45] M.J. Llorente, B.H. Nguyen, C.P. Kubiak, K.D. Moeller, Paired electrolysis in the simultaneous production of synthetic intermediates and substrates, J. Am. Chem. Soc. 138 (46) (2016) 15110–15113.

[46] Y.L. Lyubchenko, L.S. Shlyakhtenko, M. Binus, C. Gaillard, F. Strauss, Visualization of hemiknot DNA structure with an atomic force microscope, Nucl. Acids Res. 30 (22) (2002) 4902–4909.

[47] J.W. Pavlicek, E.A. Oussatcheva, R.R. Sinden, V.N. Potaman, O.F. Sankey, Y.L. Lyubchenko, Supercoiling-induced DNA bending, Biochemistry 43 (33) (2004) 10664–10668.

[48] Y.-T. Tseng, H.-Y. Lu, J.-R. Li, W.-J. Tun, W.-H. Chen, L.-K. Chau, Facile functionalization of polymer surfaces in aqueous and polar organic solvents via 3-mercaptopropylsilatrane, ACS Appl. Mater. Interfaces 8 (49) (2016) 34159–34169.

[49] W.-H. Chen, Y.-T. Tseng, S. Hsieh, W.-C. Liu, C.-W. Hsieh, C.-W. Wu, et al., Silanization of solid surfaces via mercaptopropylsilatrane: a new approach of constructing gold colloid monolayers, RSC Adv. 4 (2014) 46527–46535.

[50] S. Hsieh, W.-J. Chao, C.-W. Hsieh, Improved performance of aminopropylsilatrane over aminopropyltriethoxysilane as an adhesive film for anchoring gold nanoparticles on silicon surfaces, J. Nanosci. Nanotechnol. 9 (5) (2009) 2894–2901.

[51] J.K. Puri, R. Singh, V.K. Chahal, Silatranes: a review on their synthesis, structure, reactivity and applications, Chem. Soc. Rev. 40 (3) (2011) 1791–1840.

[52] H. Lee, S.M. Dellatore, W.M. Miller, P.B. Messersmith, Mussel-inspired surface chemistry for multifunctional coatings, Science 318 (5849) (2007) 426–430.

[53] E. Faure, C. Falentin-Daudre, C. Jerome, J. Lyskawa, D. Fournier, P. Woisel, et al., Catechols as versatile platforms in polymer chemistry, Prog. Polym. Sci. 38 (1) (2013) 236–270.

[54] C.L. Gao, G.Z. Li, H. Xue, W. Yang, F.B. Zhang, S.Y. Jiang, Functionalizable and ultra-low fouling zwitterionic surfaces via adhesive mussel mimetic linkages, Biomaterials. 31 (7) (2010) 1486–1492.

[55] G.Z. Li, H. Xue, G. Cheng, S.F. Chen, F.B. Zhang, S.Y. Jiang, Ultralow fouling zwitterionic polymers grafted from surfaces covered with an initiator via an adhesive mussel mimetic linkage, J. Phys. Chem. B 112 (48) (2008) 15269–15274.

[56] Y.H. Zhu, H.S. Sundaram, S.J. Liu, L. Zhang, X.W. Xu, Q.M. Yu, et al., A robust graft-to strategy to form multifunctional and stealth zwitterionic polymer-coated mesoporous silica nanoparticles, Biomacromolecules 15 (5) (2014) 1845–1851.
[57] H. Wei, N. Insin, J. Lee, H.S. Han, J.M. Cordero, W.H. Liu, et al., Compact zwitterion-coated iron oxide nanoparticles for biological applications, Nano. Lett. 12 (1) (2012) 22–25.
[58] Z.J. Zhou, L.R. Wang, X.Q. Chi, J.F. Bao, L.J. Yang, W.X. Zhao, et al., Engineered iron-oxide-based nanoparticles as enhanced T-1 contrast agents for efficient tumor imaging, ACS Nano 7 (4) (2013) 3287–3296.
[59] H. Wei, O.T. Bruns, O. Chen, M.G. Bawendi, Compact zwitterion-coated iron oxide nanoparticles for in vitro and in vivo imaging, Integrat. Biol. 5 (1) (2013) 108–114.
[60] B.V. Slaughter, S.S. Khurshid, O.Z. Fisher, A. Khademhosseini, N.A. Peppas, Hydrogels in regenerative medicine, Adv. Mater. 21 (32–33) (2009) 3307–3329.
[61] M.A.C. Stuart, W.T.S. Huck, J. Genzer, M. Muller, C. Ober, M. Stamm, et al., Emerging applications of stimuli-responsive polymer materials, Nat. Mater. 9 (2) (2010) 101–113.
[62] S.M. Murphy, C.J. Hamilton, M.L. Davies, B.J. Tighe, Polymer membranes in clinical sensor applications. 2. The design and fabrication of permselective hydrogels for electrochemical devices, Biomaterials. 13 (14) (1992) 979–990.
[63] D. Buenger, F. Topuz, J. Groll, Hydrogels in sensing applications, Prog. Polym. Sci. 37 (12) (2012) 1678–1719.
[64] Y. Wang, C.J. Huang, U. Jonas, T.X. Wei, J. Dostalek, W. Knoll, Biosensor based on hydrogel optical waveguide spectroscopy, Biosens. Bioelectron. 25 (7) (2010) 1663–1668.
[65] C.J. Huang, J. Dostalek, W. Knoll, Long range surface plasmon and hydrogel optical waveguide field-enhanced fluorescence biosensor with 3D hydrogel binding matrix: on the role of diffusion mass transfer, Biosens. Bioelectron. 26 (4) (2010) 1425–1431.
[66] O. Andersson, A. Larsson, T. Ekblad, B. Liedberg, Gradient hydrogel matrix for microarray and biosensor applications: an imaging SPR study, Biomacromolecules 10 (1) (2009) 142–148.
[67] N. Yang, X.D. Su, V. Tjong, W. Knoll, Evaluation of two- and three-dimensional streptavidin binding platforms for surface plasmon resonance spectroscopy studies of DNA hybridization and protein-DNA binding, Biosens. Bioelectron. 22 (11) (2007) 2700–2706.
[68] N. Li, Z.Z. Liu, S.G. Xu, Dynamically formed poly (vinyl alcohol) ultrafiltration membranes with good anti-fouling characteristics, J. Membr. Sci. 169 (1) (2000) 17–28.
[69] M. Guvendiren, S. Yang, J.A. Burdick, Swelling-induced surface patterns in hydrogels with gradient crosslinking density, Adv. Funct. Mater. 19 (19) (2009) 3038–3045.
[70] I. Tokarev, I. Tokareva, S. Minko, Gold-nanoparticle-enhanced plasmonic effects in a responsive polymer gel, Adv. Mater. 20 (14) (2008) 2730.
[71] D. Puchberger-Enengl, C. Krutzler, F. Keplinger, M.J. Vellekoop, Single-step design of hydrogel-based microfluidic assays for rapid diagnostics, Lab. Chip. 14 (2) (2014) 378–383.
[72] H.J. van der Linden, S. Herber, W. Olthuis, P. Bergveld, Stimulus-sensitive hydrogels and their applications in chemical (micro)analysis, Analyst 128 (4) (2003) 325–331.
[73] A.J. Marshall, J. Blyth, C.A.B. Davidson, C.R. Lowe, PH-sensitive holographic sensors, Anal. Chem. 75 (17) (2003) 4423–4431.

[74] A. Richter, A. Bund, M. Keller, K.F. Arndt, Characterization of a microgravimetric sensor based on pH sensitive hydrogels, Sens. Actuator B Chem. 99 (2–3) (2004) 579–585.

[75] D. Kuckling, J. Hoffmann, M. Plotner, D. Ferse, K. Kretschmer, H.J.P. Adler, et al., Photo cross-linkable poly(*N*-isopropylacrylamide) copolymers III: micro-fabricated temperature responsive hydrogels, Polymer (Guildf). 44 (16) (2003) 4455–4462.

[76] C.E. Reese, M.E. Baltusavich, J.P. Keim, S.A. Asher, Development of an intelligent polymerized crystalline colloidal array colorimetric reagent, Anal. Chem. 73 (21) (2001) 5038–5042.

[77] G. Gerlach, M. Guenther, G. Suchaneck, J. Sorber, K.F. Arndt, A. Richter, Application of sensitive hydrogels in chemical and pH sensors, Macromol. Symp. 210 (2004) 403–410.

[78] S.A. Asher, S.F. Peteu, C.E. Reese, M.X. Lin, D. Finegold, Polymerized crystalline colloidal array chemical-sensing materials for detection of lead in body fluids, Anal. Bioanal. Chem. 373 (7) (2002) 632–638.

[79] Y.F. Zhang, H.F. Ji, G.M. Brown, T. Thundat, Detection of CrO_4^{2-} using a hydrogel swelling microcantilever sensor, Anal. Chem. 75 (18) (2003) 4773–4777.

[80] B.K. Oh, M.E. Robbins, M.H. Schoenfisch, Planar nitric oxide (NO)-selective ultramicroelectrode sensor for measuring localized NO surface concentrations at xerogel microarrays, Analyst 131 (1) (2006) 48–54.

[81] E. Jang, W.G. Koh, Multiplexed enzyme-based bioassay within microfluidic devices using shape-coded hydrogel microparticles, Sens. Actuator B Chem. 143 (2) (2010) 681–688.

[82] A. Vagias, K. Sergelen, K. Koynov, P. Kosovan, J. Dostalek, U. Jonas, et al., Diffusion and permeation of labeled IgG in grafted hydrogels, Macromolecules 50 (12) (2017) 4770–4779.

[83] A. Aulasevich, R.F. Roskamp, U. Jonas, B. Menges, J. Dostalek, W. Knoll, Optical waveguide spectroscopy for the investigation of protein-functionalized hydrogel films, Macromol. Rapid Commun. 30 (9–10) (2009) 872–877.

[84] M. Toma, U. Jonas, A. Mateescu, W. Knoll, J. Dostalek, Active control of SPR by thermoresponsive hydrogels for biosensor applications, J. Phys. Chem. C 117 (22) (2013) 11705–11712.

[85] Q.W. Zhang, Y. Wang, A. Mateescu, K. Sergelen, A. Kibrom, U. Jonas, et al., Biosensor based on hydrogel optical waveguide spectroscopy for the detection of 17 beta-estradiol, Talanta 104 (2013) 149–154.

[86] J. Lahiri, L. Isaacs, J. Tien, G.M. Whitesides, A strategy for the generation of surfaces presenting ligands for studies of binding based on an active ester as a common reactive intermediate: a surface plasmon resonance study, Anal. Chem. 71 (4) (1999) 777–790.

[87] L. Yan, C. Marzolin, A. Terfort, G.M. Whitesides, Formation and reaction of interchain carboxylic anhydride groups on self-assembled monolayers on gold, Langmuir 13 (25) (1997) 6704–6712.

[88] L. Betancor, F. Lopez-Gallego, A. Hidalgo, N. Alonso-Morales, G. Mateo, R. Fernandez-Lafuente, et al., Different mechanisms of protein immobilization on glutaraldehyde activated supports: effect of support activation and immobilization conditions, Enzyme Microb. Technol. 39 (4) (2006) 877–882.

[89] V. Gauvreau, P. Chevallier, K. Vallieres, E. Petitclerc, R.C. Gaudreault, G. Laroche, Engineering surfaces for bioconjugation: developing strategies and quantifying the extent of the reactions, Bioconjugate Chem. 15 (5) (2004) 1146–1156.

[90] T. Viitala, I. Vikholm, J. Peltonen, Protein immobilization to a partially cross-linked organic monolayer, Langmuir 16 (11) (2000) 4953–4961.
[91] K. Kunath, T. Merdan, O. Hegener, H. Haberlein, T. Kissel, Integrin targeting using RGD-PEI conjugates for in vitro gene transfer, J. Gene Med. 5 (7) (2003) 588–599.
[92] S.C. Rizzi, J.A. Hubbell, Recombinant protein-co-PEG networks as cell-adhesive and proteolytically degradable hydrogel matrixes. Part 1: development and physicochemical characteristics, Biomacromolecules 6 (3) (2005) 1226–1238.
[93] R. Fernandezlafuente, C.M. Rosell, V. Rodriguez, C. Santana, G. Soler, A. Bastida, et al., Preparation of activated supports containing low pK amino-groups—a new tool for protein immobilization via the carboxyl coupling method, Enzyme Microb. Technol. 15 (7) (1993) 546–550.
[94] W. Tischer, F. Wedekind, Immobilized enzymes: methods and applications, Biocatal. Discov. Appl. 200 (1999) 95–126.
[95] C. Mateo, O. Abian, G. Fernandez-Lorente, J. Pedroche, R. Fernandez-Lafuente, J.M. Guisan, Epoxy sepabeads: a novel epoxy support for stabilization of industrial enzymes via very intense multipoint covalent attachment, Biotechnol. Prog. 18 (3) (2002) 629–634.
[96] J.E. Moses, A.D. Moorhouse, The growing applications of click chemistry, Chem. Soc. Rev. 36 (8) (2007) 1249–1262.
[97] J.C. Jewett, C.R. Bertozzi, Cu-free click cycloaddition reactions in chemical biology, Chem. Soc. Rev. 39 (4) (2010) 1272–1279.
[98] S. Ulrich, D. Boturyn, A. Marra, O. Renaudet, P. Dumy, Oxime ligation: a chemoselective click-type reaction for accessing multifunctional biomolecular constructs, Chem. Eur. J. 20 (1) (2014) 34–41.
[99] O. Melnyk, X. Duburcq, C. Olivier, F. Urbes, C. Auriault, H. Gras-Masse, Peptide arrays for highly sensitive and specific antibody-binding fluorescence assays, Bioconjugate Chem. 13 (4) (2002) 713–720.
[100] T. Doussineau, J.O. Durand, O. El-Mahdi, C. Maillet, O. Melnyk, C. Olivier, et al., Semicarbazide-functionalized silicate nanoparticles for peptide ligation, Eur. J. Inorg. Chem. 14 (2006) 2766–2772.
[101] C. Olivier, D. Hot, L. Huot, N. Ollivier, O. El-Mahdi, C. Gouyette, et al., Alpha-oxo semicarbazone peptide or oligodeoxynucleotide microarrays, Bioconjugate Chem. 14 (2) (2003) 430–439.
[102] R. Merkel, P. Nassoy, A. Leung, K. Ritchie, E. Evans, Energy landscapes of receptor-ligand bonds explored with dynamic force spectroscopy, Nature 397 (6714) (1999) 50–53.
[103] J. Turkova, Oriented immobilization of biologically active proteins as a tool for revealing protein interactions and function, J. Chromatogr. B 722 (1–2) (1999) 11–31.
[104] E. Ostuni, R.G. Chapman, R.E. Holmlin, S. Takayama, G.M. Whitesides, A survey of structure-property relationships of surfaces that resist the adsorption of protein, Langmuir 17 (18) (2001) 5605–5620.
[105] E. Ostuni, R.G. Chapman, M.N. Liang, G. Meluleni, G. Pier, D.E. Ingber, et al., Self-assembled monolayers that resist the adsorption of proteins and the adhesion of bacterial and mammalian cells, Langmuir 17 (20) (2001) 6336–6343.
[106] R.E. Holmlin, X.X. Chen, R.G. Chapman, S. Takayama, G.M. Whitesides, Zwitterionic SAMs that resist nonspecific adsorption of protein from aqueous buffer, Langmuir 17 (9) (2001) 2841–2850.

[107] G. Pasut, F.M. Veronese, State of the art in PEGylation: the great versatility achieved after forty years of research, J. Control Release 161 (2) (2012) 461–472.

[108] H. Otsuka, Y. Nagasaki, K. Kataoka, Self-assembly of poly(ethylene glycol)-based block copolymers for biomedical applications, Curr. Opin. Colloid Interface Sci. 6 (1) (2001) 3–10.

[109] S. Krishnan, C.J. Weinman, C.K. Ober, Advances in polymers for anti-biofouling surfaces, J. Mater. Chem. 18 (29) (2008) 3405–3413.

[110] R.L.C. Wang, H.J. Kreuzer, M. Grunze, Molecular conformation and solvation of oligo(ethylene glycol)-terminated self-assembled monolayers and their resistance to protein adsorption, J. Phys. Chem. B 101 (47) (1997) 9767–9773.

[111] P. Harder, M. Grunze, R. Dahint, G.M. Whitesides, P.E. Laibinis, Molecular conformation in oligo(ethylene glycol)-terminated self-assembled monolayers on gold and silver surfaces determines their ability to resist protein adsorption, J. Phys. Chem. B 102 (2) (1998) 426–436.

[112] H.S. Choi, W. Liu, P. Misra, E. Tanaka, J.P. Zimmer, B.I. Ipe, et al., Renal clearance of quantum dots, Nat. Biotechnol. 25 (10) (2007) 1165–1170.

[113] Z.G. Estephan, P.S. Schlenoff, J.B. Schlenoff, Zwitteration as an alternative to PEGylation, Langmuir 27 (11) (2011) 6794–6800.

[114] S. Han, C. Kim, D. Kwon, Thermal-degradation of poly(ethylene glycol), Polym. Degrad. Stabil. 47 (2) (1995) 203–208.

[115] S. Han, C. Kim, D. Kwon, Thermal/oxidative degradation and stabilization of polyethylene glycol, Polymer. (Guildf). 38 (2) (1997) 317–323.

[116] L.Y. Li, S.F. Chen, S.Y. Jiang, Protein interactions with oligo(ethylene glycol) (OEG) self-assembled monolayers: OEG stability, surface packing density and protein adsorption, J. Biomater. Sci. Polym. Ed. 18 (11) (2007) 1415–1427.

[117] J.B. Schlenoff, Zwitteration: coating surfaces with zwitterionic functionality to reduce nonspecific adsorption, Langmuir 30 (32) (2014) 9625–9636.

[118] R.R. Choudhury, J.M. Gohil, S. Mohanty, S.K. Nayak, Antifouling, fouling release and antimicrobial materials for surface modification of reverse osmosis and nanofiltration membranes, J. Mater. Chem. A 6 (2) (2018) 313–333.

[119] S.F. Chen, L.Y. Li, C. Zhao, J. Zheng, Surface hydration: principles and applications toward low-fouling/nonfouling biomaterials, Polymer. (Guildf). 51 (23) (2010) 5283–5293.

[120] R.F.A. Zwaal, P. Comfurius, L.L.M. Van Deenen, Membrane asymmetry and blood coagulation, Nature 268 (1977) 358–360.

Development of portable immunoassay device for future Internet of Things applications

6

Ryoji Kurita[1], *Osamu Niwa*[2] *and Yuko Ueno*[3]
[1]National Institute of Advanced Industrial Science and Technology, Tsukuba, Japan,
[2]Saitama Institute of Technology, Fukaya, Japan, [3]NTT Basic Research Laboratories, NTT Corporation, Atsugi, Japan

6.1 Introduction

Immunoassay has been utilized for various kinds of analysis including clinical, environmental, and food analysis. This is because immunological methods are highly selective since target-specific antibodies can be prepared. A low detection limit and high sensitivity can be achieved by utilizing a monoclonal antibody and a highly sensitive detection technology such as laser-induced fluorescence and chemiluminescence. However, highly sensitive immunoassays have mainly been employed in clinical laboratories using automatic analyzers. Immunoassay techniques such as enzyme-linked immunosorbent assay (ELISA) require multiple steps and reagents and so are complicated for unskilled users. In addition, high-performance analyzers are large and expensive. More recently, small analyzers have been developed that can be used for point-of-care testing (POCT) in small clinics. One typical application is the analysis of cardiac disease markers. Cardiac diseases such as myocardial infarction require rapid diagnosis, and this means that portable analyzers that operate rapidly are more important in these cases than for many other diseases [1].

In contrast, lateral flow immunoassay including immunochromatography is widely employed for such purposes as influenza diagnosis in small clinics, and as home pregnancy testing kits [2]. The immunochromatography detection limit has recently been improved by chemical amplification. One typical example is highly sensitive Ag amplification by employing a photographic development technology, which consists of depositing a large Ag particle on individual gold nanoparticles (AuNPs) [3].

Immunochromatography can be employed simply by introducing the sample into the device, and no extra reagents are required. However, immunochromatography is not quantitative and requires extra detection equipment to realize a quantitative improvement, but this increases its total size. Although the immunoassay technique

Chemical, Gas, and Biosensors for the Internet of Things and Related Applications.
DOI: https://doi.org/10.1016/B978-0-12-815409-0.00006-1

has many drawbacks when applied as a simple and inexpensive sensing technique for personal or in-home uses, it is highly advantageous compared with other biosensing methods. This is because it can deal with a very wide range of samples with high sensitivity and selectivity, which makes it possible to obtain a more accurate health check and provide preemptive medical treatment by working in combination with Internet of Things (IoT) technology. If we are to realize such a system, the following requirements must be met for future immunoassay methods with the IoT.

1. The entire device must be downsized while maintaining a very low detection limit.
2. The device must be inexpensive and simple (and label-free detection is desirable).
3. The process for one assay must involve one or only a few steps without any extra reagents.

In addition, a reduction in the total assay time is very important, and it will be achieved by reducing the sample and detection cell volumes thanks to the very short diffusion time of the analytes.

This chapter introduces some examples of portable or chip based immunosensing devices including a surface plasmon resonance (SPR) based portable immunosensor combined with microfluidic devices, a one-chip immunosensing device fabricated with a nanoimprint process, and a microfluidic immunosensor based on fluorescence detection with one measurement step, which could be improved for future IoT immunosensing devices.

6.2 Portable immunoassay system based on surface plasmon resonance for urinary immunoassay

It is very important to detect various biological components contained in body fluids such as blood and urine for the early detection and diagnosis of diseases, and also to determine therapeutic strategy [4,5]. Since the instruments used for medical analysis are generally large and expensive, the biomolecules that can be measured at the bedside or at home are currently limited. Therefore, biofluids are generally sent to an analytical institution where various biomolecular concentrations in the samples are measured by specialized technicians. So, it can take several days to obtain assay results.

In recent years, a clinical examination method called POCT has been attracting attention [6–8]. This is aimed at quickly obtaining an assay result on-site after collecting blood or urine using portable. By obtaining the results quickly on-site, it will be possible to make rapid disease diagnoses and treatment decisions, and we can expect the treatment outcome to improve and medical expenses to be reduced. A simple enzyme-modified electrochemical biosensor has already been commercialized for biomolecules that exist at relatively high concentrations in biofluids, for example, blood sugar [9]. With this sensor, it is possible to obtain a result within 1 minute of applying blood to the sensor chip. This sensor is considered to be a typical POCT device that allows the patient to measure and manage his/her own blood glucose level. The ability to perform measurements with a portable device makes it possible to use different diagnostic methods, so new markets are being developed. The DNA analysis market was formerly expected to become huge; however, it has become saturated. We think that this market saturation is because the development of the DNA analyzer has been mainly for use in comprehensive analysis for basic research, and a simple and inexpensive detection system that focuses on specific diseases is still under development. It is an important issue not only from a medical standpoint but also from an industrial and economic point of view, that is, that it is possible to detect disease markers even with a small POCT device.

We developed a portable surface plasmon resonance device (Handy-SPR) as a future POCT platform in collaboration with a Japanese company [10]. We also developed some microfluidic devices for the Handy-SPR, which detect disease markers in blood and urine samples. We realized the quick, easy, and on-site measurement of low concentration disease markers with the Handy-SPR as described below.

Fig. 6.1 shows a picture of the Handy-SPR device that we developed in collaboration with NTT Advanced Technology Corporation. A feature of the Handy-SPR is that it increases the precision of the incident angle, which is important when measuring the SPR phenomenon by using a point light source type LED to increase the focusing precision on the measurement surface, and the SPR phenomenon can be measured with high sensitivity. The Handy-SPR allows the real-time display of the SPR curve and SPR angle on a personal computer via a USB cable. The pixel number of the line CCD is plotted on the horizontal axis, the luminance obtained at each pixel is plotted on the vertical axis, and only the components below the threshold are extracted and fitted to calculate the angle of the minimum luminance, that is, the SPR angle. The Handy-SPR is 16 cm high, 6 cm deep, 9.5 cm high, and weighs about 770 g. Although the device achieves improved compactness and weight reduction for use in bedside measurement, a shift in the small SPR angle (around 10^{-3} degrees) can also be detected.

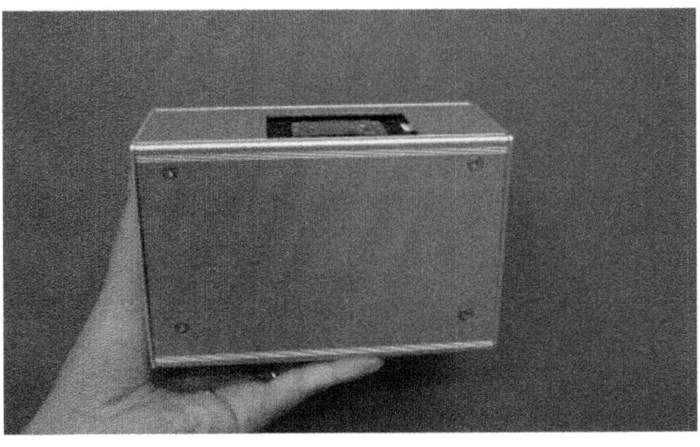

Figure 6.1 Photograph of our Handy-SPR device, which is 16 cm wide, 6 cm deep, 9.5 cm high, and weighs about 770 g [10].

For the detection of tumor markers, we focus on urinary samples that can be collected noninvasively. However, it is difficult to obtain accurate results because of the fluctuation in urine concentration due to hydration and sweating. The creatinine present in urine is measured simultaneously in urinary marker analysis to correct the urine concentration. However, the SPR method has a problem of poor sensitivity to small molecules such as creatinine (molecular weight = 113). To measure creatinine and a tumor marker (transferrin, molecular weight = 75 kDa) simultaneously with the SPR method, it is necessary to enhance the signal for creatinine. Therefore, we developed a microanalytical device for unlabeled, real-time urine analysis independent of molecular size using the SPR phenomenon as described below.

Fig. 6.2 shows a photograph and a schematic diagram of a microanalytical device designed for analyzing urine samples. The microdevice is composed of a glass substrate with thin gold films and a poly-dimethyl siloxane (PDMS) substrate with a flow channel 20 μm deep and 2 mm wide. Two circular patterns (2 mm diameter, 50 nm thick) are patterned with a photolithographic technique on the glass substrate for SPR measurement. One of the gold films was used for performing a transferrin measurement, and it was modified with a self-assembled monolayer (SAM) of decanethiol carboxylate and protein A with a carbodiimide coupling reaction. We performed a sandwich immunoassay by first introducing an antitransferrin antibody and then a mixture solution of target transferrin and a detection antibody into our device. The shift in the SPR angle caused by the immune reaction was monitored, and the transferrin concentration was estimated from the value of the SPR angle shift. On the other gold surface, we immobilized osmium bipyridine complex containing horseradish peroxidase [Os (bpy)-HRP]. Furthermore, three kinds of enzymes (creatininase, creatinase, and sarcosine oxidase) were immobilized upstream of the Os (bpy)-HRP region. The Os (bpy)-HRP was oxidized with hydrogen peroxide that was produced by sequentially reacting creatinine with the three enzymes, and the SPR angle decreased. The creatinine concentration was calculated by obtaining the change rate of the SPR angle [11].

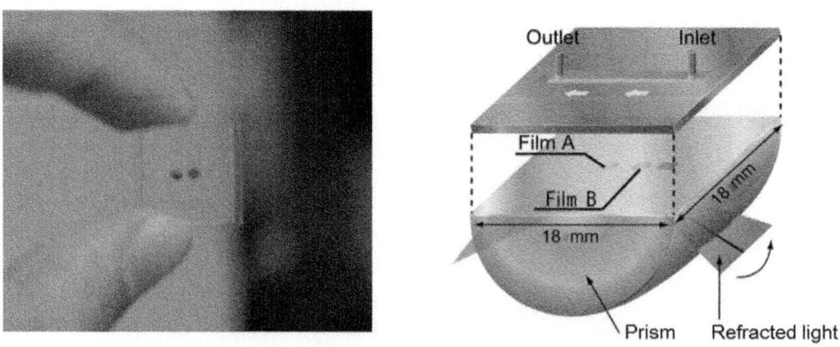

Figure 6.2 Photograph and schematic of a microanalytical device for urine samples. The device has two gold circular patterns (2 mm diameter, 50 nm thick) in a flow channel (20 μm deep and 2 mm wide) for SPR measurement [11].

Fig. 6.3 shows SPR sensorgrams when a mixed solution of target transferrin and creatinine is introduced into our device. After the introduction of the sample, the SPR angle gradually increased due to the antigen–antibody reaction on the transferrin detection surface and reached a saturation state in about 15 minutes. On the other hand, on the creatinine detection surface, the SPR angle sharply decreased due to oxidation of the osmium complex, and after 2 minutes an almost constant SPR angle was maintained.

Fig. 6.4 shows calibration curves for creatinine and transferrin. Since the concentration of creatinine normally present in urine is 2–30 mM, it can be used for correcting the concentration of transferrin in urine since the creatinine concentration can be sufficiently quantified even when diluted about 10 times. The detectable range in transferrin was 100 ng/mL–10 μg/mL and the limit of detection

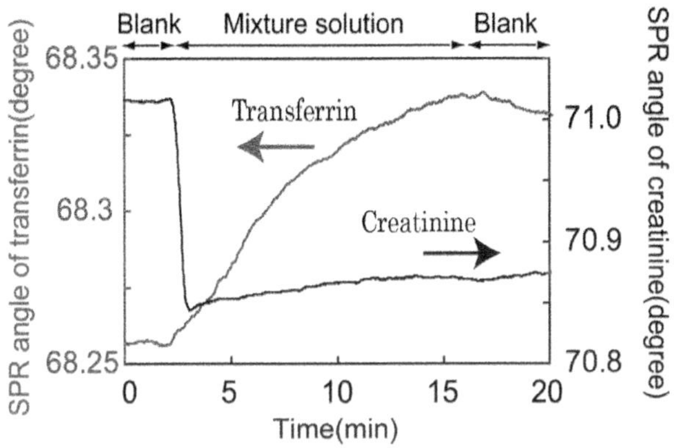

Figure 6.3 Variations in SPR angles when injecting a mixture solution of 10 μg/mL transferrin (left axis) and 1 mM creatinine (right axis).

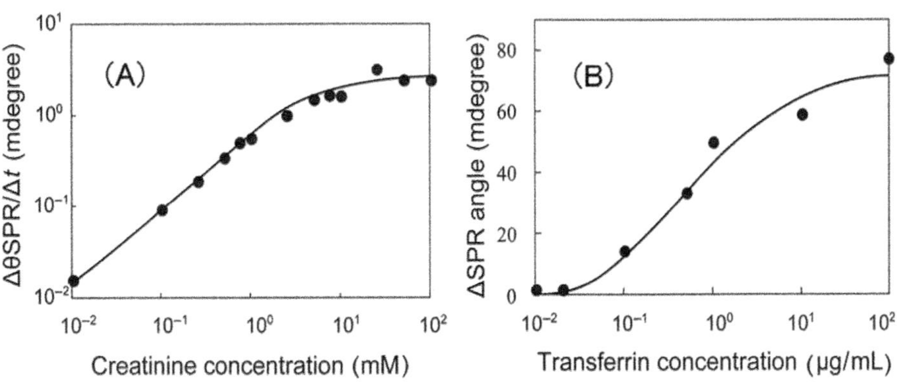

Figure 6.4 Calibration curves for (A) creatinine and (B) transferrin. The concentration range for creatinine and transferrin are from 10 μM to 100 mM and 10 ng/mL to 100 μg/mL, respectively.

was 20 ng/mL. The required concentration for a medical examination is around 800 ng/mL, and a satisfactory measurement can be performed with our device. The required sample volume is about 50 μL and the measurement is completed within 15 minutes, thus outperforming conventional ELISA. Two components with significantly different molecular sizes can be measured simultaneously, and a urinary immunoassay can be carried out more quickly and easily using our Handy-SPR equipment and microfluidic device [12].

6.3 One-chip immunosensing fabricated with nanoimprinting technique

6.3.1 Fabrication of local plasmon resonance devices with various processes

Miniaturized SPR sensing equipment has been developed for the on-site measurement of biomolecules, which is particularly important for POCT and personal healthcare. To realize IoT sensors, a simple and inexpensive device must be developed. However, the traditional SPR system is limited as regards miniaturization because it has an optical system that includes a prism, a lens, and a polarizer. The portable SPR system shown in Fig. 6.1 also employs a conventional Kretschmann configuration [10], but it is much smaller than the GE healthcare system. A highly miniaturized one-chip type SPR sensor called Spreeta was reported by Texas Instruments as shown in Fig. 6.5A [13]. This sensor consists of a transparent plastic block in which an LED light source and a linear CCD are embedded. One face of

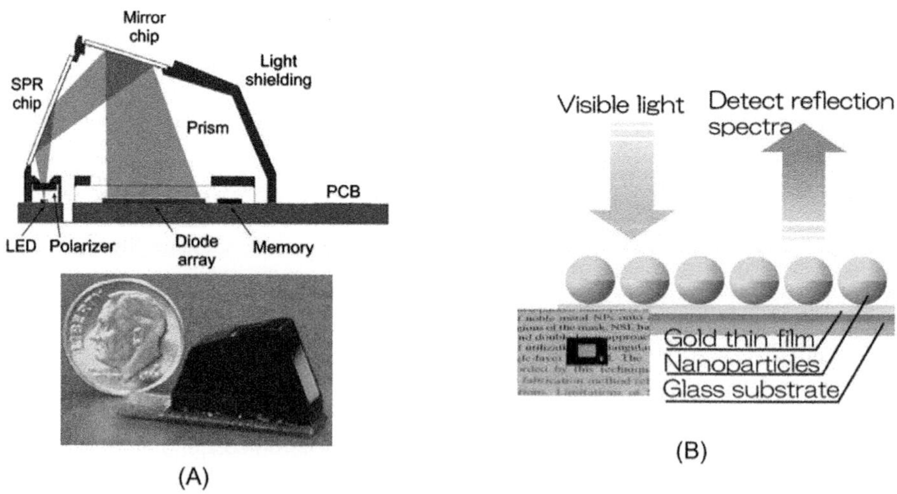

Figure 6.5 Miniaturization of SPR sensors. (A) One-chip SPR sensor Spreeta (Texas Instruments [13]). (B) Plasmonic device fabricated by arranging Au nanoparticles.

the block is coated with gold film, which can be used as the SPR sensing surface. However, the cost per chip will remain high since all the required devices such as the LED and CCD are disposable because they are embedded in a plastic block.

Instead of an SPR sensor with a Kretschmann configuration, plasmonic devices without any prism or lens for plasmon coupling and changing the incident angle have been demonstrated as shown in Fig. 6.5B. These kinds of devices are well known as local surface plasmon resonance (LSPR) sensors that employ nanostructured surfaces including nanoparticle [14] and nanorod [15] structures. Their group used transmitted light images and were able to image submonolayer quantities of alkanethiols.

However, unlike the conventional SPR sensor, for LSPR sensing devices the peak wavelength shift induced by surface refractive index change is measured. The key to realizing LSPR devices that provide highly sensitive and accurate measurements is to fabricate uniform nanostructures with a large area. Table 6.1 compares nanostructure fabrication processes for realizing LSPR sensing devices.

Electron beam (EB) lithography or the focused ion beam technique can realize nanostructured devices with excellent accuracy. However, the low throughput of the EB fabrication process greatly increases the cost, particularly when fabricating large area devices. LSPR sensors have been fabricated by arranging AuNPs on a substrate in a solution containing AuNPs. However, it is very difficult to arrange AuNPs in good order without defects. Moreover, it is particularly difficult to fabricate them with large surface areas. In addition, the arranged nanoparticles should be immobilized to suppress particle detachment. A more practical and inexpensive nanofabrication method is colloidal lithography, which employs colloids as an etching mask [16,17]. For example, Sannomiya et al. reported a short-range ordered hole array that was essentially defect free [18]. A long-range nanohole structure was also fabricated by changing the colloid size [19]. However, an etching process is needed to obtain a nanohole array structure, and the colloid size should be adjusted to control the nanohole size. Compared with the three techniques shown in Table 6.1, the nanoimprint technique has advantages for fabricating a nanostructured device. This process allows us to replicate a defect-free nanostructured device precisely from an original mold with a size of several tens of nanometers over a large area. By changing the mold design, a wide range of, for example, hole diameters, depths, and distances between each hole can be realized to obtain the desired optical parameters for LSPR sensors [20]. Imprint lithography achieved 25 nm resolution about 20 years ago [21]. The process used to make the device is

Table 6.1 Comparison of processes for fabricating LSPR devices

Method	Pattern accuracy	Throughput	Cost
EB lithography	Excellent	Poor	High
Arrangement in solution	Poor	Fair	Low
Colloidal mask	Poor[a]	Good	Low
Nanoimprint	Good	Good	Very low

[a]Particularly for large areas.

inexpensive once the master molding has been completed, although the master mold, which is commonly fabricated from metal or carbon, is rather expensive. However, molds can be used repeatedly, and the technique is suitable for biosensing devices thanks to the inexpensive fabrication process. For the above reasons, the nanoimprint process is suitable for fabricating biochemical sensors, which should generally be disposable and inexpensive. There have been various studies using nanohole array—based devices including LSPR sensors as described in a recent review [22].

6.3.2 Surface plasmon resonance biosensors fabricated by nanoimprint technique

Thermoplastic or UV-curable material has been employed to fabricate nanohole arrays for plasmonic devices. With thermoplastic polymers, the temperature should be raised above the glass transition temperature (Tg) prior to imprint and reduced below Tg to enable the mold to be detached without deformation [22]. In contrast, the nanoimprint process using UV-curable resin can be performed without elevating the temperature thus eliminating any thermal size change. Nuzzo and Roger's group fabricated periodical patterns consisting of submicron order nanohole arrays by using a soft nanoimprint technique that employed a flexible PDMS mold and UV-curable polyurethane resin [23,24]. They employed transmission spectra for measuring the peak wavelength and obtained an image of the alkanethiol submonolayer. However, the diameter and depth of the nanohole are both relatively large. It will be difficult to fabricate a PDMS mold with a smaller structure than hard molds such as nickel and GC. Yu et al. fabricated a 70 nm sputter-deposited polycarbonate substrate and employed a thermal imprint process to fabricate a submicrometer grating, which they used it for cysteine detection [25]. A Au nanodisk array was also fabricated with thermal nanoimprint lithography using a Si mold and applied to an enzyme immunoassay to detect enzymatic reaction products adsorbed on the device surface [26]. In contrast, a UV imprint process was employed to fabricate POCT devices for detecting prostate-specific antigen (PSA) [27]. Fig. 6.6 describes the process for fabricating a simple plasmonic nanohole array device that can be employed to measure an immune reaction using UV nanoimprint as reported by the authors [28]. Glassy carbon molds with six different patterns were employed because a GC mold is less expensive than a nickel mold but more robust than a PDMS or Si mold. Nanohole arrays with different hole depths (50, 100, 150 nm) and periodicities (400, 500, 600, 900 nm) have been successfully fabricated with no observable defects. The nanohole shape is almost unchanged after the sputter deposition of Au film.

The effect of hole depth and periodicity on the optical properties was evaluated with the system shown in Fig. 6.7. The system consists of a white light source, a portable spectrophotometer, and a custom-made bundle fiber, which illuminates the device surface and also collects reflected light and introduces it into the spectrophotometer. The wavelength shift per refractive index unit and shape of the spectra were changed by changing the hole depth. The spectra agreed very well with

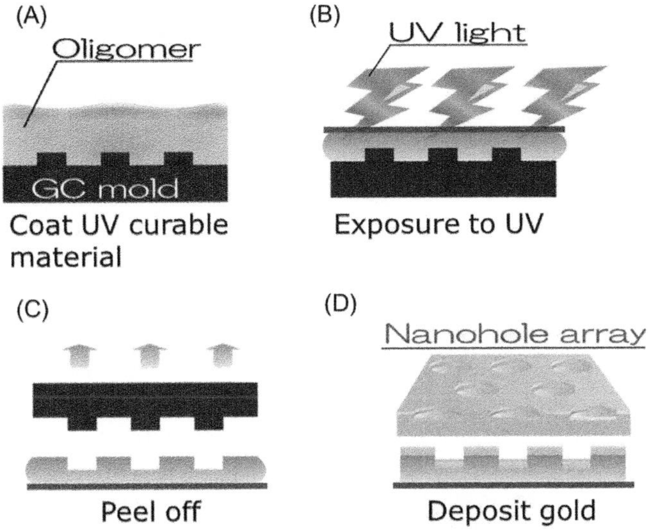

Figure 6.6 Procedure for fabricating a gold nanohole array using the nanoimprint technique and gold film deposition. **(A)** A UV-curable oligomer was dropped into a glassy carbon mold. (B) PET (poly-ethylene telephtalete) film was placed on the oligomer solution and then subjected to UV irradiation. (C) The mold was detached from the fabricated nanohole array. (D) Gold film was deposited on the nanohole array.

simulated spectra. In addition, the wavelength shift increased when the periodicity was increased from 400 to 600 nm. An optimized nanohole array (300 nm diameter, 30 nm depth, and a periodicity of 600 nm) was employed for detecting the tumor necrosis factor (TNF-α). Fig. 6.8A shows a schematic representation of an immunoassay on the nanohole array sensor. The capture antibody was immobilized by a carbodiimide coupling reaction on a gold nanohole array via a SAM membrane. The PDMS microchannel was covered on the nanohole array and TNF-α was introduced. Calibration curves with and without a gold labeled secondary antibody were obtained to compare the sensitivity and detection limit as shown in Fig. 6.8A (2) and (3). From the calibration curves, we estimated a detection limit of 21 ng/mL with labeling and 45 ng/mL without labeling. The assay protocol without labeling is more suitable for realizing an IoT immunosensor. A more sophisticated nanostructure design for improving the detection limit should be studied by considering the current detection limit of several tens of nanomolar/mL.

6.4 Microfluidic biosensor with one-step optical detection

We created an on-chip graphene biosensor to detect biologically important proteins such as cancer markers simply by adding a sample smaller than 1 μL to the chip

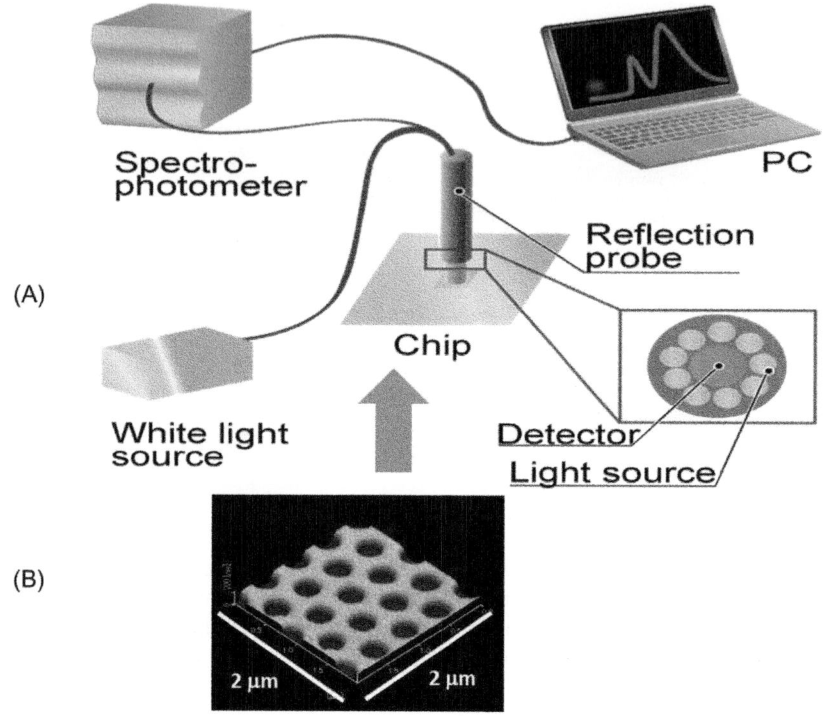

Figure 6.7 (A) Measurement system for detecting the reflection spectrum from an LSPR device. (B) Atomic force microscopy images of a gold nanohole array with a hole depth of 149 nm.

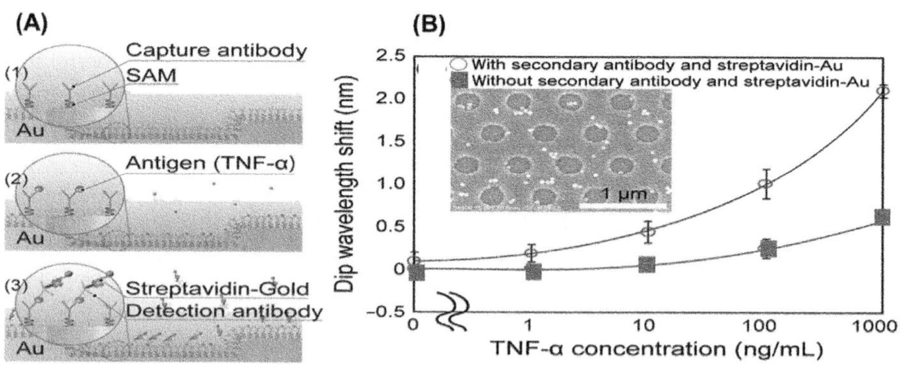

Figure 6.8 (A) Schematic representation of immunoassay process. (1) The capture antibody was immobilized by a carbodiimide coupling reaction on the Au nanohole array. (2) Sample solution containing TNF-α was injected into the nanohole array incorporated microchannel. (3) Streptavidin-AuNP was injected after the detection antibody was injected. (B) Calibration curve for immunoassay between anti-TNF-α antibody and TNF-α with (circle) or without (square) streptavidin–gold. All flow rates were 5 mL/min. (Inset) SEM image of the gold nanohole array with streptavidin gold.

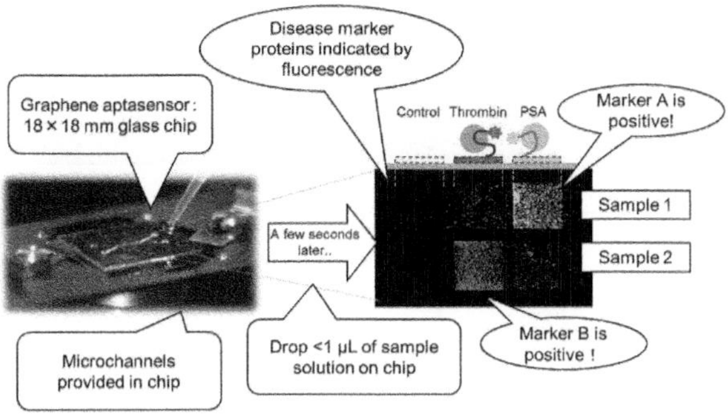

Figure 6.9 Photograph taken when using the on-chip graphene aptasensor (left) and color-composite fluorescence image of graphene aptasensor after injecting PSA solution and thrombin solution into the top and bottom channels (right), respectively.

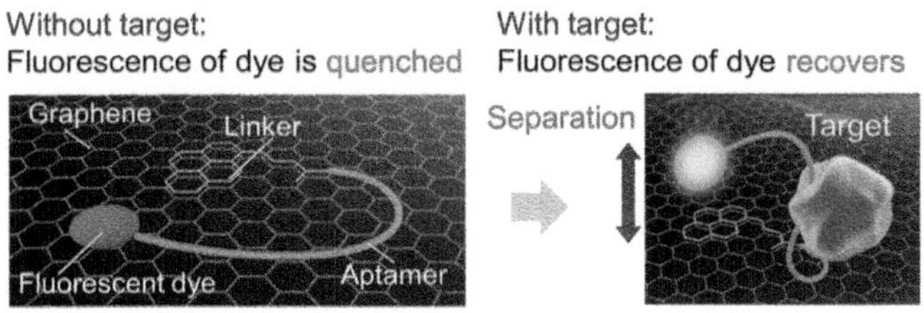

Figure 6.10 Schematic illustration of design and operating mechanism of GO FRET aptasensor.

(Fig. 6.9) [29]. Reducing the number of operation steps is important for IoT sensors, especially when they are to be operated by unskilled patients. It brings our sensors closer to such practical applications as home health checks, which are useful in helping to avoid congestion in hospitals and for easing the stress of a patient who would normally have to wait for the results of a hospital health check. In this section, we review our recent achievements using the on-chip graphene aptasensor for protein detection. These achievements include the simultaneous detection of multiple target molecules on a single chip and the molecular design of a probe for enhancing the sensitivity of the graphene aptasensor.

6.4.1 Mechanism of graphene aptasensor

In our graphene or graphene oxide (GO) sensor, the graphene surface is modified with a pyrene-aptamer-dye probe. The components work as a linker to the graphene surface, a protein recognition part, and a fluorescence detection tag. The system allows us to detect molecules on a solid surface, and so is a powerful tool for realizing an on-chip sensor (Fig. 6.10) [30,31].

It is noteworthy that a PDMS microchannel requires no external power or expensive equipment to induce a liquid flow, because the flow in the microchannel is

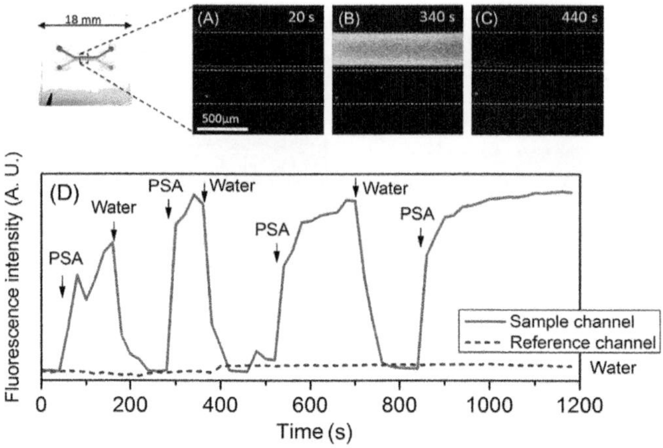

Figure 6.11 Fluorescence images of an on-chip graphene aptasensor under microfluidic conditions observed at (A) 20, (B) 340, and (C) 440 s, respectively. The sensor has two channels, both of which are filled with DI water at the initial stage, where only the top channel is replaced by 100 μg/mL PSA solution and DI water. (D) The plot shows the average fluorescence intensities of the top and bottom channel areas over time.

driven by capillary force [32]. We fabricated an on-chip graphene aptasensor by placing a sheet with microchannels on a solid substrate, and we formed the graphene aptasensor on its surface. We can fabricate a microchannel with the desired design by using photolithographic techniques. The number of channels is also variable. Thus, if we form multiple microchannels on a single chip, we can perform a high-throughput analysis by performing several different measurements in parallel. Moreover, if we use one of the multichannels as an internal standard to eliminate the effect of fluorescence degradation caused by laser exposure and other noises, a consistent reference-to-sample comparison is possible that allows us to realize an accurate quantitative analysis. Thus, the target protein can be detected simply by adding a sample smaller than 1 μL to the on-chip sensor, and the simultaneous measurement of the fluorescence emitted from the graphene surfaces located in each microchannel is also possible. The reaction that forms the aptamer–protein complex is completed in about a minute. Real-time detection is possible by tracking the fluorescence intensity at the microchannels by adding different sample solutions sequentially (Fig. 6.11). For comparison, we used an ELISA, which is one of the most widely-used protein detection methods [33]. A commercially obtained ELISA kit requires a sample volume of at least 100 μL and an assay time of three hours.

6.4.2 Multichannel linear array for multiple protein detection

A merit of using an on-chip sensor is that it enables us to realize a parallel analysis system such as array sensors [34]. This allows us to undertake a quantitative comparison of different samples by forming a multichannel configuration and/or a micropattern with the probes. We fabricated a 2×3 linear-array graphene aptasensor by using two different aptamers for different targets, namely thrombin and PSA, which were labeled with TAMRA (λ_{max}(abs)/λ_{max} (em) = 555 nm/580 nm) and FAM (λ_{max} (abs)/λ_{max} (em) = 494 nm/518 nm), respectively. (Fig. 6.9, right). Before making the array sensor, we confirmed the versatility of the on-chip graphene aptasensor by the detection of three different proteins, namely, thrombin, PSA, and hemagglutinin, simply by changing the aptamers but retaining the same sensor composition. We used a confocal laser scanning microscope to measure the fluorescence image of this sensor after injecting thrombin and PSA solutions into the top and bottom microchannels, respectively. Bright fluorescence was only observed in the areas where we assumed that the correct aptamer–target pair had been formed. The simultaneous detection of multiple target molecules on a single chip was successfully demonstrated [29].

6.4.3 Molecular design for enhanced sensitivity

The most interesting feature of aptasensors is that we can design and construct various kinds of biomolecular probes by incorporating additional functions with an aptamer. We can improve the sensitivity of an on-chip graphene aptasensor by modifying an aptamer with an ssDNA spacer (Fig. 6.12). The strategy was to

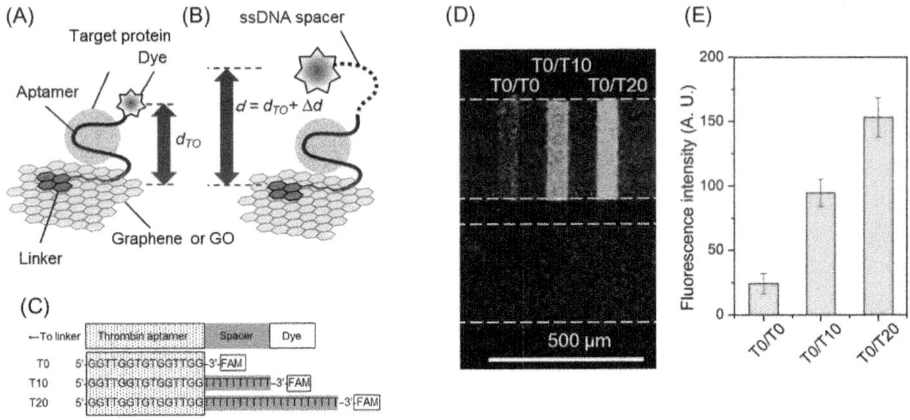

Figure 6.12 Schematic representation of a technique for increasing fluorescence intensity by modifying an aptamer with an ssDNA spacer. (A) When the aptamer recognizes the target protein, the fluorescence quenching of the dye becomes less efficient as the distance between the dye and the graphene/GO surface (D) increases. (B) The increase in (D) becomes more potent by introducing a spacer between the aptamer sequence and the dye. (C) The probes consist of a linker, a thrombin aptamer, a spacer, and a dye. They are denoted as T0 by using the numbers of thymines (*N*) for the spacer. (D) Fluorescence image of the 2 × 3 multichannel linear-array aptasensor patterned with T0/T0, T0/T10, and T0/T20, from left to right. Thrombin solution (100 μg/mL) and water were injected into the top and bottom microchannels. (E) Average fluorescence intensity of the patterned area in the top microchannel.
Source: Reprinted with permission from Y. Ueno, K. Furukawa, K. Matsuo, S. Inoue, K. Hayashi, H. Hibino, Chem. Commun. 49 (2013) 10346–10348. Copyright 2013 Royal Society of Chemistry.

increase the distance between the fluorescence dye and the graphene surface, which is crucial for FRET-based sensors, when forming a complex with the target protein.

We fabricated a 2×3 linear-array GO aptasensor by using three different probes and introducing ssDNA spacers with 0, 10, and 20 thymine segments between the aptamer and the dye. The fluorescence intensity increased significantly with increases in the spacer length [35,36]. The limit for thrombin detection was less than 10 nM, which corresponds to the in vivo concentration range during blood clotting. The results show that introducing an ssDNA spacer at the correct position is an effective way of enhancing sensor sensitivity.

In general, aptasensors are versatile because they can be extended to the detection of many different targets by replacing the aptamers [37]. We confirmed that several DNA and RNA aptamers are sufficiently active to realize an on-chip graphene/GO aptasensor through the detection of corresponding target proteins [29]. The versatility of the on-chip graphene/GO aptasensor was successfully demonstrated simply by changing the aptamers but with the sensor composition remaining the same. Moreover, aptamers are chemically stable [38]. We confirmed that the aptasensors that had been stored at a normal temperature and pressure for more than a month operated as normally as if they had just been made. Stability and robustness are also important factors for biosensors in IoT applications. We believe that aptasensors can provide one possible solution.

6.5 Future trend

This chapter introduced simple and rapid immunosensing devices. The specifications of each sensing device are insufficient for use as IoT sensors. However, an aspect of the performance of each sensor is important for IoT application. For example, SPR based BNP sensors (first example) exhibit a very low detection limit in the pg/mL range with a very small volume sample. This means these sensors can be used with many kinds of biologically important samples, although their size is still large and some steps are required to complete the assay. A one-chip immunosensing device (second example) will be inexpensive because it can be fabricated with molding technology. However, the detection system must be much smaller and the detection limit must be improved. The microfluidic fluorescence sensing device (third example) can be operated with a one-step process with a relatively good detection limit, but the entire system, and particularly the optical detection system, must be miniaturized to a portable device size.

In the near future immunosensing devices that meet the above requirements will be needed so that we can achieve an advanced healthcare system connected to the IoT. Collaboration between various fields including material science, biotechnology, micromachining, and electrical engineering will be necessary if we are to realize these immunosensing devices of the future.

References

[1] V. Cheng, R. Kazanagra, A. Garcia, L. Lenert, P. Krishnaswamy, N. Gardetto, et al., A rapid bedside test for B-type peptide predicts treatment outcomes in patients admitted for decompensated heart failure: a pilot study, J. Am. Coll. Cardiol. 37 (2001) 386–391.

[2] A. Chen, S. Yang, Replacing antibodies with aptamers in lateral flow immunoassay, Biosens. Bioelectron. 71 (2015) 230–242.

[3] K. Mitamura, H. Shimizu, M. Yamazaki, M. Ichikawa, K. Nagai, J. Katada, et al., Clinical evaluation of highly sensitive silver amplification immunochromatography systems for rapid diagnosis of influenza, J. Virol. Meth. 194 (2013) 123–128.

[4] P. von Lode, Point-of-care immunotesting: approaching the analytical performance of central laboratory methods, Clin. Biochem. 38 (2005) 591–606.

[5] J. Homola, Surface plasmon resonance sensors for detection of chemical and biological species, Chem. Rev. 108 (2008) 462–493.

[6] V. Gubala, L.F. Harris, A.J. Ricco, M.X. Tan, D.E. Williams, Point of care diagnostics: status and future, Anal. Chem. 84 (2012) 487–515.

[7] T. Endo, K. Kerman, N. Nagatani, H.M. Hiepa, D.K. Kim, Y. Yonezawa, et al., Multiple label-free detection of antigen – antibody reaction using localized surface plasmon resonance-based core – shell structured nanoparticle layer nanochip, Anal. Chem. 78 (2006) 6465–6475.

[8] A. Posthuma-Trumpie, J. Korf, A. van Amerongen, Lateral flow (immuno)assay: its strengths, weaknesses, opportunities and threats. A literature survey, Anal. Bioanal. Chem. 393 (2009) 569–582.

[9] J.D. Newman, A.P.F. Turner, Home blood glucose biosensors: a commercial perspective, Biosens. Bioelectron. 20 (2005) 2435–2453.

[10] R. Kurita, Y. Yokota, Y. Sato, F. Mizutani, O. Niwa, On-chip enzyme immunoassay of a cardiac marker using a microfluidic device combined with a portable surface plasmon resonance system, Anal. Chem. 78 (2015) 5525–5531.

[11] K. Nakamoto, R. Kurita, N. Sekioka, O. Niwa, Simultaneous on-chip surface plasmon resonance measurement of disease marker protein and small metabolite combined with immuno- and enzymatic reactions, Chem. Lett. 37 (2008) 698–699.

[12] K. Nakamoto, R. Kurita, O. Niwa, One-chip biosensor for simultaneous disease marker/calibration substance measurement in human urine by electrochemical surface plasmon resonance method, Biosens. Bioelectron. 26 (2010) 1536–1542.

[13] B.D. Spangler, E.A. Wilkinson, J.T. Murphy, B.J. Tyler, Comparison of the Spreeta® surface plasmon resonancesensor and a quartz crystal microbalance for detection of Escherichia coli heat-labile enterotoxin, Anal. Chim. Acta 444 (2001) 149–161.

[14] N. Elahi, N. Kamali, M.H. Baghersad, Recent biomedical applications of gold nanoparticles: a review, Talanta 184 (2018) 537–556.

[15] J. Zhou, Z. Cao, N. Panwar, R. Hu, X. Wang, J. Qu, et al., Functionalized gold nanorods for nanomedicine: past, present and future, Coord. Chem. 352 (2017) 15–66.

[16] A. Dahlin, M. Zach, T. Rindzevicius, M. Kall, D.S. Sutherlamd, F. Hook, Localized surface plasmon resonance sensing of lipid-membrane-mediated biorecognition events, J. Am. Chem. Soc. 127 (2005) 5043–5048.

[17] N. Roca, N.H. Pandya, S. Nath, A.J. Haes, Linear assembly of gold nanoparticle clusters via centrifugation, Langmuir 26 (2010) 2035–2041.

[18] T. Sannomiya, O. Scholder, K. Jefimovs, C. Hafner, A.B. Dahlin, Investigation of plasmon resonances in metal films with nanohole arrays for biosensing applications, Small 7 (2011) 1653–1663.
[19] J.F. Masson, M.P. Murray-Methot, L.S. Live, Nanohole arrays in chemical analysis: manufacturing methods and applications, investigation of plasmon resonances in metal films with nanohole arrays for biosensing applications, Analyst 135 (2010) 1483–1489.
[20] B.D. Lucas, J.S. Kim, C. Chin, L.J. Guo, Nanoimprint lithography based approach for the fabrication of large-area, uniformly-oriented plasmonic arrays, Adv. Mater. 20 (2008) 1129–1134.
[21] S.Y. Chou, P.R. Krauss, P.J. Renstrom, Imprint lithography with 25-nanometer resolution, Science 272 (1996) 85–87.
[22] C. Escobedo, On-chip nanohole array based sensing: a review, Lab. Chip 13 (2013) 2445–2463.
[23] M.E. Stewart, N.H. Mack, V. Malyarchuk, J.A.N.T. Soares, T.-W. Lee, S.K. Gray, et al., Quantitative multispectral biosensing and 1D imaging using quasi-3D plasmonic crystals, Proc. Natl. Acad. Sci. 103 (2006) 17143–17148.
[24] J. Yao, M.E. Stewart, J. Maria, T.-W. Lee, S.K. Gray, J.A. Rogers, et al., Seeing molecules by eye: surface plasmon resonance imaging at visible wavelengths with high spatial resolution and submonolayer sensitivity, Angew. Chem., Int. Ed. 47 (2008) 5013–5017.
[25] C.C. Yu, K.H. Ho, H.L. Chen, S.-Y. Chuang, S.-C. Tseng, W.-F. Su, Using the nanoimprint-in-metal method to prepare corrugated metal structures for plasmonic biosensors through both surface plasmon resonance and index-matching effects, Biosens. Bioelectron. 33 (2012) 267–273.
[26] S.-W. Lee, K.-S. Lee, J. Ahn, J.-J. Lee, M.-G. Kim, Y.-B. Shin, Highly sensitive biosensing using arrays of plasmonic Au nanodisks realized by nanoimprint lithography, ACS Nano 5 (2011) 897–904.
[27] X. Zhou, T.I. Wong, H.Y. Song, L. Wu, Y. Wang, P. Bai, et al., Development of localized surface plasmon resonance-based point-of-care system, Plasmonics 9 (2014) 835–844.
[28] K. Nakamoto, R. Kurita, O. Niwa, T. Fujii, M. Nishida, Development of a mass-producible on-chip plasmonic nanohole array biosensor, Nanoscale 3 (2011) 5067–5075.
[29] Y. Ueno, K. Furukawa, K. Matsuo, S. Inoue, K. Hayashi, H. Hibino, On-chip graphene oxide aptasensor for multiple protein detection, Anal. Chim. Acta 866 (2015) 1–9.
[30] K. Furukawa, Y. Ueno, E. Tamechika, H. Hibino, Protein recognition on a single graphene oxide surface fixed on a solid support, J. Mater Chem. B 1 (2013) 1119–1124.
[31] K. Furukawa, Y. Ueno, M. Takamura, H. Hibino, Graphene FRET Aptasensor, ACS Sens. 1 (2016) 710–716.
[32] G.M. Walker, D.J. Beebe, A passive pumping method for microfluidic devices, Lab. Chip 2 (2002) 131–134.
[33] A. Voller, A. Bartlett, D.E. Bidwell, Enzyme immunoassays with special reference to ELISA techniques, J. Clinical Pathology 31 (1978) 507–520.
[34] M. Schena, D. Shalon, R.W. Davis, P.O. Brown, Quantitative monitoring of gene expression patterns with a complementary DNA microarray, Science 270 (1995) 467–470.
[35] Y. Ueno, K. Furukawa, K. Matsuo, S. Inoue, K. Hayashi, H. Hibino, Molecular design for enhanced sensitivity of a FRET aptasensor built on the graphene oxide surface, Chem. Commun. 49 (2013) 10346–10348.

[36] Y. Ueno, K. Furukawa, A. Tin, H. Hibino, On-chip FRET graphene oxide aptasensor: quantitative evaluation of enhanced sensitivity by aptamer with a double-stranded DNA spacer, Anal., Sci. 31 (2015) 875–879.

[37] P. Hong, W. Li, J. Li, Applications of aptasensors in clinical diagnostics, Sensors 12 (2012) 1181–1193.

[38] S. Klussmann, The Aptamer Handbook: Functional Oligonucleotides and Their Applications, first ed., Wiley-VCH, Weinheim, 2006.

Sensitive and reusable surface acoustic wave immunosensor for monitoring of airborne mite allergens

7

Koji Toma, Takahiro Arakawa and Kohji Mitsubayashi
Department of Biomedical Devices and Instrumentation, Institute of Biomaterials and Bioengineering, Tokyo Medical and Dental University, Tokyo, Japan

7.1 Introduction

Allergic diseases are disorders of the immune system and are accompanied by symptoms such as asthma, rhinitis, or itching [1–5]. The prevalence of the allergic diseases is becoming a global health issue. For example, it is reported that more than 40% of children are diagnosed with allergic rhinitis, and about 300 million people suffer from asthma in the world [3,6]. Among many causes of allergies, environmental allergens such as pollens and house dust mites (HDMs) are one of the popular substances because they exist in a living space, and it is difficult to escape from them. Especially they easily become airborne, which increases the risk of allergen inhalation (Fig. 7.1A). The concentration of airborne allergens is known to fluctuate over time by the influence of temperature, humidity, and human activities [7,8]. It makes a situation in which people are exposed to highly concentrated environmental allergens without being aware of it and increases the risk of sensitization spontaneously. Up to now, few studies have focused on airborne allergens because there is no system able to continuously measure airborne allergen concentration [9–12]. Recently we introduced a concept of an airborne allergen monitoring system that consists of an airborne allergen sampler and an allergen detector. It will collect airborne allergens and measure them rapidly and repeatedly—semicontinuously—and alert people to the environmental quality (Fig. 7.1B). This chapter focuses on the allergen detector, which is necessary to measure allergens selectively and semicontinuously. In general, high selectivity to allergens can be obtained by employing an immunoassay, which relies on the specificity of an antibody to an antigen. However, conventional immunoassays, such as enzyme-linked immunosorbent assays (ELISA) or lateral flow immunoassays, are designed to be used only for one-time measurement [13–15]. A major challenge in the use of immunoassays for repeated measurement is surface regeneration, in which a formed immunocomplex must be broken down and removed from a sensor surface. The regeneration often

Chemical, Gas, and Biosensors for the Internet of Things and Related Applications.
DOI: https://doi.org/10.1016/B978-0-12-815409-0.00007-3

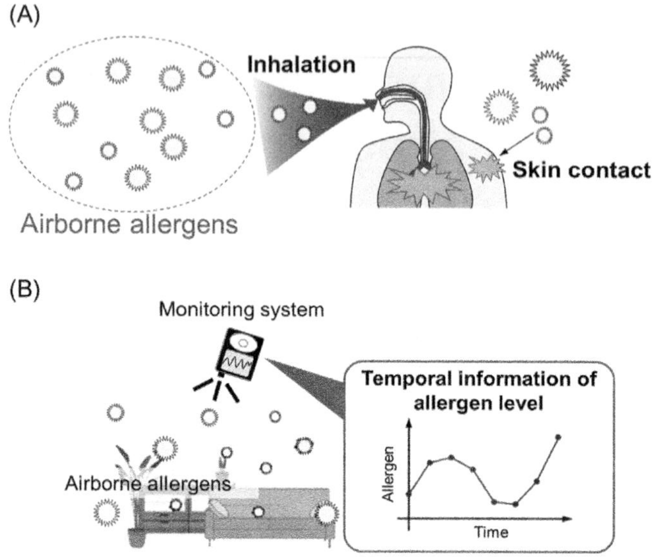

Figure 7.1 (A) Main route of sensitization to allergens. (B) Airborne allergen monitoring system for prevention of exposure to airborne allergens.

damages proteins and deteriorates sensor performance. To overcome this challenge, we have studied and developed an immunosensor that does not lose sensor function during a repeated measurement using a pH-tolerant protein scaffold for sensor surface modification. This chapter describes such an immunosensor for HDM allergen, and an approach to improve the sensor's sensitivity for the advanced semicontinuous measurement.

7.2 Surface acoustic wave immunosensor for repeated measurement of house dust mite allergens

A surface acoustic wave (SAW) device (Japan Radio Co., Ltd.) was used as a transducer. The device consists of an interdigitated transducer for both input and output, a SAW propagating area (sensing area), and a reflector (Fig. 7.2A). These parts were fabricated by sputtering a 90-nm-thick gold layer and the lift-off process on a 36Y-X quartz substrate. In the device, shear-horizontal (SH-) SAW, which is more suitable in the liquid media than Rayleigh SAW due to less damping, excitation at the input transducer at the center frequency of 250 MHz, and propagation back and forth along the sensing area surface. Molecules binding to the sensing area induces a viscoelastic change in vicinity of the surface, which alters the SAW velocity and results in a phase shift of the alternating voltage between the input and output transducers.

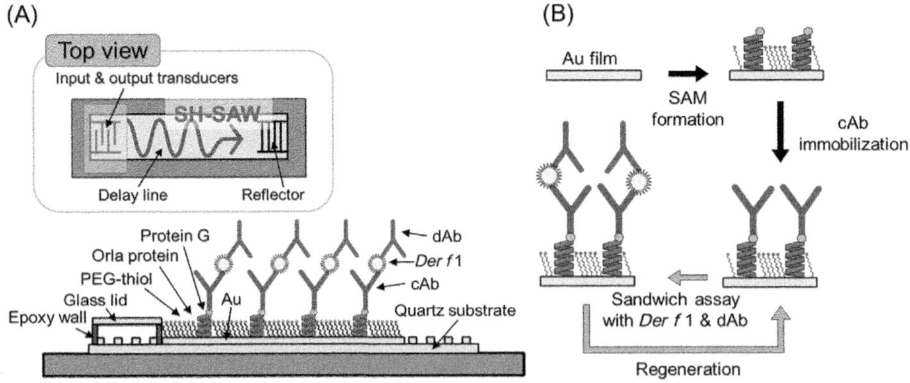

Figure 7.2 (A) Schematic diagram of a SAW device (top view) and an immunosensor (side view). (B) Processes of the sensor surface preparation and repeated measurement of *Der f* 1. Source: Reprinted with permission from K. Toma, D. Miki, C. Kishikawa, N. Yoshimura, K. Miyajima, T. Arakawa, et al., Repetitive immunoassay with a surface acoustic wave device and a highly stable protein monolayer for on-site monitoring of airborne dust mite allergens, Anal. Chem. 87 (2015) 10470–10474. doi:10.1021/acs.analchem.5b02594 [16]. Copyright (2015) American Chemical Society

For repeated measurement, the sensing area surface of the SAW device was modified with a mixed self-assembled monolayer of pH-tolerant protein scaffold, ORLA85 protein (Orla Protein Technologies), and (11-mercaptoundecyl)hexa(ethylene glycol) (6PEG-thiol). ORLA85 protein is a protein G-fused membrane protein and able to anchor its dithiol on a gold surface via gold-sulfur bond (Fig. 7.2B). Since protein G allows oriented antibody immobilization, high surface density of capture antibodies (cAbs) on a SAW device was expected. 6PEG-thiol was used as a filler molecule between ORLA85 proteins on a surface to prevent nonspecific binding of nontargeted substances. Semicontinuous measurement of *Dermatophagoides farinae* group 1 (*Der f* 1) was demonstrated by sandwich assay with *Der f* 1 and detection antibody (dAb) and surface regeneration with hydrochloride (HCl, pH 1) to remove *Der f* 1 and dAb for the next measurement.

To maximize cAb density on the sensor surface, influence of pH and concentration of cAb solution on the cAb binding efficiency was investigated. Fig. 7.3A shows a phase shift occurred by the binding of cAb (ΔP_{cAb}) with various solution pH, which was determined as a difference between an averaged phase for the last 30 seconds and before applying cAb—baseline. The cAb solution was prepared using acetate buffered saline (ABS, pH 4.0–5.0) and phosphate buffered saline with Tween 20 (PBS-T, pH 6.0–8.0). The maximum binding was observed with pH 6.0 PBS-T solution, and it was about fourfold larger amount than that with pH 4.0 ABS solution. This result well agreed to the previous report that binding between an antibody and a protein G was the strongest under acidic conditions [17]. Similarly, the influence of cAb concentration was investigated. As the result shows in Fig. 7.3B, the phase shift ΔP_{cAb} dramatically increased at the cAb

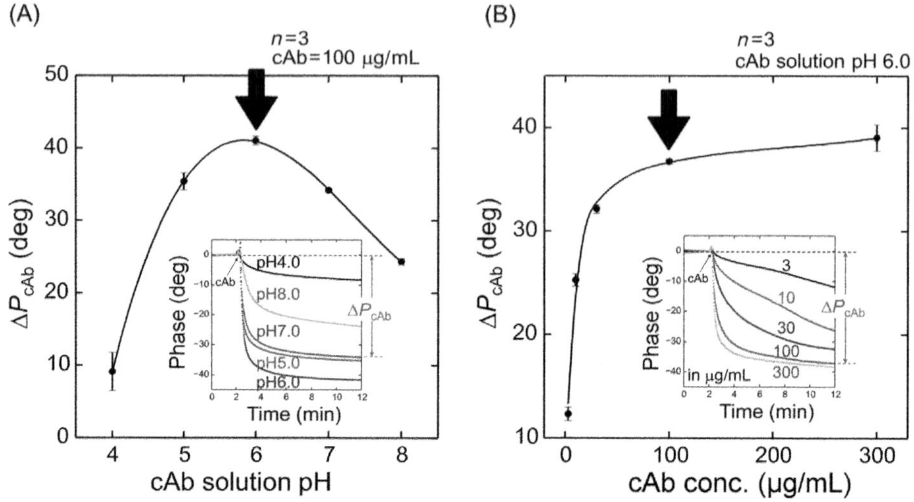

Figure 7.3 Influences of (A) buffer solution pH dissolving cAb and (B) cAb concentration on the efficiency of cAb binding to protein G. The insets show the binding kinetics of cAb under various conditions.
Source: Adapted with permission from K. Toma, D. Miki, C. Kishikawa, N. Yoshimura, K. Miyajima, T. Arakawa, et al., Repetitive immunoassay with a surface acoustic wave device and a highly stable protein monolayer for on-site monitoring of airborne dust mite allergens, Anal. Chem. 87 (2015) 10470–10474. doi:10.1021/acs.analchem.5b02594 [16]. Copyright (2015) American Chemical Society.

concentration less than 50 μg/mL and reached a saturation at the concentration over 100 μg/mL. It was probably due to two reasons: at the concentration less than 100 μg/mL, the binding rate was diffusion-limited, and the cAb concentration dominantly influences the binding rate; at the concentration over 100 μg/mL, the number of binding sites on the sensor surface rather limited the binding rate.

After binding of cAb to ORLA85, cAb was immobilized on the sensor surface by crosslinking between cAb and protein G because the immobilization enables to the skip rebinding process of cAb in the repeated measurement and shortens the total measurement time. Fig. 7.4 shows a phase shift during the crosslinking of cAb. After binding of cAb, a crosslinker, 0.5 mM PEGylated bis(sulfosuccinimidyl)-suberate [BS(PEG)$_5$], was applied to the sensor surface and reacted for 10 minutes. After rinsing, noncrosslinked cAb was removed by applying HCl pH 1.0 to the sensor surface. At the end of the crosslinking, unreacted active esters of BS (PEG)$_5$ were passivated by 1 M ethanolamine pH 8.4. To evaluate the amount of crosslinked cAb, the immobilization rate was calculated as a ratio between the phase shifts ΔP_{cAb} before (Δa) and after (Δb) applying HCl. The result showed that the immobilization rate of the first crosslinking was of 85.5%, and it was improved up to 99.5% by repeating the crosslinking process three times. This result indicates that almost all bound cAb was successfully immobilized on the sensor surface.

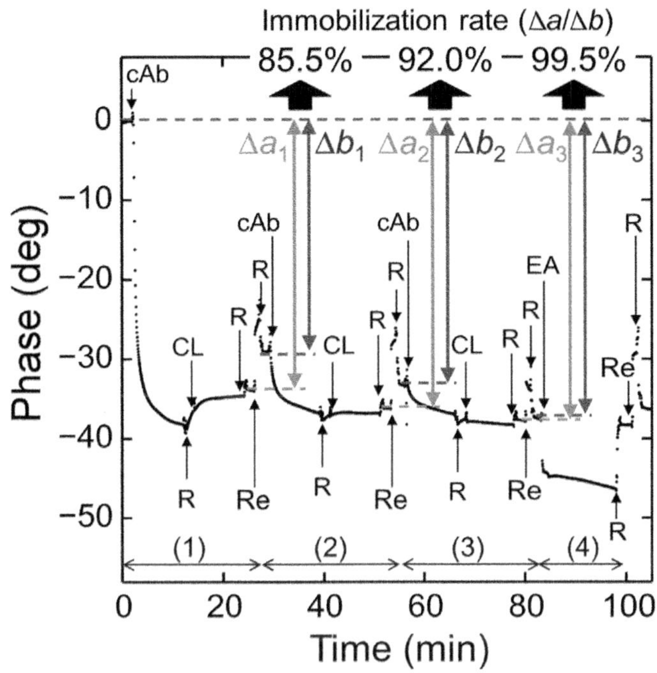

Figure 7.4 Phase change during crosslinking between cAb and protein G for the immobilization. CL, crosslinking with BS(PEG)$_5$; R, rinsing; Re, regeneration with HCl; EA, passivation with ethanolamine.
Source: Adapted with permission from K. Toma, D. Miki, C. Kishikawa, N. Yoshimura, K. Miyajima, T. Arakawa, et al., Repetitive immunoassay with a surface acoustic wave device and a highly stable protein monolayer for on-site monitoring of airborne dust mite allergens, Anal. Chem. 87 (2015) 10470–10474. doi:10.1021/acs.analchem.5b02594 [16]. Copyright (2015) American Chemical Society.

7.3 Sensor characteristics and semicontinuous measurement of *Der f* 1

Measurement of *Der f* 1 was demonstrated by the developed SAW immunosensor. Fig. 7.5A shows a phase shift during measurement of *Der f* 1. After a slight phase change by *Der f* 1 binding, the large phase change by dAb binding was observed. It was attributed to a molecular weight difference between two molecules. After the measurement, the sensor surface was regenerated by 100 mM HCl, and the phase recovered to the baseline as bound *Der f* 1 and dAb were almost completely removed. In addition, owing to the cAb immobilization, the measurement time was 30 minutes including the regeneration, which is about 1/6 of the measurement time of ELISA. The limit of detection (LOD) was revealed by measuring various concentrations of *Der f* 1 and determined to be 6.1 ng/mL. Although the LOD of the SAW immunosensor was not as good as that of ELISA (0.58 ng/mL), which

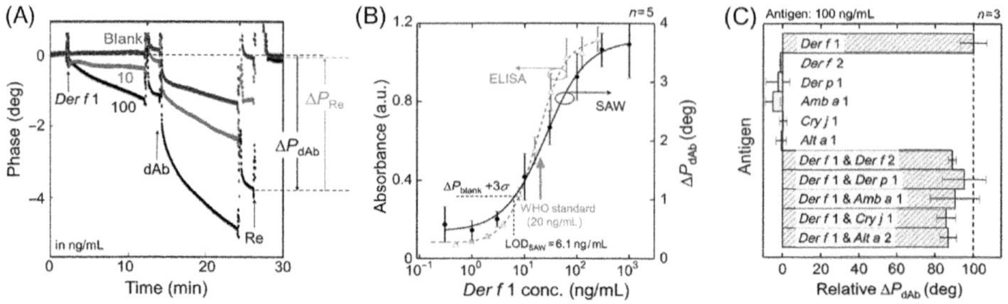

Figure 7.5 (A) Phase change during measurement of *Der f* 1 at various concentrations (blank, 10 ng/mL, and 100 ng/mL) via sandwich assay. Re, regeneration with HCl. (B) Calibration curves of the SAW immunosensor and ELISA (control) to *Der f* 1. (C) Sensor's selectivity to *Der f* 1.
Source: Adapted with permission from K. Toma, D. Miki, C. Kishikawa, N. Yoshimura, K. Miyajima, T. Arakawa, et al., Repetitive immunoassay with a surface acoustic wave device and a highly stable protein monolayer for on-site monitoring of airborne dust mite allergens, Anal. Chem. 87 (2015) 10470–10474. doi:10.1021/acs.analchem.5b02594 [16]. Copyright (2015) American Chemical Society.

encompassed the standard (20 ng/mL) set by the World Health Organization (WHO) (Fig. 7.5B). Here, the LOD was determined as the intersection where the phase shift from a blank sample ($\Delta P_{\mathrm{blank}}$) plus three times a standard deviation (σ) equals the phase shift ΔP_{dAb}.

The selectivity of the SAW immunosensor was also evaluated with six different environmental allergens—*Der f* 1, *Der f* 2, *Der p* 1, *Amb a* 1, *Cry j* 1, and *Alt a* 1—by comparing the sensor output ΔP_{dAb} with each sample and the mixture. Fig. 7.5C shows the relative sensor outputs to that of *Der f* 1, which was obtained by subtracting the phase shift of a blank sample from that of *Der f* 1 ($\Delta P_{\mathrm{dAb}} - \Delta P_{\mathrm{blank}}$). It shows clear sensor outputs only from the samples containing *Der f* 1, which validates high selectivity of the SAW immunosensor, especially from the fact that no sensor output was observed from *Der f* 2 and *Der p* 1, which tend to result in the cross-reactivity due to their similar protein conformations to *Der f* 1.

Before demonstrating repeated measurement of *Der f* 1, pH of the regeneration reagent was optimized. HCl solution at various pH (1.0–3.0) was used to regenerate the surface after measurement of *Der f* 1, and the regeneration rate were compared. Here, the regeneration rate was defined as a ratio of the phase recovery (ΔP_{Re}) to the sensor output (ΔP_{dAb}). The result showed that the regeneration rate was larger than 90% in the pH range of 1.0–2.0 and decreased in the pH value greater than 2.0 (Fig. 7.6A). Taking the reproducibility of the regeneration rate into account, HCl at pH 1.0 showing the smallest error bar was chosen for repeated measurement of *Der f* 1.

Now, 10 repeated measurements of *Der f* 1 (100 ng/mL) are demonstrated. The phase change represented in Fig. 7.6B shows high reproducibility although the baseline gradually decreased. This decrease of the baseline was probably due to imperfect surface regeneration, and some *Der f* 1 or dAb remained on the surface.

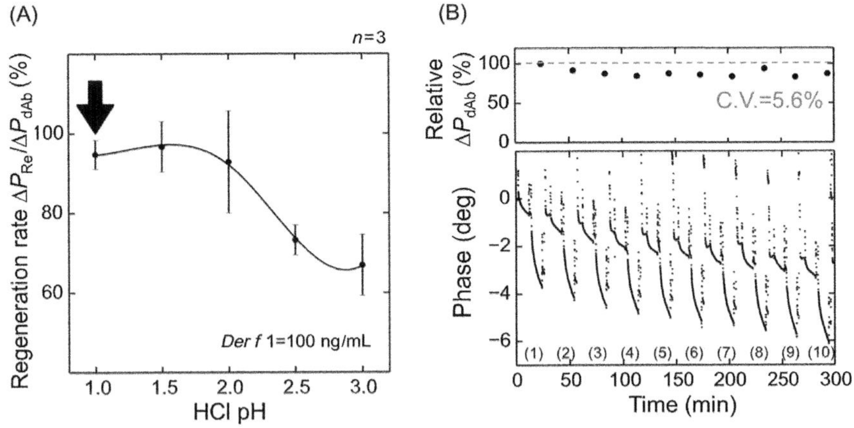

Figure 7.6 (A) Influence of regeneration solution pH on the regeneration rate. (B) Phase change during 10 repeated measurement of *Der f* 1.
Source: Adapted with permission from K. Toma, D. Miki, C. Kishikawa, N. Yoshimura, K. Miyajima, T. Arakawa, et al., Repetitive immunoassay with a surface acoustic wave device and a highly stable protein monolayer for on-site monitoring of airborne dust mite allergens, Anal. Chem. 87 (2015) 10470–10474. doi:10.1021/acs.analchem.5b02594 [16]. Copyright (2015) American Chemical Society.

However, there were still enough binding sites on the sensor surface, and the influence on the sensor performance was negligible. As a result, the coefficient of variation (C.V.) of the sensor outputs during 10 repeated measurement was 5.6%, indicating a high potential for semicontinuous measurement.

7.4 Sensitivity improvement via gold nanoparticles

Although the LOD of the SAW immunosensor met the WHO's standard, more improvement in the sensitivity is required because a final concentration of the HDM allergens at the sensing part in the monitoring system is likely to be lower than that of the originally sampled due to loss through transportation of the allergens, especially at a gas–liquid interface, a border at the collecting and the sensing parts in the monitoring system. To enhance the sensitivity of the SAW immunosensor, gold nanoparticles (AuNPs) were employed because an AuNP has larger mass than proteins such as *Der f* 1 and dAb. In the SAW immunosensor, AuNP binds to dAb via streptavidin–biotin interaction, which are conjugated to AuNP and dAb, respectively, after the above described sandwich assay with *Der f* 1 and dAb (Fig. 7.7).

Fig. 7.8A shows a sensor response during *Der f* 1 measurement using AuNP without any capping molecule. Loading AuNPs on the sensor induced a larger phase change (ΔP_{AuNP}) than that of *Der f* 1 (ΔP_{Ag}) or dAb (ΔP_{dAb}) because of the mass difference between AuNP and the proteins. However, the phase did not

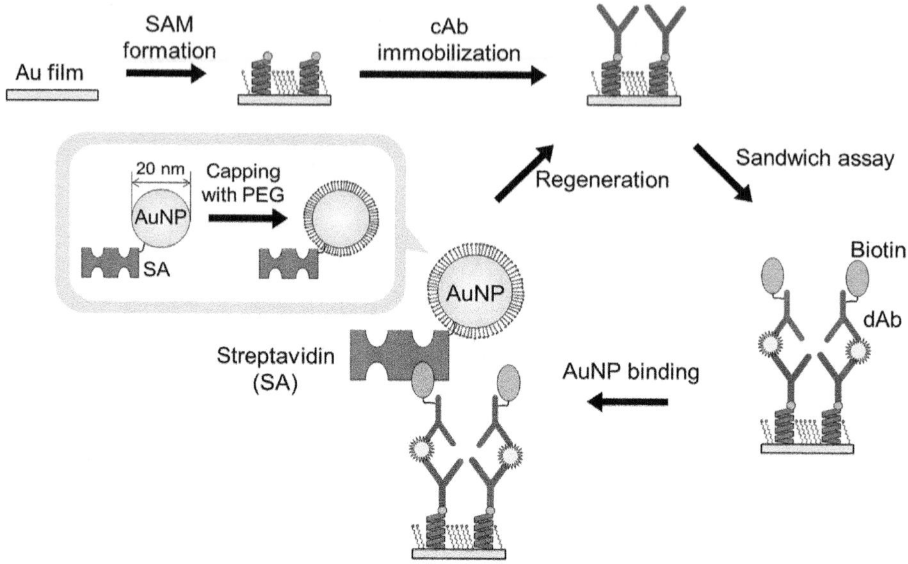

Figure 7.7 Process of sandwich assay with AuNPs for sensitive measurement of *Der f* 1.
Source: Adapted from K. Toma, D. Miki, N. Yoshimura, T. Arakawa, H. Yatsuda, K. Mitsubayashi, A gold nanoparticle-assisted sensitive SAW (surface acoustic wave) immunosensor with a regeneratable surface for monitoring of dust mite allergens, Sens. Actuators B Chem. 249 (2017) 685–690. doi:10.1016/j.snb.2017.04.073 [18].

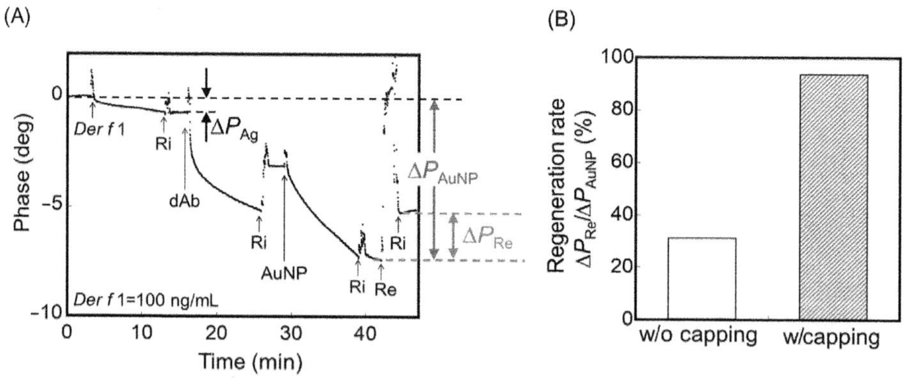

Figure 7.8 (A) Phase change during measurement of *Der f* 1 using AuNPs without a capping layer. (B) Regeneration rate without and with the capping layer on AuNPs. Ri, rinsing; Re, regeneration with HCl; ΔP_{Ag}, phase shift from *Der f* 1; ΔP_{dAb}, additional phase shift from dAb; ΔP_{AuNP}, additional phase shift from AuNP; ΔP_{Re}, phase recovery through sensor regeneration.
Source: Adapted from K. Toma, D. Miki, N. Yoshimura, T. Arakawa, H. Yatsuda, K. Mitsubayashi, A gold nanoparticle-assisted sensitive SAW (surface acoustic wave) immunosensor with a regeneratable surface for monitoring of dust mite allergens, Sens. Actuators B Chem. 249 (2017) 685–690. doi:10.1016/j.snb.2017.04.073 [18].

recover to the baseline after the surface regeneration with HCl pH 1.0 probably due to aggregation or nonspecific binding of uncapped AuNPs, which were hardly removed. For repeated measurement with AuNP, it was necessary to minimize the nonspecific binding of AuNPs on the surface, and therefore, a surface of AuNP was modified with a capping monolayer of (11-mercaptoundecyl)tri(ethylene glycol) (3PEG-thiol). Fig. 7.8B represents the regeneration rate of the measurement using AuNPs without or with 3PEG-thiol capping layer. Here another regeneration rate was determined to be a ratio the phase recovery after the regeneration (ΔP_{Re}) and ΔP_{AuNP}. As a result of the capping layer, the regeneration rate was improved from 31% to 93%, and the effectiveness of the capping layer for prevention of nonspecific binding was confirmed.

The AuNP concentration is also an important parameter that appears to influence the signal amplification and the regeneration rate. The signal amplification was calculated as a ratio of the sensor output signal with AuNPs to that without them ($\Delta P_{AuNP}/\Delta P_{dAb}$). It was found that both the signal amplification and the regeneration rate have linear relations to the AuNP concentration, but there is a trade-off relation between them (Fig. 7.9). Since the regeneration rate was also important for repeated use in the monitoring system, we chose 1.7×10^{11} particles/mL (about threefold higher concentration than that of dAb) AuNPs solution in successive experiments with which the signal amplification of about 200%, and the regeneration rate of about 70% were expected.

Now, the sensitivity was evaluated. The sensor outputs from various concentrations of *Der f* 1 were plotted in Fig. 7.10. The plots were fitted by the following equation with the correlation coefficients of 0.999:

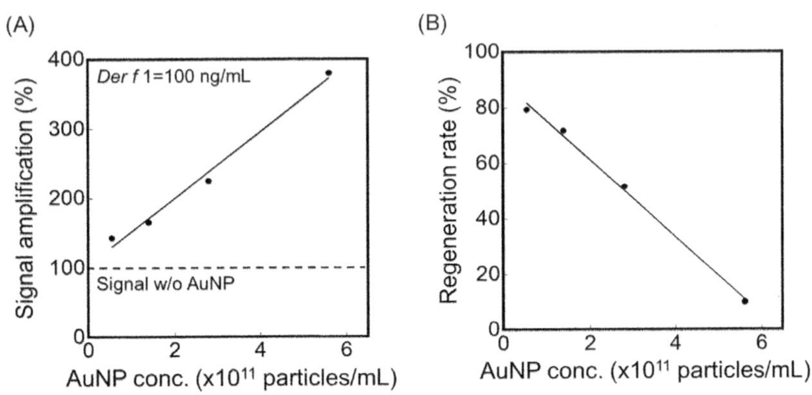

Figure 7.9 Influences of AuNP concentration on (A) signal amplification and (B) regeneration rate.
Source: Reprinted from K. Toma, D. Miki, N. Yoshimura, T. Arakawa, H. Yatsuda, K. Mitsubayashi, A gold nanoparticle-assisted sensitive SAW (surface acoustic wave) immunosensor with a regeneratable surface for monitoring of dust mite allergens, Sens. Actuators B Chem. 249 (2017) 685–690. doi:10.1016/j.snb.2017.04.073 [18].

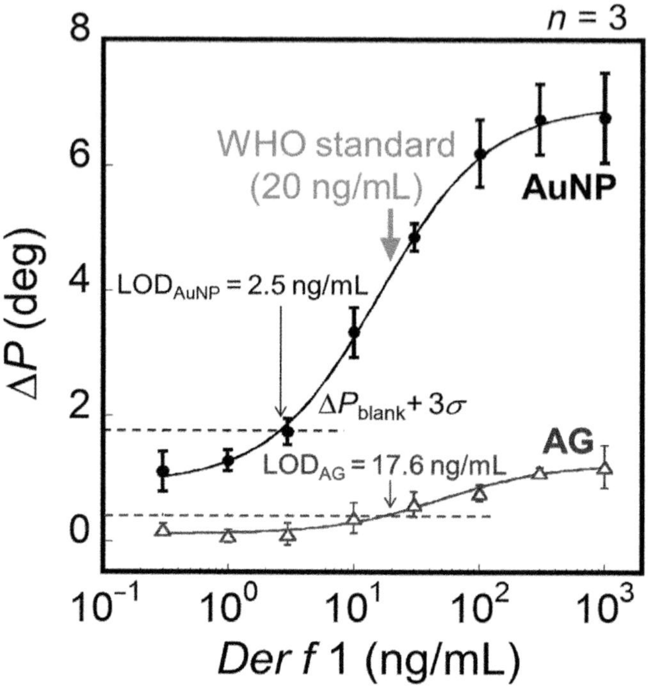

Figure 7.10 Calibration curves of the SAW immunosensor for *Der f* 1 (•) with AuNP-assisted signal amplification and (△) from only *Der f* 1 antigens.
Source: Reprinted from K. Toma, D. Miki, N. Yoshimura, T. Arakawa, H. Yatsuda, K. Mitsubayashi, A gold nanoparticle-assisted sensitive SAW (surface acoustic wave) immunosensor with a regeneratable surface for monitoring of dust mite allergens, Sens. Actuators B Chem. 249 (2017) 685–690. doi:10.1016/j.snb.2017.04.073 [18].

$$\Delta P_{\mathrm{AuNP}}(\mathrm{deg}) = A + (B - A)/\{1 + [Der\,f\ 1]/C^{D}\}, \quad (7.1)$$

where $A = 7.0$, $B = 0.91$, $C = 18$, and $D = 0.95$ are the coefficients, and [*Der f* 1] is the concentration of loaded *Der f* 1 in ng/mL. The LOD of the AuNP-assisted method was determined to be 2.5 ng/mL (100 pM). This is 2.5-fold improvement from the above described sandwich assay method [*Der f* 1, 6.1 ng/mL (0.24 nM)]. For reference, the LOD was also evaluated for the method with only *Der f* 1, that is, without dAb and AuNP, and was 17.6 ng/mL (704 pM). The sensitivity was not improved as expected from a mass of AuNP, which is 300-fold larger than that of *Der f* 1. However, optimal concentration of AuNPs and length of capping PEG-thiol may improve the sensitivity because the signal amplification and deterioration of regeneration rate have a trade-off relation to the concentration. Considering that the sensor outputs were highly reproducible through a successive measurement of *Der f* 1 during the evaluation of the sensitivity, the AuNP-assisted method can also be applied to the repeated measurement.

7.5 Conclusion

This chapter has described a SAW immunosensor for the repeated measurement of HDM allergen, *Der f* 1. The results indicated the sensor's advantages, such as high sensitivity emcompassing the WHO's standard, high selectivity, short measurement time, and reusability over 10 measurements. In addition, the sensitivity can be further improved using AuNP while maintaining the reusability. From these characteristics, the SAW immunosensor is potentially able to lead to the monitoring system for prevention of environmental allergen-caused diseases. In principle, the SAW immunosensor can be applied to any other target molecules by exchanging an antibody, and therefore a monitoring system that is able to monitor various environmental allergens simultaneously is also expected by arraying the SAW immunosensors.

References

[1] A. Togias, Rhinitis and asthma: evidence for respiratory system integration, J. Allergy Clin. Immunol. 111 (2003) 1171–1183. Available from: https://doi.org/10.1067/mai.2003.1592.

[2] R. Beasley, U. Keil, E. Von Mutius, N. Pearce, Worldwide variation in prevalence of symptoms of asthma, allergic rhinoconjunctivitis, and atopic eczema: ISAAC. The International Study of Asthma and Allergies in Childhood (ISAAC) Steering Committee, Lancet (London, England) 351 (1998) 1225–1232. Available from: https://doi.org/10.1016/S0140-6736(97)07302-9.

[3] R. Pawankar, G.W. Canonica, R.F. Lockey, S.T. Holgate (Eds.), WAO White Book on Allergy 2011-2012: Executive Summary, 2011.

[4] S.J. Arbes, P.J. Gergen, L. Elliott, D.C. Zeldin, T. Park, Prevalences of positive skin test responses to 10 common allergens in the US population: results from the Third National Health and Nutrition Examination Survey, J. Allergy Clin. Immunol. 116 (2005) 377–383. Available from: https://doi.org/10.1016/j.jaci.2005.05.017.

[5] A.P. Jackson, A.P. Foster, B.J. Hart, C.R. Helps, S.E. Shaw, Prevalence of house dust mites and dermatophagoides group 1 antigens collected from bedding, skin and hair coat of dogs in south-west England, Vet. Dermatol. 16 (2005) 32–38. Available from: https://doi.org/10.1111/j.1365-3164.2005.00427.x.

[6] A.L. Wright, C.J. Holberg, F.D. Martinez, M. Halonen, W. Morgan, L.M. Taussig, Epidemiology of physician-diagnosed allergic rhinitis in childhood, Pediatrics 94 (1994) 895–901.

[7] L.D. Knibbs, C. He, C. Duchaine, L. Morawska, Vacuum cleaner emissions as a source of indoor exposure to airborne particles and bacteria, Environ. Sci. Technol. 46 (2012) 534–542. Available from: https://doi.org/10.1021/es202946w.

[8] M. Sakaguchi, M. Hashimoto, N.S. Hospital, S. Inouye, R. Sasaki, M. Hashimoto, et al., Measurement of airborne mite allergen exposure in individual subjects, J. Allergy Clin. Immunol. 97 (1996) 1040–1044. Available from: https://doi.org/10.1016/S0091-6749(96)70255-5.

[9] T. Plattsmills, D. Vervloet, W. Thomas, R. Aalberse, M. Chapman, Indoor allergens and asthma: report of the Third International Workshop, J. Allergy Clin. Immunol. 100 (1997) S2–S24. Available from: https://doi.org/10.1016/S0091-6749(97)70292-6.

[10] T. Takai, T. Kato, H. Yasueda, K. Okumura, H. Ogawa, Analysis of the structure and allergenicity of recombinant pro- and mature Der p 1 an*d Der* f 1: major conformational IgE epitopes blocked by prodomains, J. Allergy Clin. Immunol. 115 (2005) 555–563. Available from: https://doi.org/10.1016/j.jaci.2004.11.024.
[11] T.J. Nuttall, J.R. Lamb, P.B. Hill, Characterisation of major and minor dermatophagoides allergens in canine atopic dermatitis, Res. Vet. Sci. 71 (2001) 51–57. Available from: https://doi.org/10.1053/rvsc.2001.0485.
[12] E.R. Tovey, C. Almqvist, Q. Li, D. Crisafulli, G.B. Marks, Nonlinear relationship of mite allergen exposure to mite sensitization and asthma in a birth cohort, J. Allergy Clin. Immunol. 122 (2008) 114–118. Available from: https://doi.org/10.1016/j.jaci.2008.05.010.
[13] U. Wahn, S. Lau, R. Bergmann, M. Kulig, J. Forster, K. Bergmann, et al., Indoor allergen exposure is a risk factor for sensitization during the first three years of life, J. Allergy Clin. Immunol. 99 (1997) 763–769.
[14] A. Tsay, L. Williams, E.B. Mitchell, M.D. Chapman, A rapid test for detection of mite allergens in homes, Clin. Exp. Allergy 32 (2002) 1596–1601.
[15] M.D. Chapman, A. Tsay, L.D. Vailes, Home allergen monitoring and control–improving clinical practice and patient benefits, Allergy 56 (2001) 604–610.
[16] K. Toma, D. Miki, C. Kishikawa, N. Yoshimura, K. Miyajima, T. Arakawa, et al., Repetitive immunoassay with a surface acoustic wave device and a highly stable protein monolayer for on-site monitoring of airborne dust mite allergens, Anal. Chem. 87 (2015) 10470–10474. Available from: https://doi.org/10.1021/acs.analchem.5b02594.
[17] B. Akerstrom, L. Bjorck, A physicochemical study of protein G, a molecule with unique immunoglobulin G-binding properties, J. Biol. Chem. 261 (1986) 10240–10247.
[18] K. Toma, D. Miki, N. Yoshimura, T. Arakawa, H. Yatsuda, K. Mitsubayashi, A gold nanoparticle-assisted sensitive SAW (surface acoustic wave) immunosensor with a regeneratable surface for monitoring of dust mite allergens, Sens. Actuat. B Chem. 249 (2017) 685–690. Available from: https://doi.org/10.1016/j.snb.2017.04.073.

Aptameric sensors utilizing its property as DNA

8

Kinuko Ueno, Kaori Tsukakoshi and Kazunori Ikebukuro
Department of Biotechnology and Life Science, Graduate School of Engineering, Tokyo University of Agriculture and Technology, Tokyo, Japan

8.1 Introduction

As one of the novel sensing elements, aptamers, the nucleic acid ligands that can bind to target molecules, have been widely studied [1]. Aptamers can bind to target molecules as strong as antibodies, which can show picomolar to nanomolar levels of K_D. Also, aptamers can bind to a target with high specificity and they can be chemically synthesized with high designability at low cost. Moreover, aptamers have the properties of nucleic acids, such as the feasibility of structure change or the capability of amplification by forming double strands. By applying these properties of aptamers, diverse kinds of biosensing systems, such as electrochemical sensors, colorimetric detection, and other composite systems have been developed [2,3].

To apply biosensors to Internet of Things (IoT), it is essential to obtain electric signals so that they can be easily converted into information suitable for a connection between several instruments. Otherwise, the sensors that can output the signals in the optical form are needed so that we can convert the signal for quantitative data using smartphone-based capturing or image recognition technique. To meet these requirements, aptameric sensors can be applicable to convert signals into information suitable for transmission and applied to a ubiquitous network or IoT. In this chapter, we would like to address the features of aptamers and how they can be used for IoT-connected biosensors and give several provide several study outcomes that have accomplished monitoring components in food, water, and blood.

8.2 Aptamer-immobilized electrochemical sensor

A great advantage of using aptamers is that their sequences can be chemically synthesized and modified easily. This feature enables us to develop several types of aptamer immobilized sensors. By immobilizing aptamers to the electrode, we could obtain an electrical signal that can be easily transferred to the network using data communication.

Chemical, Gas, and Biosensors for the Internet of Things and Related Applications.
DOI: https://doi.org/10.1016/B978-0-12-815409-0.00008-5

As representative immobilization methods, there are three representative methods. One of the simple methods is the adsorption process, which immobilizes aptamers to the carrier directly using absorption. A typical absorption process uses the electrostatic adsorption of DNA, which has a negative charge derived from a phosphoric acid, and on a carrier charged positively by treatment with polylysine, nitrocellulose, aminophenyltrimethoxylene [4–6]. Although the absorption process can immobilize aptamers without labor, the amounts of immobilized aptamers are dependent on the area of the carrier.

To immobilize large amounts of aptamers, an entrapment process is widely used. This is a method of entrapping and immobilizing aptamer by incorporating DNA when forming a polymer such as polyaniline or polypyrrole by electropolymerization on a carrier [7,8]. It is also possible to immobilize avidin on the carrier to entrap biotin-modified DNA. In addition to polymers, a porous material such as dendrion can be used for the entrapment process [9]. When the entrapment method is used, the large amounts of oligonucleotides can be immobilized, however, it is difficult to maintain the immobilized state and the effect of leakage after immobilization should be a concern. Moreover, when the polymer itself is positively charged there is also an increase in nonspecific adsorption [10].

As another approach to immobilize the aptamers onto the electrode, a covalent binding method is also widely used. This method can immobilize an amino group–modified aptamer on the carrier, which is treated with functional groups, such as carbonyl group or a thiol group via covalent binding. For example, when immobilizing to a graphite electrode or carbon nanotube by covalent bonding method, *N*-Hydroxysuccinimide (NHS) and 1-ethyl-3-(3-dimethylaminopropyl) carbodiimide hydrochloride (EDC) can be used. By adding NHS or EDC, it is possible to convert the carboxylic acid on the surface into an active ester named NHS form, which can trap the amino group–modified aptamers through the coupling reaction [11]. In the case where the carrier is a conductive polymer, it can likewise be immobilized by introducing a functional group such as a carboxyl group or an amino group into the conductive polymer. Additionally, on the same principle, thiol modified aptamers can be immobilized on a gold electrode [12,13]. Also, self-monomolecular structured film (SAM) is often formed using alkyl SH in expectation of orientation control of the aptamer and suppression of nonspecific adsorption to the electrode [14,15]. Because this method can be applied for various carriers, including gold nanoparticles, we can say that this gold electrode modification is the most widely used method among the covalent bonding methods [16–18].

8.3 Detection using complementary chain formation

One of the major features of DNA is the ability to specifically bind to DNA strands that have complementary sequences. Therefore, it is possible to immobilize single-stranded DNA on the carrier using the immobilization method described above and immobilize the aptamer via complementary chain formation by linking the complementary chain to the aptamer. Since this technique can immobilize the aptamer without destroying its structure, it is applied in combination with various immobilization methods. In detail, we can easily fix the single-stranded DNA via complementary chain formation to single-stranded DNA that were immobilized by covalent bonding on the carrier.

8.3.1 Strand displacement assay

For the application of an aptamer sensor, it is possible to electrochemically detect the structural change of DNA by immobilization of the complementary strand, which was modified with redox probes such as methylene blue and ferrocene. Heeger and coworkers immobilized the thrombin aptamer and its complementary chain on the electrode and methylene blue labeling on the complementary side [19]. When the aptamer binds to thrombin, the double-stranded structure collapses and methylene blue of the complementary strand comes close to the electrode surface, so it is detected as an increase in the response-current value. This detection method, which utilizes complementary chain formation and strand displacement reaction of DNA, is called the *Strand displacement assay* and it has been applied to various methods [20,21].

To amplify the signal of the strand displacement assay, another approach to utilize the double aptamers was reported. Xiang and colleagues developed a thrombin detection method using a strand displacement assay through this complementary chain forming part, by designing a sequence in which complementary chain forming sequences are ligated to each of two kinds of aptamers binding to thrombin [22]. Previously, two different thrombin aptamers have been reported for thrombin, and it was revealed that their binding sites are different. In this report, the authors designed a sequence, using a strand displacement reaction, that occurs to a portion of the DNA duplex immobilized on the gold electrode by adding a complementary chain forming sequence to each thrombin aptamer [22] (Fig. 8.1). At this time, since the sequence linked to the aptamer partially forms a complementary chain with the sequence immobilized on the electrode, a part of the immobilized single-stranded DNA is exposed. The methylene blue–labeled single strand binds to this exposed sequence, and the strand displacement reaction occurs again, whereby electron transfer via methylene blue close to the electrode surface occurs. In the end, thrombin is detected by an increase in current value. A feature of this method is that, even if the complementary chain-forming sequence linked to the aptamer is dissociated by the second strand displacement reaction, the strand displacement reaction is repeated by binding to the DNA complementary chain in the vicinity. As a result, the signal obtained from one molecule of thrombin is repeatedly amplified.

8.3.2 Bound/Free separation using complementary chain formation

One of the advantages of immobilizing the recognition element in the target molecule detection system is that bound/free separation (BF separation) can be

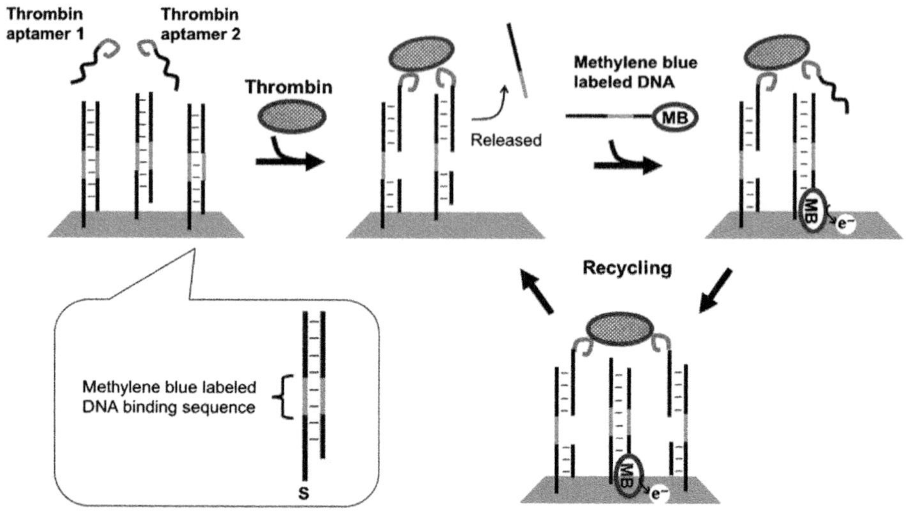

Figure 8.1 Thrombin detection using two thrombin aptamers and the strand displacement assay. Two thrombin aptamers cause strand displacement and the methylene blue–labeled single strand binds to the immobilized DNA chain, which resulted in the change of current value.

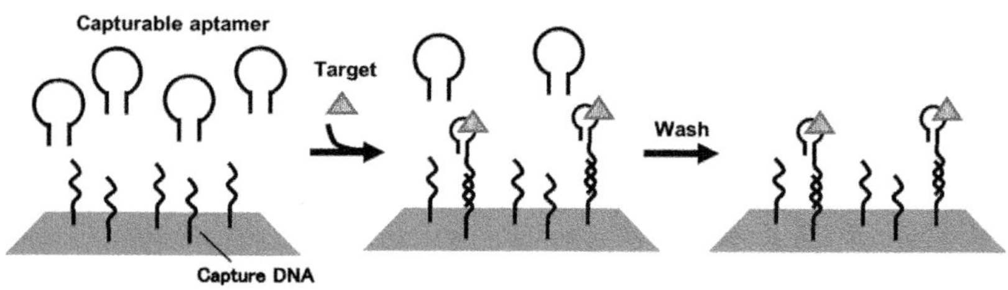

Figure 8.2 Scheme of detection using capturable aptamer and capture DNA. The single stranded part of capturable aptamer can form a complementary chain with the capture DNA so that it can detect the target specifically.

performed. BF separation refers to removal of molecules other than molecules bound to a target substance by a washing technique, and it is possible to greatly suppress the background value in detection of a target molecule. We have reported a method to make BF separation easier by taking this advantage [23]. A single-stranded DNA sequence was immobilized on a carrier as capture DNA. On the other hand, an aptamer sequence and its complementary strand forming sequence were ligated and designed as a capturable aptamer (Fig. 8.2). A capturable aptamer causes a structural change when the target substance is present so that its single stranded part is exposed. The single stranded part of the capturable aptamer can form a complementary chain with the capture DNA, so it is possible to detect only the signal from the aptamer bound to the target substance by performing the washing technique after adding the target substance.

8.4 Aptamer sensor combined with enzymes

As another biosensor element, enzymes are powerful molecular recognition tools because of their high specificity, sensitivity, and various enzymatic reactions, which can be used for broad ranges of applications. Especially, some enzymes that can perform oxidoreduction are suitable for biosensing elements using electrochemical detection. Additionally, aptamers can be easily modified with enzymes by chemically synthesizing or biotin-avidin bonding. Focusing on these features, there are several efforts to detect the target substance by the structural change of aptamers and signal from the enzyme activity [24,25].

Lu and coworkers reported a detection method using invertase that was modified with a complementary strand of single-stranded DNA [26]. This single-stranded DNA can bind to the aptamer immobilized on magnetic beads. When this bead is added to the solution containing the target substance, the aptamer portion immobilized on the beads is peeled off from the complementary chain and binds to the target substance, so that the invertase immobilized by the complementary chain is liberated. At this time, by adding sucrose, which is a substrate of invertase, into the solution, it is possible to detect the target substance by measuring glucose produced by invertase using a commercially available glucose sensor when the target substance is present. In this report, the detection system of cocaine or adenosine was constructed using this principle.

In addition, Tan and colleagues conjugated the aptamer binding to the target substance and its complementary chain to two different polyacrylamide polymers, respectively, and glucoamylase was immobilized in the gel formed by this DNA and polymer [27]. In the presence of the target substance, the aptamer sequence contained in the polymer binds to the target, resulting in a structural change, and the glucoamylase captured in the gel is eluted. At the same time, by adding amylose as a substrate, it becomes possible to detect the target substance by measuring glucose decomposed by glucoamylase with a commercially available glucose sensor. In this report, the authors have succeeded in developing a detection method for cocaine, benzouregonin, and ecogynin methyl ester using this principle.

Also, we reported a method for electrochemically detecting a target substance by selecting an enzyme having direct electron transfer capability with an electrode as an enzyme to be immobilized [28] (Fig. 8.3). Focusing on FAD (flavin adenine dinucleotide) glucose dehydrogenase (FADGDH) having direct electron transfer capability, a gold electrode was prepared in which a sequence was obtained by fusing an aptamer capable of binding to FADGDH and the adenosine aptamer was immobilized. When adenosine, which is the target substance, is present, the guanine quadruplex structure of the aptamer is folded so that FADGDH is brought close to the electrode and the response current value rises directly by electron transfer, so electrochemically adenosine can be detected.

8.5 Utilizing structural change of aptamers to biosensor

One of the features of aptamer is its drastic structure changes under the presence of target molecules. Many aptamers form a specific structure when binding to a target

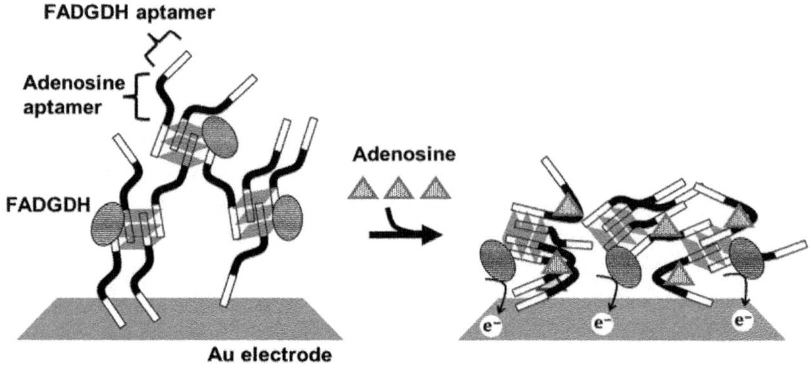

Figure 8.3 Scheme of detection using FADGDH and adenosine aptamer. The structural change of the immobilized DNA sequence containing adenosine aptamer and FADGDH aptamer make FADGDH closer to the electrode so that the target can be detected.

molecule and this feasibility of structural change has been applied for biosensors. In particular, it has been reported that in the detection system, including a field effect transistor, microgravimetric quartz crystal microbalance sensors, and impedance measurements, it is possible to detect a structural change of a negatively charged aptamer and obtain it as an electrical signal [29].

Furthermore, we reported the aptameric enzyme subunit (AES) as a new detection element combining aptamer and enzyme [30]. AES is composed of the enzyme, thrombin, and an aptamer, which is capable of controlling the thrombin activity. This aptamer sequence contains a combination of an aptamer that binds to a target molecule and an aptamer that can inhibit thrombin activity. When a target molecule is present, the aptamer forms a specific structure and binds to thrombin, resulting in a decrease in thrombin activity. On the other hand, in the absence of the target substance, the structure of the aptamer collapses so that the thrombin activity can be observed. By developing a sensor based on this combination of an enzyme and its inhibitory aptamer, we succeeded in the development of a sensing element that does not require a washing operation and can be detected with high sensitivity [31,32]. Furthermore, by combining multiple aptamer sequences or combining with an appropriate electrochemical detection system, development of a sensing device making use of the feature of aptamer called structural change can be said to be useful for the development of a wide variety of sensors [33–35].

8.6 Utilizing structural change of aptamers to biosensor

As described above, a number of aptamers have been reported as novel recognition elements and applied to biosensors. However, among the aptamers, there are several DNA sequences that retain activity that resembles enzyme activity when forming a specific structure [36,37]. Such a DNA sequence is called a DNAzyme or aptazyme and has been applied to various detection mechanisms. A major feature of DNAzyme is that it has both enzymatic activity and characteristics of DNA, such as complementary chain formation and various structural changes. For this reason, development of various detection systems has been attempted by combining with DNAzymes and other designed DNA sequences.

There are several efforts to develop a detection method using DNAzyme that has DNA cleavage activity. For example, Brown and coworkers focused on the fact that the activity of DNAzyme was increased under the presence of specific metal ions and applied this feature to Pb^{2+} detection [38]. As a specific detection mechanism, fluorescence modification is performed on a DNA fragment that is a substrate to be cleaved, whereas a DNAzyme is modified with a quencher (Fig. 8.4). In this case, when the concentration of the metal ion is low, cleavage does not occur even if the substrate DNA forms a complementary strand with the DNAzyme, so that the fluorescent group is quenched by the quencher and the fluorescence is not observed. On the other hand, when the metal ion concentration is high, the activity of DNAzyme increases and substrate DNA fragment will be cleaved. After the digestion of substrate DNA, the quencher and the fluorescent group are physically separated so that the fluorescence is observed. Based on this detection principle, this report succeeded in detecting lead ions contained in Lake Michigan. This lead detection method was recently combined with a microchannel to improve its sensitivity to the subfemtomolar order [39]. Also, combining Pb^{2+} sensitive DNAzyme and copper sensitive DNAzyme, multiple metal ions can be detected [40]. Furthermore, a detection system capable of colorimetric determination by modifying a thiol group to a DNAzyme or a substrate DNA fragment to be combined with gold nanoparticles has also been developed [41].

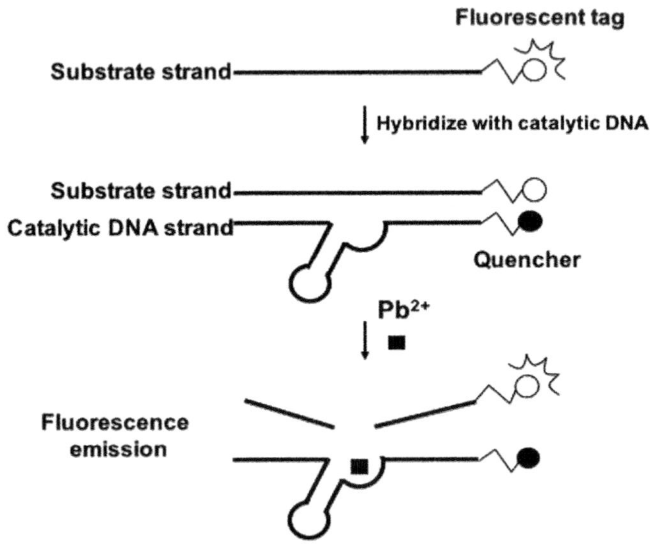

Figure 8.4 Scheme of detection using fluorescent tagged substrate strand and catalytic DNA strand.

Another example of a detection mechanism using a DNAzyme is a detection system using a DNA sequence having peroxidase activity. It has been reported that hemin aptamer has a peroxidase activity [42]. By immobilizing this hemin aptamer sequence on the gold electrode after thiol modification, it is possible to electrochemically detect hemin. In addition, a method for detecting a change in color using 2,2′-azino-bis (3-ethyl-benzothiazoline-6-sulfonic acid, ABTS) as a substrate has also been widely applied [43].

For example, Wang and coworkers reported a mercury detection system that focused on the fact that thymine and mercury present in the DNA sequence bind to each other and destroy the aptamer structure [44]. In this report, a DNA fragment having both a thymine-rich sequence and a hemin aptamer sequence was designed. In the presence of mercury, the structure of the aptamer was broken and the activity of the DNAzyme was not observed. Conversely, in the absence of mercury, the aptamer forms a structure and binds to hemin, and the peroxidase activity can be observed by colorimetric determination of ABTS.

In addition, Li and coworkers combine a DNA fragment having an aptamer sequence capable of binding to a target substance and a primer sequence, which can ligate to a cyclic DNA harboring a hemin aptamer sequence (Fig. 8.5) [45]. In the presence of the target, the cyclic DNA can bind to the target-DNA fragment complex and the sequence of cyclic DNA can be amplified by using DNA polymerase. As a result, huge signals from numbers of hemine aptamer sequences can be obtained from one target molecule, which resulted in high sensitivity. In this report, they succeeded in detecting growth factor B-chain (PDGF-BB), a platelet-derived of 2 pg/mL within 2 hours.

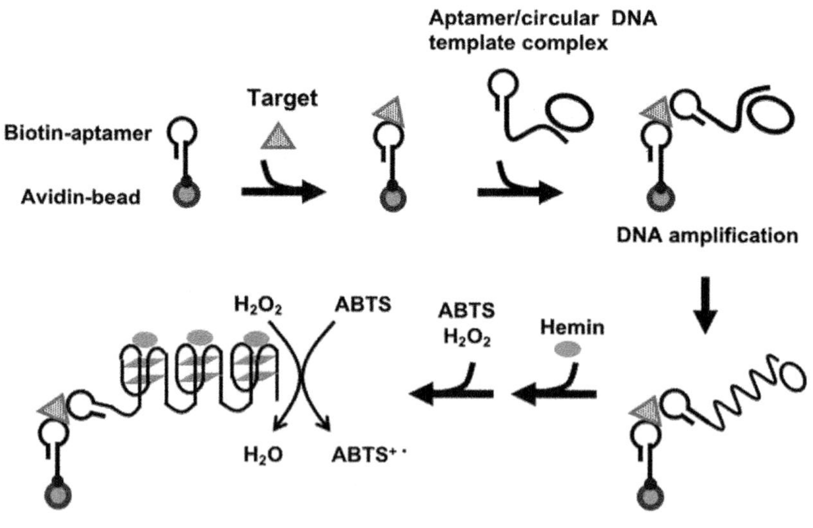

Figure 8.5 Detection method utilizing hemin aptamer and circular DNA amplification. Aptamer/circular DNA template complex can amplify the hemin aptamer sequence so that the target can be detected by DNAzyme activity.

Thus, in the detection method using DNAzyme, it is possible to output various signals such as electrochemistry, fluorescence, and colorimetric quantification by designing the detection method. In addition, the signal can be easily enhanced simply by amplifying the active DNA sequence with a polymerase.

8.7 Development of highly sensitive sensors by amplifying DNA strands

A major advantage of using an aptamer as a molecular recognition element is that the detection signal can be easily enhanced. The DNA sequence can be extended easily by using DNA polymerase, and also it is possible to combine and amplify fragmented DNA sequences by repeating complementary chain formation without using polymerase.

For example, we would like to introduce the detection system using rolling cycle amplification (RCA). Although RCA is one of the DNA amplification reactions using DNA polymerase, it is possible to synthesize a long DNA sequence containing a specific sequence periodically by using a circular DNA strand as a template [46,47]. An amplified signal can be obtained by binding a DNA fragment obtained by modifying a quantum dot, a fluorescent group, or an enzyme itself for detection to a DNA strand obtained by elongating a template strand.

As another example, hybridization chain reaction (HCR) can be mentioned as a method for amplifying DNA fragments without requiring enzymes such as DNA polymerase. HCR is an amplification reaction using two DNA fragments having complementary chain forming sequences, and is a technique in which two kinds of sequences alternately repeat complementary chain formation to form a long DNA fragment.

For the application of HCR, Zuo and coworkers reported a microRNA detection method combining HCR-capable DNA fragments with a tetrapod-type structure composed of DNA (Fig. 8.6) [48]. In this method, when the targeted microRNA

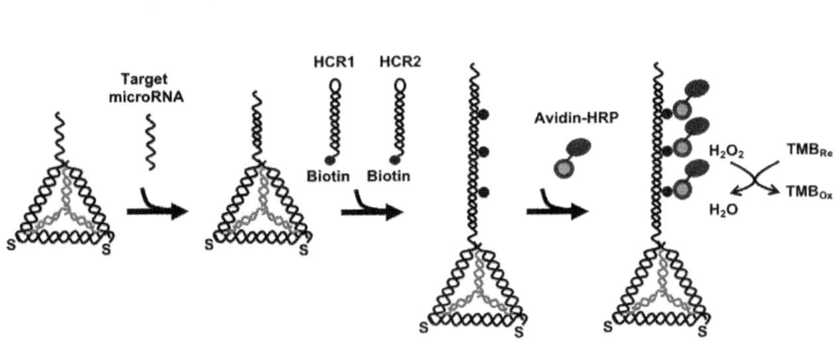

Figure 8.6 Detection method utilizing HCR mediated amplification. The target microRNA can bind to the tetrapod-type DNA structure so that the hybridization chain reaction sequences (HCR1, HCR2) and avidin fused HRP can bind to the DNA structure.

binds to the tetrapod-type DNA structure, two types of biotin-modified HCR sequences bind to the microRNA and extend the length by HCR. By linking avidin-fused horseradish peroxidase (HRP) to the HCR sequences and detecting the reaction of HRP, it is possible to amplify the signal obtained from one microRNA. As a result, this report succeeded in detecting 10 aM microRNA and it can be said that it is a highly sensitive detection method.

8.8 Colorimetric detection using aptameric sensor and smart devices

Development of biosensors using aptamers and DNA fragments has been widely performed so far, but in recent years, it has been attempted to develop a biosensor combining with smart devices. In particular, some efforts were reported to apply colorimetric quantification using small cameras attached to smartphones.

For example, Chen and colleagues have developed a colorimetric method using gold nanoparticles and DNA fragments for the purpose of detection of mercury contained in the environment (Fig. 8.7) [49]. It was known that mercury can bind with thymine in the DNA fragment to form a hairpin structure. In this method, the DNA fragment, which was fixed to the beads, is released from the beads when the mercury binds to the DNA fragments because of its structural change. After release of

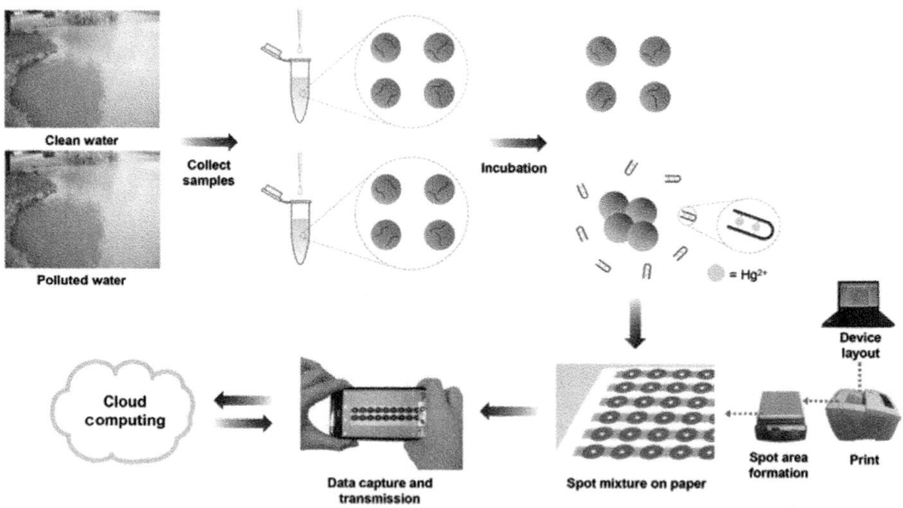

Figure 8.7 Detection of mercury in polluted water using smartphone. The colorimetric detection can be performed by the change of the gold nanoparticles and smartphone. Source: Reprinted with permission from G.H. Chen, W.Y. Chen, Y.C. Yen, C.W. Wang, H. T. Chang, C.F. Chen, Detection of mercury(II) ions using colorimetric gold nanoparticles on paper-based analytical devices, Anal. Chem. 86 (14) (2014) 6843–6849. Copyright 2014, American Chemical Society.

the DNA fragments, the gold nanoparticles, which removes its negative charge, can form aggregation and surface plasmon coupling, resulting in a blue coloration. By using this scheme, it is possible to detect mercury by its colorimetric determination. The characteristic point of this method is that the wax coated cellulose paper, named μPAD, is used for detection. To use this μPAD, a circle for a spot is printed to μPAD using a commercially available printer. After spotting the sample on the circle drawn on μPAD, colorimetric determination is performed by photographing the circular pattern with the smartphone and analyzing the picture with application software. As a result, they succeeded in detecting mercury above 25 nM in this report.

In addition, a detection system combining a smartphone with a complementary strand of DNA has also been developed. Zheng and colleagues developed a method of streptomycin detection by using a streptomycin aptamer and a sequence capable of forming a complementary chain with the aptamer (Fig. 8.8) [50]. Under the presence of streptomycin, the aptamer sequences become DNA single strands and bind to streptomycin, resulting in the decrease of double stranded DNA (dsDNA). When SYBR Green, one of the dsDNA binding fluorescence dyes, was added to the sample, its fluorescence was decreased. On the other hand, when streptomycin is not present in the sample, since the aptamer does not bind to the target and most of the DNA fragments are in a state of forming complementary strands, the fluorescence of SYBR Green will be increased. In short, we can detect the target component just by mixing samples and DNA sequences and observe its fluorescence. This detection principle is classical, but the feature of this method is that the authors succeeded in detecting this change in fluorescence intensity using a smartphone. However, this method employs fluorescence for detection and the signals cannot be detected unless excitation light is applied to the SYBR Green reagent, so it is necessary to prepare a black box and UV lamp in addition to the smartphone.

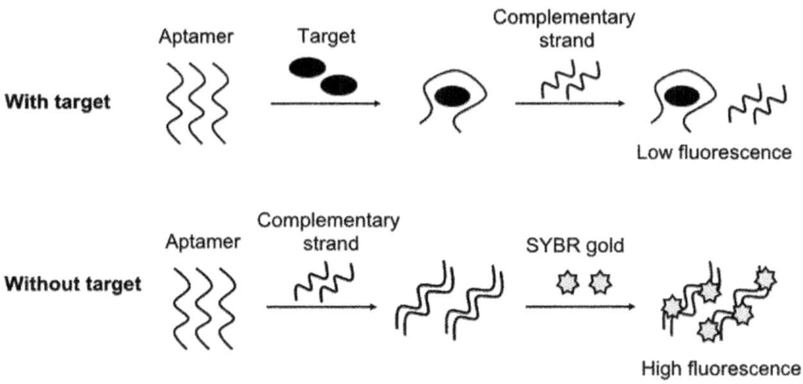

Figure 8.8 Fluorescence detection scheme using aptamer and its complementary strand. With target, the complementary strand cannot bind to the aptamer sequence, which caused low fluorescence, while complementary strand can emit high fluorescence without target.

8.9 Conclusion

In this chapter, we introduced several remarkable features of aptamers and basic principles of aptameric sensors. Because it is easy to chemically modify the aptamer's sequence, an aptamer can be immobilized to various carriers, which enables us to develop a detection method using electrochemical sensing, colorimetric sensing, and fluorescence. Also, the aptamer and other DNA sequences can be easily amplified with or without enzymes, so we can sensitize the aptameric sensors. Additionally, as described above, aptamer sensors are also applied to the detection using smart devices recently. Especially, colorimetric determination can be easily applied to smartphone-based detection because we can obtain and analyze the results just by taking a photograph. In the case of fluorescence detection, some additional equipment may be necessary; however, various detection principles can be applied for the sensor because DNA can be modified easily. Taking advantage of the high image analysis technology of smartphones and the great merit of being able to digitize immediately leads us to share information on the Internet quickly. It is considered to be a powerful means to directly link biosensors to IoT. For that reason, aptamer sensors capable of various output including colorimetric quantification and electrochemical detection can be expected to be applied to detection systems directly connected to IoT.

References

[1] T. Mairal, V.C. Ozalp, P. Lozano Sánchez, M. Mir, I. Katakis, C.K. O'Sullivan, Aptamers: molecular tools for analytical applications, Anal. Bioanal. Chem. 390 (4) (2008) 989–1007.

[2] K.S. Park, Nucleic acid aptamer-based methods for diagnosis of infections, Biosens. Bioelectron. 102 (2018) 179–188.

[3] K.M. Song, S. Lee, C. Ban, Aptamers and their biological applications, Sensors (Basel) 12 (1) (2012) 612–631.

[4] R.C. Williams, M.J. Chamberlin, Electron microscope studies of transient complexes formed between Escherichia coli RNA polymerase holoenzyme and T7 DNA, Proc. Natl. Acad. Sci. U.S.A. 74 (9) (1977) 3740–3744.

[5] P.L. Dolan, Y. Wu, L.K. Ista, R.L. Metzenberg, M.A. Nelson, G.P. Lopez, Robust and efficient synthetic method for forming DNA microarrays, Nucleic Acids Res. 29 (21) (2001) E107.

[6] B.A. Stillman, J.L. Tonkinson, FAST slides: a novel surface for microarrays, Biotechniques 29 (3) (2000) 630–635.

[7] P.P. Sengupta, J.N. Gloria, M.K. Parker, A.S. Flynt, A polyaniline-based sensor of nucleic acids, J. Vis. Exp. (117) (2016).

[8] W. Joseph, M. Jiang, A. Fortes, B. Mukherijee, New label-free DNA recognition based on doping nucleic-acid probes within conducting polymer films, Anal. Chim. Acta 402 (1999).

[9] B.J. Hong, S.J. Oh, T.O. Youn, S.H. Kwon, J.W. Park, Nanoscale-controlled spacing provides DNA microarrays with the SNP discrimination efficiency in solution phase, Langmuir 21 (10) (2005) 4257–4261.

[10] S.B. Nimse, K. Song, M.D. Sonawane, D.R. Sayyed, T. Kim, Immobilization techniques for microarray: challenges and applications, Sensors (Basel) 14 (12) (2014) 22208–22229.
[11] K. Maehashi, T. Katsura, K. Kerman, Y. Takamura, K. Matsumoto, E. Tamiya, Label-free protein biosensor based on aptamer-modified carbon nanotube field-effect transistors, Anal. Chem. 79 (2) (2007) 782–787.
[12] A.B. Steel, T.M. Herne, M.J. Tarlov, Electrochemical quantitation of DNA immobilized on gold, Anal. Chem. 70 (22) (1998) 4670–4677.
[13] J.A. Wood, S. Basray, S. Bok, Enhanced DNA detection through the incorporation of nanocones and cavities into a plasmonic grating sensor platform, IEEE Send. J. 16 (2015) 3403–3408.
[14] J.C. Love, L.A. Estroff, J.K. Kriebel, R.G. Nuzzo, G.M. Whitesides, Self-assembled monolayers of thiolates on metals as a form of nanotechnology, Chem. Rev. 105 (4) (2005) 1103–1169.
[15] C. Nicosia, J. Huskens, Reactive self-assembled monolayers: from surface functionalization to gradient formation, Mater Horizons 1 (2014) 32.
[16] A. Miodek, E.M. Regan, N. Bhalla, N.A. Hopkins, S.A. Goodchild, P. Estrela, Optimisation and characterisation of anti-fouling ternary SAM layers for impedance-based aptasensors, Sensors (Basel) 15 (10) (2015) 25015–25032.
[17] W. Xiao, M. Xiao, Q. Fu, S. Yu, H. Shen, H. Bian, et al., A portable smart-phone readout device for the detection of mercury contamination based on an aptamer-assay nanosensor, Sensors 16 (11) (2016) 1871.
[18] Q. Wei, R. Nagi, K. Sadeghi, S. Feng, E. Yan, S.J. Ki, et al., Detection and spatial mapping of mercury contamination in water samples using a smart-phone, ACS Nano 8 (2) (2014) 1121–1129.
[19] Y. Xiao, B.D. Piorek, K.W. Plaxco, A.J. Heeger, A reagentless signal-on architecture for electronic, aptamer-based sensors via target-induced strand displacement, J. Am. Chem. Soc. 127 (51) (2005) 17990–17991.
[20] C.P. Vary, F.J. McMahon, F.P. Barbone, S.E. Diamond, Nonisotopic detection methods for strand displacement assays of nucleic acids, Clin. Chem. 32 (9) (1986) 1696–1701.
[21] T.G. Walker, J.G. Nadeau, C.P. Linn, A DNA probe assay using strand displacement amplification (SDA) and filtration to separate reacted and unreacted detector probes, Mol. Cell. Probes 9 (1995) 399–403.
[22] J. Yang, B. Dou, R. Yuan, Y. Xiang, Aptamer/protein proximity binding-triggered molecular machine for amplified electrochemical sensing of thrombin, Anal. Chem. 89 (9) (2017) 5138–5143.
[23] D. Ogasawara, N.S. Hachiya, K. Kaneko, K. Sode, K. Ikebukuro, Detection system based on the conformational change in an aptamer and its application to simple bound/free separation, Biosens. Bioelectron. 24 (5) (2009) 1372–1376.
[24] W. Yoshida, K. Abe, K. Ikebukuro, Emerging techniques employed in aptamer-based diagnostic tests, Expert Rev. Mol. Diagn. 14 (2) (2014) 143–151.
[25] K. Abe, W. Yoshida, K. Ikebukuro, Electrochemical biosensors using aptamers for theranostics, Adv. Biochem. Eng. Biotechnol. 140 (2014) 183–202.
[26] Y. Xiang, Y. Lu, Using personal glucose meters and functional DNA sensors to quantify a variety of analytical targets, Nat. Chem. 3 (9) (2011) 697–703.
[27] Z. Zhu, C. Wu, H. Liu, Y. Zou, X. Zhang, H. Kang, et al., An aptamer cross-linked hydrogel as a colorimetric platform for visual detection, Angew. Chem. Int. Ed. Engl. 49 (6) (2010) 1052–1056.

[28] Y. Morita, W. Yoshida, N. Savory, S.W. Han, M. Tera, K. Nagasawa, et al., Development of a novel biosensing system based on the structural change of a polymerized guanine-quadruplex DNA nanostructure, Biosens. Bioelectron. 26 (12) (2011) 4837–4841.
[29] O. Willner, M. Zyats, Electronic aptamer-based sensors, Angew. Chem. 46 (34) (2007) 6408–6418.
[30] W. Yoshida, K. Sode, K. Ikebukuro, Aptameric enzyme subunit for biosensing based on enzymatic activity measurement, Anal. Chem. 78 (10) (2006) 3296–3303.
[31] W. Yoshida, K. Sode, K. Ikebukuro, Homogeneous DNA sensing using enzyme-inhibiting DNA aptamers, Biochem. Biophys. Res. Commun. 348 (1) (2006) 245–252.
[32] W. Yoshida, E. Mochizuki, M. Takase, H. Hasegawa, Y. Morita, H. Yamazaki, et al., Selection of DNA aptamers against insulin and construction of an aptameric enzyme subunit for insulin sensing, Biosens. Bioelectron. 24 (5) (2009) 1116–1120.
[33] K. Ikebukuro, W. Yoshida, K. Sode, Aptameric enzyme subunit for homogeneous DNA sensing, Biotechnol. Lett. 30 (2) (2008) 243–252.
[34] W. Yoshida, K. Sode, K. Ikebukuro, Label-free homogeneous detection of immunoglobulin E by an aptameric enzyme subunit, Biotechnol. Lett. 30 (3) (2008) 421–425.
[35] W. Yoshida, K. Sode, K. Ikebukuro, Aptameric enzyme subunit for homogeneous protein sensing, Nucleic Acids Symp. Ser. (Oxf). (51) (2007) 99–100.
[36] Y. Li, D. Sen, Toward an efficient DNAzyme, Biochemistry 36 (18) (1997) 5589.
[37] C.R. Geyer, D. Sen, Evidence for the metal-cofactor independence of an RNA phosphodiester-cleaving DNA enzyme, Chem. Biol. (8) (1997) 579.
[38] Y. Lu, J. Liu, P.J. Bruesehoff, C.M. Pavot, A.K. Brown, New highly sensitive and selective catalytic DNA biosensors for metal ions, Biosens. Bioelectron. 5-6 (2003) 529.
[39] K.A. Shaikh, K.S. Ryu, E.D. Goluch, J.M. Nam, J. Liu, C.S. Thaxton, et al., A modular microfluidic architecture for integrated biochemical analysis, Proc. Natl. Acad. Sci. U. S.A. 102 (28) (2005) 9745–9750.
[40] P. Zuo, B.C. Yin, B.C. Ye, DNAzyme-based microarray for highly sensitive determination of metal ions, Biosens. Bioelectron. 25 (4) (2009) 935–939.
[41] D. Mazumdar, J. Liu, G. Lu, J. Zhou, Y. Lu, Easy-to-use dipstick tests for detection of lead in paints using non-cross-linked gold nanoparticle-DNAzyme conjugates, Chem. Commun. (Camb). 46 (9) (2010) 1416–1418.
[42] P. Travascio, Y. Li, D. Sen, DNA-enhanced peroxidase activity of a DNA-aptamer-hemin complex, Chem. Biol. 5 (9) (1998) 505–517.
[43] G. Pelossof, R. Tel-Vered, J. Elbaz, I. Willner, Amplified biosensing using the horseradish peroxidase-mimicking DNAzyme as an electrocatalyst, Anal. Chem. 82 (11) (2010) 4396–4402.
[44] T. Li, S. Dong, E. Yang, Label-free colorimetric detection of aqueous mercury ion (Hg^{2+}) using Hg^{2+}-modulated G-quadruplex-based DNAzymes, Anal. Chem. 81 (2009) 2144.
[45] L. Tang, Y. Liu, M.M. Ali, D.K. Kang, W. Zhao, J. Li, Colorimetric and ultrasensitive bioassay based on a dual-amplification system using aptamer and DNAzyme, Anal. Chem. 84 (11) (2012) 4711–4717.
[46] H. Ji, F. Yan, J. Lei, H. Ju, Ultrasensitive electrochemical detection of nucleic acids by template enhanced hybridization followed with rolling circle amplification, Anal. Chem. 84 (16) (2012) 7166–7171.
[47] C. Feng, X. Mao, Y. Yang, X. Zhu, Y. Yin, G. Li, Rolling circle amplification in electrochemical biosensor with biomedical applications, J. Electroanal. Chem. 781 (2016) 223–232.

[48] Z. Ge, M. Lin, P. Wang, H. Pei, J. Yan, J. Shi, et al., Hybridization chain reaction amplification of microRNA detection with a tetrahedral DNA nanostructure-based electrochemical biosensor, Anal. Chem. 86 (4) (2014) 2124–2130.
[49] G.H. Chen, W.Y. Chen, Y.C. Yen, C.W. Wang, H.T. Chang, C.F. Chen, Detection of mercury(II) ions using colorimetric gold nanoparticles on paper-based analytical devices, Anal. Chem. 86 (14) (2014) 6843–6849.
[50] B. Lin, Y. Yu, Y. Cao, M. Guo, D. Zhu, J. Dai, et al., Point-of-care testing for streptomycin based on aptamer recognizing and digital image colorimetry by smartphone, Biosens. Bioelectron. 100 (2018) 482–489.

Electrochemical sensing techniques using carbon electrodes prepared by electrolysis toward environmental Internet of Things sensor

9

Hiroaki Matsuura
Department of Life Science & Green Chemistry, Faculty of Engineering, Saitama Institute of Technology, Fukaya, Japan

9.1 Introduction

9.1.1 Electrochemical monitoring support Internet of Things services

Internet of Things (IoT) is an emerging technology that is experiencing rapid developmental growth. This technology is embodied in a wide field of networked products, systems, and chemical sensors. In particular, chemical sensor applications supported by IoT services have been developed for use in smart home and smart health applications, as well as the environment, medicine, food, and industry. For environmental sensors, IoT technology is particularly important because remote sensing with number of sensors dispersed in the fields enables to detect local environmental information such as local pollution, and gas leaking in real-time without human labor as illustrated in Fig. 9.1. To realize the above systems, there are some requirements for developing future environmental sensors. The sensors should be miniaturized and inexpensive while maintaining their performance in terms of sensitivity, detection limit, and long-term stability. The operation of the sensors should not be complicated and human-free operation is preferable. Electrochemical sensors are one of the candidates for realizing such systems. However, novel metals including gold and platinum are expensive, although such electrodes show good electrochemical performance.

Carbon materials have been widely used as a sensing electrode of electrochemical detection owing to a low background current, a wide potential window, low cost, and chemical inertness as compared with metal electrodes [1–5]. In addition, the chemically modified carbon surfaces have been of significant interest in various fields including electroanalytical chemistry and electrocatalysis [6,7]. Recently, it

Chemical, Gas, and Biosensors for the Internet of Things and Related Applications.
DOI: https://doi.org/10.1016/B978-0-12-815409-0.00009-7

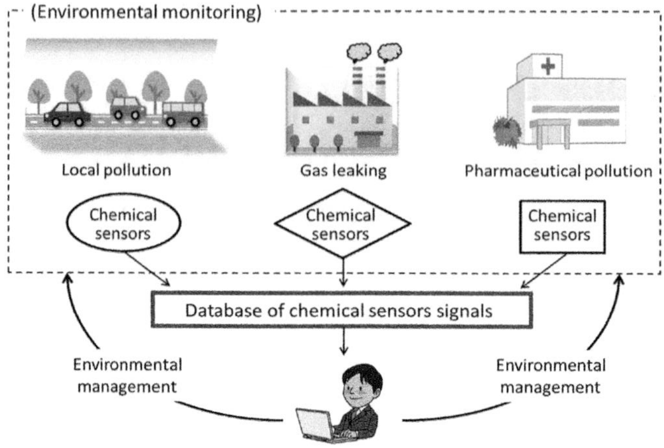

Figure 9.1 Chemical sensor applications supported by Internet of Things services.

has been clarified that nitrogen atoms containing functional groups can be easily introduced onto the glassy carbon (GC) and carbon felt (CF) based-electrode (as shown in Fig. 9.2) surface by the electrochemical oxidation of ammonium carbamate in an aqueous solution at a highly positive electrode potential [8].

Especially, it has been revealed by us that nitrogen-containing functional groups such as amino group can easily be introduced to the surfaces of the carbon surface by the electrochemical oxidation of ammonium carbamate [9,10]. In this electrochemical surface modification process, not only the primary amine group (i.e., aniline-like aromatic amine moiety) but also other nitrogen-containing functional groups (i.e., the secondary amine-like moieties containing pyrrole-type and quaternary amine-like moieties containing graphitic quaternary nitrogen) can be introduced onto the carbon electrode surface [11,12].

In contrast, electrocatalytic redox waves between the hydrogen ion (H^+) and hydrogen molecule (H_2) were observed after the long-term electrochemical reduction of the nitrogen atoms containing functional groups introduced carbon electrode in a strong acid electrolyte. During the electrode reduction of the electrooxidized carbon electrode in sulfuric acid electrolyte, platinum ion dissolved from platinum wire counter electrode is electrodeposited on the surface of nitrogen atoms containing functional groups introduced carbon electrode surface [13]. The consumption of platinum can be reduced by this electrode fabrication method, and the high electrocatalytic activity of hydrogen oxidation can be achieved. Since amperometric detection, which is most widely employed for electroanalysis, requires a calibration curve, this prevents realization of IoT sensors that require maintenance free and remote sensing. The authors have developed calibration free coulometric sensors, which can be achieved using electrodes with high electrocatalytic activity. In addition, the nitrogen atoms functional groups introduced carbon surface is advantageous for such applications.

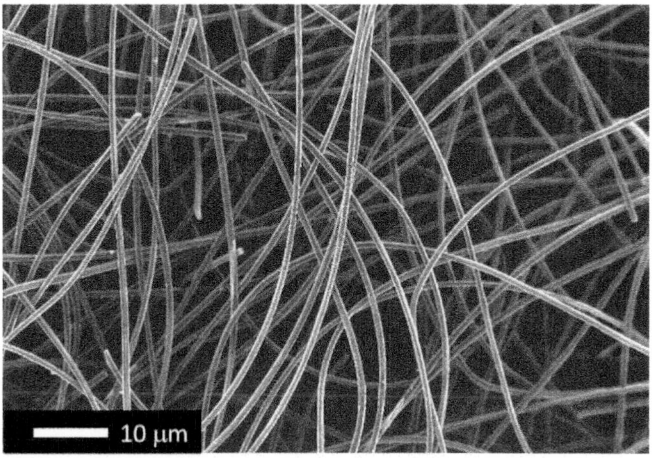

Figure 9.2 SEM image of carbon felt surface.

This chapter discusses the electrochemical modification of carbon surface and the development of electrochemical sensor using the above modified electrodes for detecting environmental species.

9.1.2 Carbon electrode surface activation

As mentioned in the exordium, carbon materials have been fabricated by various procedures. The term of activation has been used to describe procedures for increasing the electrocatalytic activity of carbon electrodes. Recently, it has been found that the oxygen reduction potential can be significantly moved to positive direction by using nitrogen-containing carbon alloy [14] and graphite with nitrogen [9]. Especially, it has been revealed that nitrogen atoms containing functional groups can simply be introduced to the surfaces of the CF and GC electrodes by the electrode oxidation of ammonium carbamate [8], and the electron transfer rates of many inorganic and organic compounds are accelerated to be able to detect excellent redox waves [15].

In addition, we have recently described an electrochemically activated platinum/carbon electrode incorporating nitrogen atom containing functional groups that was fabricated by stepwise electrolysis. This electrode exhibited specific electrocatalytic activity in relation to hydrogen molecule oxidation reaction as determined by hydrodynamic voltammetry. In spite of relatively large platinum particles, they are tightly immobilized on the activated carbon electrode compared with electrodeposited platinum on unmodified GC electrode. Moreover, we first successfully developed an amperometric sensor for dissolved hydrogen molecule based on flow injection analysis (FIA) using the electrochemically activated carbon electrode modified with platinum particles as the working electrode. Although platinum particles modified nitrogen-terminated carbon electrodes using carbon powder materials have been reported, our nitrogen-containing functional groups introduced GC electrode is the first reported to use a solid electrode for electrodeposition of platinum particles and applied for oxidizing hydrogen molecule. The solid electrodes are more advantageous for developing hydrogen sensors because they are freestanding and do not require other processes for electrode fabrication such as high temperature treatment and modification of Nafion.

9.2 Chemical sensors using electrochemical activated carbon electrodes

9.2.1 *Electrochemical activated techniques for aminated electrode preparation*

For introducing the amino group onto the GC electrode, a bare GC electrode was electrooxidized in 0.1 M ammonium carbamate aqueous solution at +1.1 V (vs Ag/AgCl) for 60 minutes. In this modification process, not only the primary amine group but also other nitrogen-containing functional groups can be introduced onto the GC electrode surface. We named this electrooxidized GC electrode as an aminated GC electrode.

9.2.2 *Electrochemical activated techniques for electrodeposited platinum particles on glassy carbon electrode modified with nitrogen-containing functional groups*

More recently, we obtained very interesting results of a novel redox waves between hydrogen ion (H^+) and hydrogen molecule (H_2) appearing at highly positive potential range after the long-term electrode reduction of the aminated GC electrode in a strong acid electrolyte, and we named this electroreduced aminated GC electrode as electrodeposited platinum particles on GC electrodes modified with nitrogen-containing functional groups (Pt-NGC). The Pt-NGC electrode was fabricated by stepwise electrolysis as follows. First, a bare GC electrode was electrooxidized in ammonium carbamate aqueous solution at +1.1 V (vs Ag/AgCl) for 60 minutes. Next, the electrooxidized GC electrode (aminated GC electrode) was electroreduced in 1.0 M sulfuric acid at −1.0 V (vs Ag/AgCl) for 20 hours at room temperature. During the electroreduction of the electrooxidized GC electrode (aminated GC electrode) in sulfuric acid electrolyte, platinum ion dissolved from platinum wire counter electrode is electrodeposited on the surface of nitrogen-containing functional groups introduced GC electrode.

9.3 Electrocatalytic activity and analytical performance

Nitrogen atoms containing functional groups such as amino groups were introduced by the electrooxidation of CF electrode in an ammonium carbamate aqueous solution. We found that the reduction wave of hypochlorite can be well separated from that of oxygen in the cyclic voltammogram (CV) obtained by using this electrooxidized CF electrode as a working electrode (Fig. 9.3). This result means that the selective reduction wave of hypochlorite appears due to an electrochemical modification of electrocatalytic active sites containing nitrogen atoms functional groups.

This electromodified CF electrode was applied for batch injection coulometric measurement of hypochlorite without dissolved oxygen interference. The cross section of the fabricated batch injection coulometric cell is shown in Fig. 9.4. A porous CF electrode has been used as a suitable working electrode for a coulometric measurement. In this coulometric cell, the analyte added to the working electrode surface is diffused into the inner part of the CF quickly in a three-dimensional direction, while undergoing the electrochemical reaction. Therefore we adopted this coulometric measurement technique to perform the rapid and simple detection. The procedure for the coulometric determination is described as follows. First, the electrode potential between a working electrode and a counter electrode was maintained at a constant value by a potentiogalvanostat. Next, an aliquot of the sample solution was added to the center of the working electrode using a micropipet after the residual current reached a constant steady-state value. The concentration of the analyte was calculated from total electrical charges measured by a coulometer with no need of calibration curve.

The reproducible current versus time curves were obtained by the repetitive measurement of hypochlorite and the relative standard deviation (RSD) for 10 successive measurements was 1.5%. Linear relationship was observed in the concentration range up to 57 ppm ($R^2 = 0.9997$) with detection limit of 0.18 ppm ($S/N = 3$). From these results, our proposed batch injection coulometric sensing technique is very

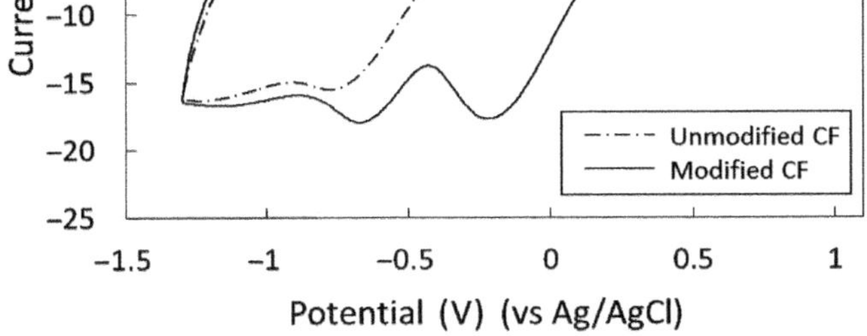

Figure 9.3 Cyclic voltammograms of carbon felt electrode modified with nitrogen-containing functional groups in 0.1 M PBS solution (pH 7.0) with 1.0 mM hypochlorite (solid line) and without hypochlorite (dotted line). Potential sweep rate: 50 mV/s.

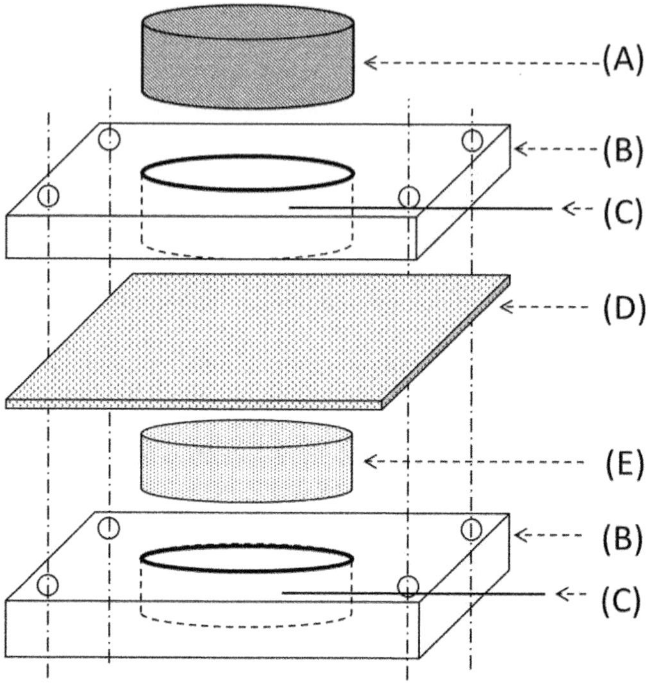

Figure 9.4 Configuration of batch injection coulometric cell: (A) working modified electrode, (B) acrylic plate, (C) Pt lead wire, (D) ion exchange membrane, and (E) counter carbon felt.

promising for hypochlorite detection without dissolved oxygen interference in practical applications.

An amperometric determination based on FIA of hydrogen peroxide was first developed using electrodeposited platinum particles on GC electrodes modified with nitrogen-containing functional groups (Pt-NGC) prepared by stepwise electrolysis as the working electrode. We demonstrated that the electrocatalytic redox waves of hydrogen peroxide can be observed by using Pt-NGC electrode fabricated by the electrode reduction of the aminated GC electrode in sulfuric acid electrolyte. The properties of the electrocatalytic redox reactions for hydrogen peroxide have been investigated by cyclic voltammetric measurements.

Fig. 9.5 shows CVs obtained using Pt-NGC electrode in a 0.1 M phosphate buffer electrolyte (pH 7.0) with or without hydrogen peroxide (1 mM) after high purity nitrogen gas bubbling. In the CV of the Pt-NGC electrode at the electrolyte with 1 mM hydrogen peroxide, the redox pair was observed (solid line in Fig. 9.5). The potentials of this redox wave are corresponding to those that electrode reaction of hydrogen peroxide using Pt-NGC electrode. This behavior presented by Pt-NGC electrode is typical as regards the electrocatalytic redox properties of hydrogen peroxide in medium. We constructed an amperometric sensor based on FIA system of

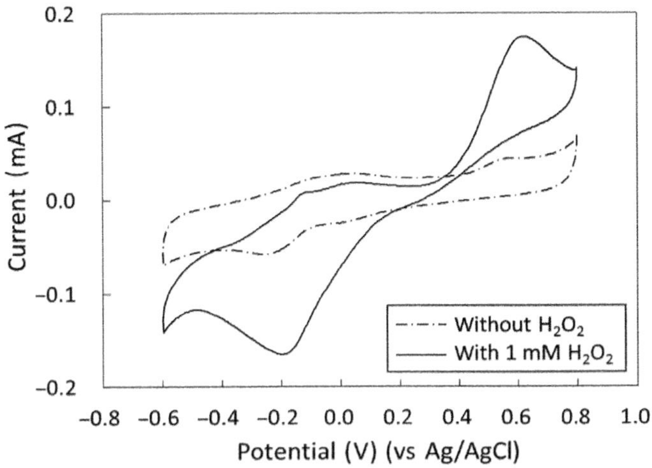

Figure 9.5 Cyclic voltammograms of Pt-NGC electrode in 0.1 M phosphate buffer solution (pH 7.0) with 1 mM H_2O_2 and without H_2O_2 after nitrogen gas bubbling. Scan rate: 50 mV/s.

hydrogen peroxide using Pt-NGC as the working electrode. To achieve the rapid measurement of hydrogen peroxide with high sensitivity, we examined the effect of applied potential to the Pt-NGC at a radial flow on the peak current. In such an extremely negative potential region, the electrocatalytic reduction of dissolved oxygen was observed by the electrode reduction, resulting in accurate determination of hydrogen peroxide. Hence, the measurement of hydrogen peroxide was performed by positive potential region. The current responses of the electrode oxidation of hydrogen peroxide increased with increasing at the positive applied potentials. On the other hand, the background current noise also increased when the electrode potential exceeded +0.8 V. Therefore the optimal potential to measure the peak current was determined to be +0.8 V. Typical current versus time curve obtained by the repetitive measurement of the hydrogen peroxide in medium are shown in Fig. 9.6. This curve indicates that the measurement of hydrogen peroxide is finished completely in a short time (~10 seconds) when flow rate is 2.0 mL/min, and no detectable residual current fluctuation appears after the electrolysis is completed. The RSD for 10 successive measurements of 20 μM was 5.6%. The peak height was actually increased with the increase in the hydrogen peroxide concentration. The current responses with good linearity ($R^2 = 0.9993$) of observed in the concentration range from 1 to 100 μM with a detection limit of approximately 0.1 μM. Then, our proposed FIA system described here must be very useful analytical technique because the simple and rapid measurement of the hydrogen peroxide concentration is easily performed.

Oxidative stress occurs from the strong cellular oxidizing potential of excessive reactive oxygen species (ROS) and free radicals [16–20]. Hydrogen molecule selectively reduced the hydroxyl radical, the most cytotoxic of ROS, and effectively protected cells. As hydrogen molecule has potential as an antioxidant in preventive

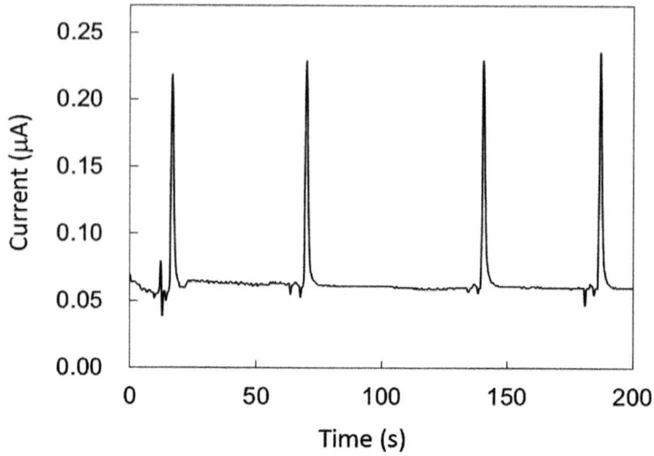

Figure 9.6 Current versus time curve obtained for the repetitive measurement of 20 μM hydrogen peroxide.

and therapeutic applications, it can be used as an effective antioxidant therapy [21]. Owing to such biological importance of hydrogen molecule, the measurement of hydrogen molecule dissolved in medium provides useful information for this purpose. We report that the electrocatalytic oxidation wave of hydrogen molecule can be observed by using Pt-NGC electrode fabricated by the electrode reduction of the aminated GC electrode in sulfuric acid electrolyte. GC electrode was covalently modified by electrochemical oxidation/reduction procedures. The properties of the electrocatalytic activity for hydrogen molecule oxidation reaction have been investigated by employing hydrodynamic voltammetric measurements with a rotating disk electrode.

Fig. 9.7 shows the hydrodynamic voltammogram (HDV) of bubbling hydrogen molecule gas in 0.10 M phosphate buffer electrolyte (pH 7.0) obtained by using Pt-NGC electrode. When the potential was scanned in the cathodic direction from +0.8 V, the oxidation wave began from +0.4 V and continued to −0.5 V. In contrast, when the potential was scanned in the anodic direction from −0.65 V, the oxidation wave began from −0.5 V and continued to +0.5 V. These results mean that an active site has an electrocatalytic activity in each abovementioned potential range. The wave heights of the oxidation/reduction processes were stable, that is, they were constant regardless of the number of potential sweeps. This behavior obtained by Pt-NGC electrode is typical as regards the electrocatalytic redox properties of hydrogen molecule dissolved in medium. As a reference, we examined the current response of hydrogen molecule using the aminated GC electrode, but no current response of hydrogen molecule dissolved was obtained. This can be considered to show that the electrodeposited platinum in collaboration with the nitrogen-containing functional groups introduced by stepwise electrolysis in ammonium carbamate aqueous solution and sulfuric acid are functioning as active sites of the

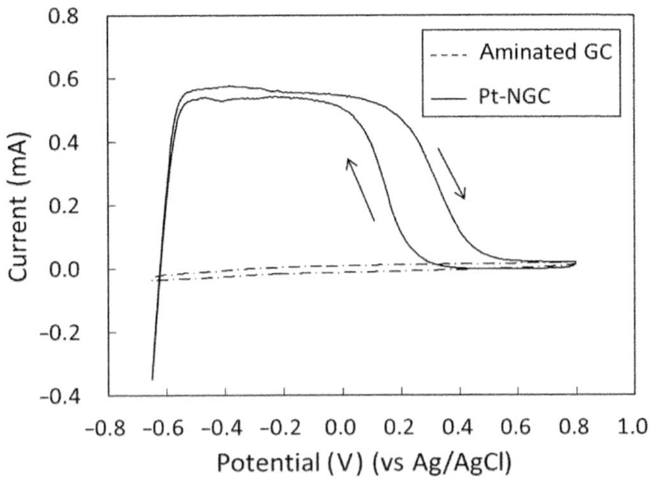

Figure 9.7 HDV of bubbling hydrogen molecule gas in 0.10 M phosphate buffer solution (pH 7.0) using aminated GC electrode (dotted line) and Pt-NGC electrode (solid line). Bubbling rate of hydrogen molecule: 50 mL/min. Sweep rate: 50 mV/s.

specific electrocatalytic current for hydrogen molecule. Furthermore, we confirmed that the electrocatalytic activity of Pt-NGC electrode for hydrogen molecule oxidation did not decrease after ultrasonication for 5 minutes. This fact indicates that platinum particles on nitrogen-containing functional groups are tightly immobilized.

We prepared an amperometric sensor based on FIA system of dissolved hydrogen molecule using Pt-NGC electrode as the sensing electrode. Owing to achieve the rapid measurement of dissolved hydrogen molecule with high sensitivity, we examined the effect of applied potential to the Pt-NGC electrode at a radial flow (-0.5 to $+0.5$ V vs Ag/AgCl) on the peak current. The current responses of the electrode oxidation of hydrogen molecule increased with increasing at the negative applied potentials. On the other hand, the background current noise also increased when the electrode potential exceeded 0 V. In such an extremely negative potential region, the electrocatalytic reduction of dissolved oxygen was observed by the electrode reduction. Therefore the optimal potential to measure the peak current was measured to be $+0.05$ V.

A typical current versus time curve obtained by the repetitive measurement of the hydrogen molecule dissolved in medium is shown in Fig. 9.8. This curve indicates that the determination of dissolved hydrogen molecule is finished completely in a short time (~ 15 seconds), and no detectable residual current fluctuation appears after the electrolysis is completed. The RSD for 11th successive measurements was 12.4%. The concentration of hydrogen molecule is so unstable because the hydrogen molecule dissolved in medium gradually escaped from the solution to air, and the improvement for the stability of dissolved hydrogen molecule concentration in solution is now in progress.

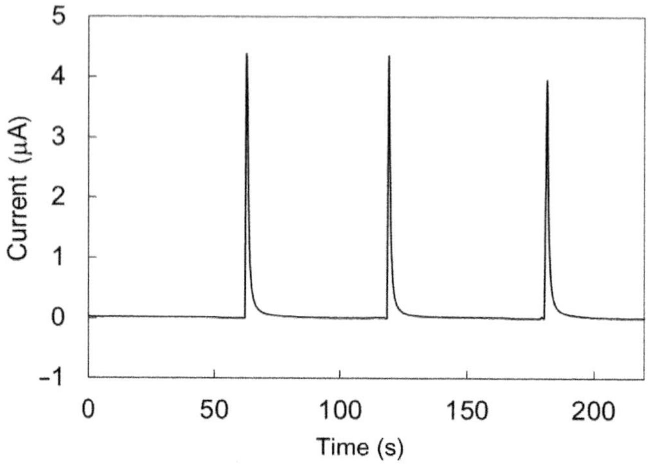

Figure 9.8 Current versus time curve obtained for the repetitive measurement of dissolved hydrogen molecule.

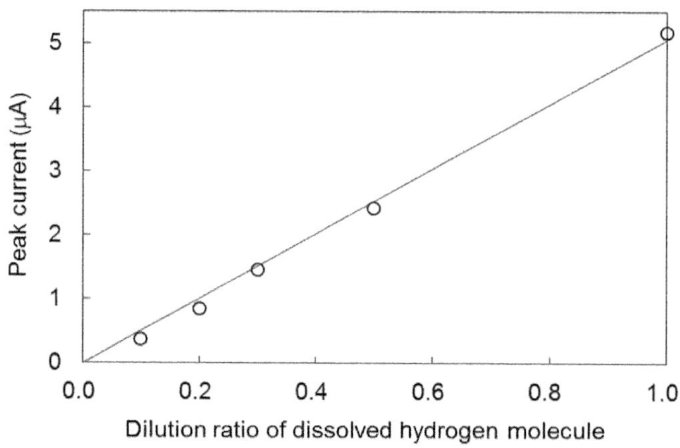

Figure 9.9 Relationship between the current response and the dilution ratio of the dissolved hydrogen molecule. The ratio of 1 corresponded to the concentration of the prepared dissolved hydrogen molecule without dilution.

Fig. 9.9 shows the relationship between the peak current for electrooxidation of hydrogen molecule dissolved in solution and the dilution ratio (DR) of the dissolved hydrogen molecule. The DR of dissolved hydrogen solution was calculated according to the formula:

$$\mathrm{DR} = \frac{C}{Co}$$

where *Co* denotes the concentration of dissolved hydrogen molecule water prepared by bubbling of hydrogen molecule gas and *C* denotes the concentration of dissolved hydrogen molecule solution diluted with arbitrary rates. As shown in Fig. 9.9, good linearity ($R^2 = 0.995$) of the current responses observed in the range of DR from 0.1 to 1. Then, our proposed FIA system described here must be very useful analytical method because the simple and rapid measurement of the dissolved hydrogen molecule concentration in solution is easily used.

9.4 Conclusion and future perspectives

Nitrogen atoms containing functional groups such as amino group were introduced by the electrooxidation of CF electrode in an ammonium carbamate solution. We found that the reduction wave of hypochlorite can be well separated from that of oxygen in the CV obtained by using this electrooxidized CF electrode. Moreover, the Pt-NGC electrode was prepared by the electrode reduction in 1.0 M sulfuric acid followed by the electrode oxidation in ammonium carbamate aqueous solution. The Pt-NGC exhibited an electrocatalytic activity of hydrogen peroxide redox reaction and hydrogen molecule oxidation reaction. The Pt-NGC electrode was prepared by the electrode reduction in 1.0 M sulfuric acid followed by the electrode oxidation in ammonium carbamate aqueous solution. In addition, the Pt-NGC electrode exhibited an electrocatalytic activity of hydrogen molecule oxidation reaction at highly positive potential range in the HDV. The oxidation wave of hydrogen molecule is observed by using the Pt-NGC electrode, which possesses electrocatalytic activity, which is strongly dependent on the electrode potential. The Pt-NGC electrode is promising for a wide range of applications related to hydrogen peroxide and hydrogen molecule detection with high sensitivity as a substitute for platinum electrode.

In a future supported by IoT services, these electrochemical modification techniques are expected to be applied for the surface modification of carbon thin layer with potential applications in flexible and stretchable wearable electronic devices. The modified carbon thin-layer materials are promising the simple and low cost fabrication of sensing electrodes for chemical sensors. In addition, the challenges exist for developing future generations of flexible and stretchable sensors for wearable technology that would be usable for the IoT.

Acknowledgments

This work was supported in part by a Grant-in-Aid for Scientific Research (No. 16K17923) from the Ministry of Education, Culture, Sports, Science and Technology (MEXT), Japan.

References

[1] G.M. Jenkins, K. Kawamura, Structure of glassy carbon, Nature 231 (1971) 175.

[2] G.M. Swain, The susceptibility to surface corrosion in acidic fluoride media: a comparison of diamond, HOPG, and glassy carbon electrodes, J. Electrochem. Soc. 141 (1994) 3382.

[3] T. Yano, D.A. Tryk, K. Hashimoto, A. Fujishima, Electrochemical behavior of highly conductive boron-doped diamond electrodes for oxygen reduction in alkaline solution, J. Electrochem. Soc. 145 (1998) 1870.

[4] O. Niwa, Electroanalytical chemistry with carbon film electrodes and micro and nano-structured carbon film-based electrodes, Bull. Chem. Soc. Jpn. 78 (2005) 559.

[5] R.L. McCreery, Advanced carbon electrode materials for molecular electrochemistry, Chem. Rev. 108 (2008) 2646.

[6] R.W. Murray, in: A.J. Bard (Ed.), Electroanalytical Chemistry, vol. 13, Marcel Dekker, New York, 1984.

[7] R.L. McCreery, in: A.J. Bard (Ed.), Electroanalytical Chemistry, vol. 17, Marcel Dekker, New York, 1991.

[8] S. Uchiyama, H. Watanabe, H. Yamazaki, A. Kanazawa, H. Hamana, Y. Okabe, Electrochemical introduction of amino group to a glassy carbon surface by the electrolysis of carbamic acid, J. Electrochem. Soc. 154 (2007) F31.

[9] H. Watanabe, H. Yamazaki, X. Wang, S. Uchiyama, Oxygen and hydrogen peroxide reduction catalyses in neutral aqueous media using copper ion loaded glassy carbon electrode electrolyzed in ammonium carbamate solution, Electrochim. Acta 54 (2009) 1362.

[10] Y. Yamawaki, K. Asaka, H. Matsuura, S. Uchiyama, Batch injection coulometry of hypochlorite using carbon felt electrodes modified with nitrogen-containing functional groups, Bunseki Kagaku 63 (2014) 411.

[11] A. Kanazawa, T. Okajima, S. Uchiyama, A. Kawauchi, T. Osaka, Characterization by electrochemical and X-ray photoelectron spectroscopic measurements and quantum chemical calculations of N-containing functional groups introduced onto glassy carbon electrode surfaces by electrooxidation of a carbamate salt in aqueous solutions, Langmuir 30 (2014) 5297.

[12] S. Uchiyama, H. Matsuura, Y. Yamawaki, Observation of hydrogen oxidation wave using glassy carbon electrode fabricated by stepwise electrolyses in ammonium carbamate aqueous solution and hydrochloric acid, Electrochim. Acta 88 (2013) 251.

[13] H. Matsuura, T. Takahashi, S. Sakamoto, T. Kitamura, S. Uchiyama, An amperometric flow injection analysis of dissolved hydrogen molecule using tightly immobilized electrodeposited platinum particles on nitrogen-containing functional groups introduced glassy carbon electrodes, Anal. Sci. 33 (2017) 703.

[14] S.M. Lyth, Y. Nabae, S. Morita, S. Kuroki, M. Kakimoto, J. Ozaki, et al., Carbon nitride as a nonprecious catalyst for electrochemical oxygen reduction, J. Phys. Chem. C 113 (47) (2009) 20148.

[15] X. Wang, S. Uchiyama, Amperometric glucose sensor fabricated by combining glucose oxidase micelle membrane and aminated glassy carbon electrode, Anal. Lett. 41 (7) (2008) 1173.

[16] D.C. Wallace, A mitochondrial paradigm of metabolic and degenerative diseases, aging, and cancer: a dawn for evolutionary medicine, Annu. Rev. Genet. 39 (2005) 359.

[17] P.H. Reddy, Amyloid precursor protein-mediated free radicals and oxidative damage: implications for the development and progression of Alzheimer's disease, J. Neurochem. 1 (2006) 96.
[18] S. Ohta, A Multi-functional organelle mitochondrion is involved in cell death, proliferation and disease, Curr. Med. Chem. 10 (2003) 2485.
[19] E. Wright Jr., J.L. Scism-Bacon, L.C. Glass, Oxidative stress in type 2 diabetes: the role of fasting and postprandial glycaemia, Int. J. Clin. Pract. 60 (2006) 308.
[20] C.C. Winterbourn, Biological reactivity and biomarkers of the neutrophil oxidant, hypochlorous acid, Toxicology 181 (2002) 223.
[21] I. Ohsawa, M. Ishikawa, K. Takahashi, M. Watanabe, K. Nishimaki, K. Yamagata, et al., Hydrogen acts as a therapeutic antioxidant by selectively reducing cytotoxic oxygen radicals, Nat. Med. 13 (2007) 688.

Chemical sensors for environmental pollutant determination

10

Hongmei Bi[1] *and Xiaojun Han*[2]
[1]College of Science, Heilongjiang Bayi Agricultural University, Daqing, P.R. China, [2]State Key Laboratory of Urban Water Resource and Environment, School of Chemistry and Chemical Engineering, Harbin Institute of Technology, Harbin, P.R. China

10.1 Introduction

The development of industry and agriculture has enhanced the living standard of human beings, however, this has also led to serious environmental pollutions from heavy metal ions (Cd^{2+}, Hg^{2+}, Pb^{2+}, etc.), toxic gases (SO_2, NO_x, etc.), volatile organic compounds, pathogenic bacteria (*Escherichia coli*, *Salmonella typhi*, etc.) or pesticides, etc. [1]. These chemicals directly or indirectly have a great impact on the ecosystem, and consequently harm environmental security and human health.

Water pollution is the major source leading to epidemic diseases or prevalent disease since it is strongly associated with daily life. To meet the urgent demand for rapid, reliable, and accurate monitoring and detecting of these pollutants, sensors are being developed to offer a user-friendly, selective, portable, and sensitive analytical platform. Sensors are devices that can analyze the target analyte/species quantitatively based on the interaction between the recognition element and the target samples [2]. They provide powerful tools to detect the toxic contaminants to protect the public environment and human health [3]. This chapter provides the underlining principles of chemical sensors and a brief review on the progress of chemical sensor fabrication and application based on a variety of transducer technologies.

10.2 Definition of a chemical sensor

Wolfeis said that "Chemical sensors are small-sized devices comprising a recognition element, a transduction element, and a signal processor capable of continuously and reversibly reporting a chemical concentration" [4]. In 1991 IUPAC proposed the definition of chemical sensor: "a chemical sensor is a device that transforms chemical information, ranging from concentration of a specific sample component to total composition analysis, into an analytically useful signal" [5]. This is

Chemical, Gas, and Biosensors for the Internet of Things and Related Applications.
DOI: https://doi.org/10.1016/B978-0-12-815409-0.00010-3

considered the official definition of chemical sensor. With the development and expansion of sensor research, nanoelements were introduced into sensor field gradually. Francia et al. defined a chemical nanosensor "as an electronic device, consisting of a transducer and a sensitive element that relies, for its operating mechanism, on at least one of the physical and chemical properties typical of the nanostate" [6]. The physical transducer and chemical recognition system are the two basic components in a typical chemical sensor system [3].

10.3 Classification of chemical sensors

Efficient signal recognition, reception, and transduction in sensors form the basis of quantitative analysis using chemical sensors. The transducer plays a vital role within a sensor because it translates the chemical interaction/signal into a physical quantity reproducibly. According to the mode of chemical signal transduction, chemical sensors are usually classified as electrochemical and optical sensors, etc.

10.3.1 Electrochemical sensors

Electrochemical sensors are normally categorized into voltammetric, amperometric, impedance spectroscopy, and potentiometric sensors.

10.3.1.1 Voltammetric sensors

Voltammetric sensors detect an analyte according to the current change against concentration as a function of applied potential. Cyclic voltammetry and differential pulse voltammetry are the primary electrochemical techniques to analyze the environmental pollutants [7–10]. Glassy carbon electrodes (GCEs) were commonly modified to detect analytes in voltammetric sensors. Porous graphene (PGR)/calcium lignosulfonate (CLS) nanocomposite modified GCE was found to be able to detect Pb^{2+} and Cd^{2+} simultaneously using differential pulse anodic stripping voltammetry [11]. Fig. 10.1A illustrates how the sensing interface was fabricated. Graphene oxide was produced using Hummers' method. CLS/PGR nanocomposites were generated through thermal reduction of graphene oxides, and were dispersed evenly into Nafion solution. The Nafion/CLS/PGR/GCE was formed by casting the solution onto a fresh GCE surface. This sensor has a wide detection range (0.05–5.0 μM) for Pb^{2+} and Cd^{2+} with detection limit of 0.01 μM for Pb^{2+} and 0.003 μM for Cd^{2+} respectively [11]. The stripping signals and calibration curves for Pb^{2+} and Cd^{2+} at different concentrations are shown in Fig. 10.1B and C respectively.

10.3.1.2 Amperometric sensors

The principle of amperometric sensors is the current changes of working electrodes against the analyte concentrations as a function of time with fixed potential. An

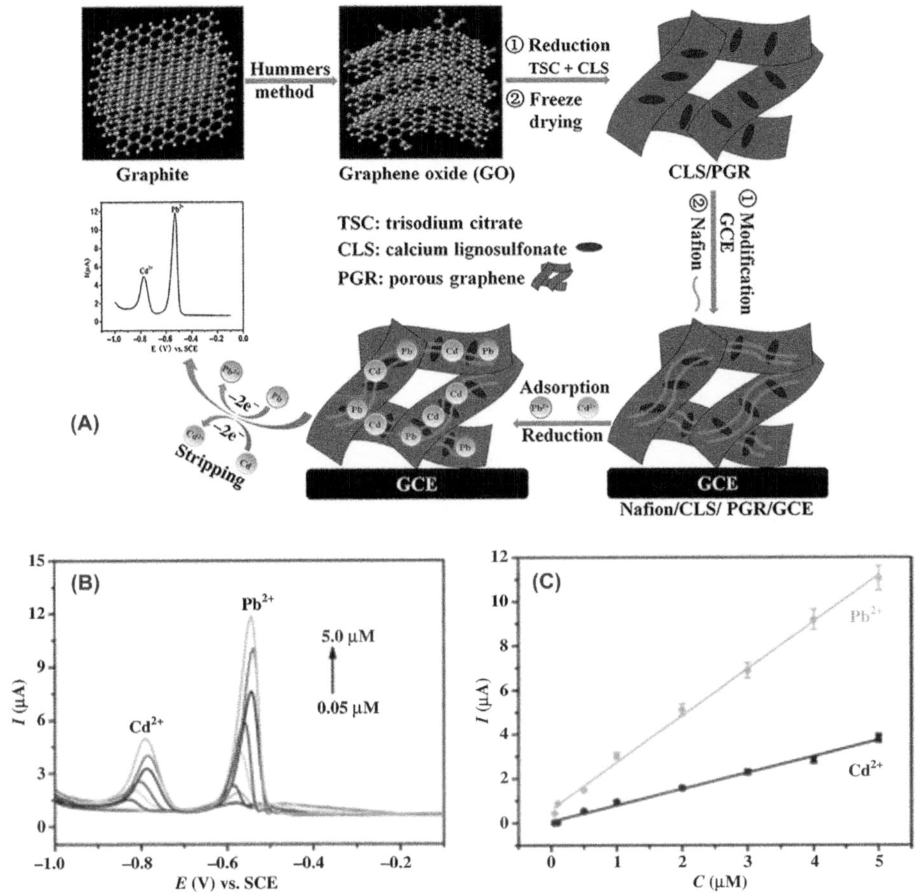

Figure 10.1 Schematic of electrochemical sensor fabrication (A). Electrochemical signals (B) and the calibration curve (C) for Pb^{2+} and Cd^{2+} at 0.05, 0.1, 0.5, 1.0, 2.0, 3.0, 4.0, 5.0 μM, respectively [11].
Source: Reproduced from L. Yu, Q. Zhang, B. Yang, Q. Xu, Q. Xu, X. Hu, Electrochemical sensor construction based on Nafion/calcium lignosulphonate functionalized porous graphene nanocomposite and its application for simultaneous detection of trace Pb^{2+} and Cd^{2+}, Sens. Actuators B Chem. 259 (2018) 540–551 with permission from Elsevier (2018).

electrochemical sensor was developed to detect *E. coli* using amperometric technique [12]. It was fabricated by the modification of bifunctional glucose oxidase (GO_x)-polydopamine (PDA)-based polymeric nanocomposites (PMNCs) onto Prussian blue modified screen-printed interdigitated microelectrodes (SP-IDMEs). The gold nanoparticles (AuNPs) at the surface of the synthesized magnetic bead-GO_x@PDA PMNCs were used to absorb antibodies and GO_x [12], which enabled the magnetic nanocomposites to capture target bacteria. After magnetical separation, the PMNCs–cell conjugates were washed and filtered through a filter paper.

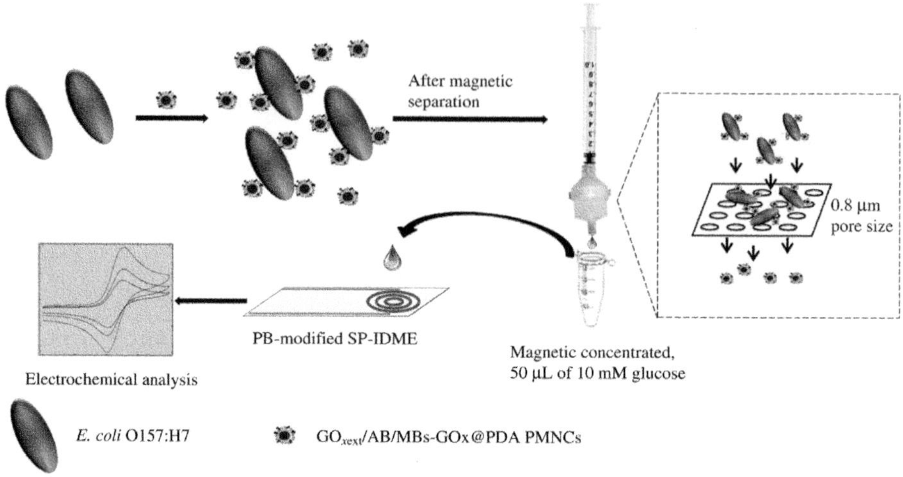

Figure 10.2 The schematic of *Escherichia coli* detection process [12].
Source: Reproduced from M. Xu, R.H. Wang, Y.B. Li, An electrochemical biosensor for rapid detection of E. coli O157:H7 with highly efficient bifunctional glucose oxidase-polydopamine nanocomposites and Prussian blue modified screen-printed interdigitated electrodes, Analyst 141 (2016) 5441–5449 with permission from the Royal Society of Chemistry (2016) (Open Access Article).

The free PMNCs penetrated the filter paper into the vial. With the fixed number of PMNCs initially, the more bacteria is in the sample solution, and the less amount of free PMNCs is in the vial. The glucose solution was mixed with the solution containing free PMNCs to allow enzymatic reaction as shown in Fig. 10.2. Finally, the sample was transferred onto SP-IDMEs to be analyzed using amperometric techniques. This sensor is sensitive to *E. coli* with the detection limit of 10^2 CFU/mL.

10.3.1.3 Electrochemical impedance spectroscopy sensors

Electrochemical impedance spectroscopy (EIS) sensors were used to detect analytes by measuring the impedance changes as a function of sample concentration. To obtain the quantitative results, equivalent circuits were often used to fit the impedance plots. Recently, an impedance-based biosensor was fabricated to detect *Listeria monocytogenes* by a two-step method, that is, first immunomagnetic separation, and second EIS detection of solution ionic strength caused by urase catalysis [13]. The fundamental principle of this method is shown in Fig. 10.3A. Magnetic nanoparticles were modified with the monoclonal antibodies by biotin-streptavidin interaction, which were used efficiently for separation of *Listeria* cells. AuNPs modified with polyclonal antibodies (PAbs) and the urease reacts with *Listeria* to form a sandwich complex. The increase of media ionic strength resulting from the hydrolysis of the urea catalyzed by urease was detected by the microelectrode.

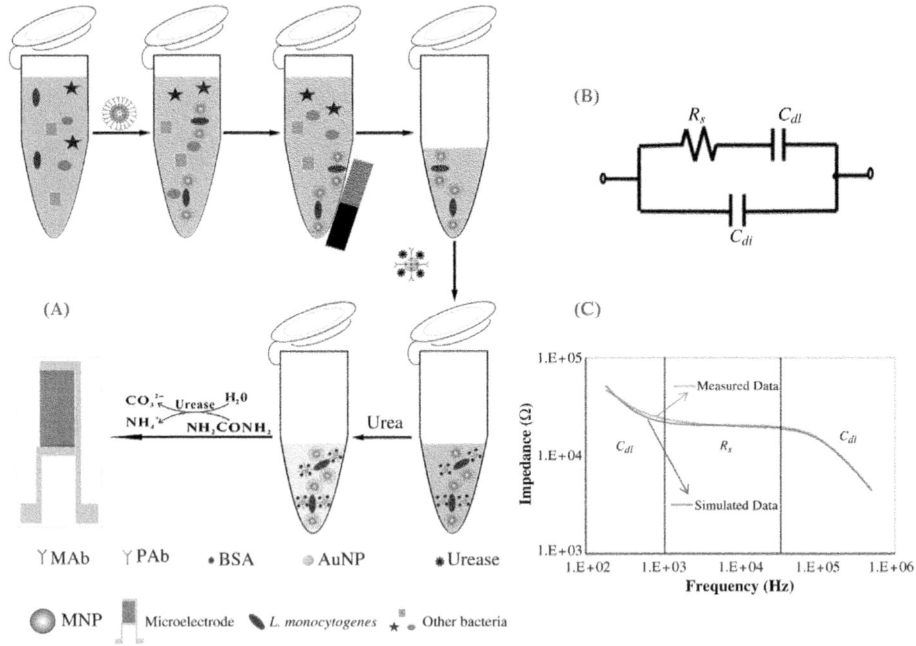

Figure 10.3 Schematic of two-step EIS biosensor working principle (A). The equivalent circuit for fitting of the impedance data (B) and bode plots of the impedance spectra of the measured and simulated data (C) [13].
Source: Reproduced from Q. Chen, J.H. Lin, C.Q. Gan, Y.H. Wang, D. Wang, Y.H. Xiong, et al., A sensitive impedance biosensor based on immunomagnetic separation and urease catalysis for rapid detection of *Listeria monocytogenes* using an immobilization-free interdigitated array microelectrode, Biosens. Bioelectron. 74 (2015) 504–511 with permission from Elsevier (2015).

The impedance change between the supernatant and deionized water was calculated using the equivalent circuit (Fig. 10.3B). 3.0×10^3 CFU/mL of *L. monocytogenes* was detected based on the fitted impedance data using this method (Fig. 10.3C). This biosensor showed a lower detection limit of 30 CFU/mL and good reusability.

10.3.1.4 Potentiometric sensors

Potentiometric sensors mainly determine the analyte concentration by measuring the variation of potential difference between working and reference electrodes at different analyte concentrations. Ion-selective electrodes belong to such sensor. The typical example is pH meter. The potentiometric sensors have been developed on pathogen detection [14–16]. Layer-by-layer technique is widely used for surface modifications. By assembling the carboxylated multiwall carbon nanotubes (CNTs), poly(diallyldimethylammonium chloride) (polycation) and aptamer (polyanion) via

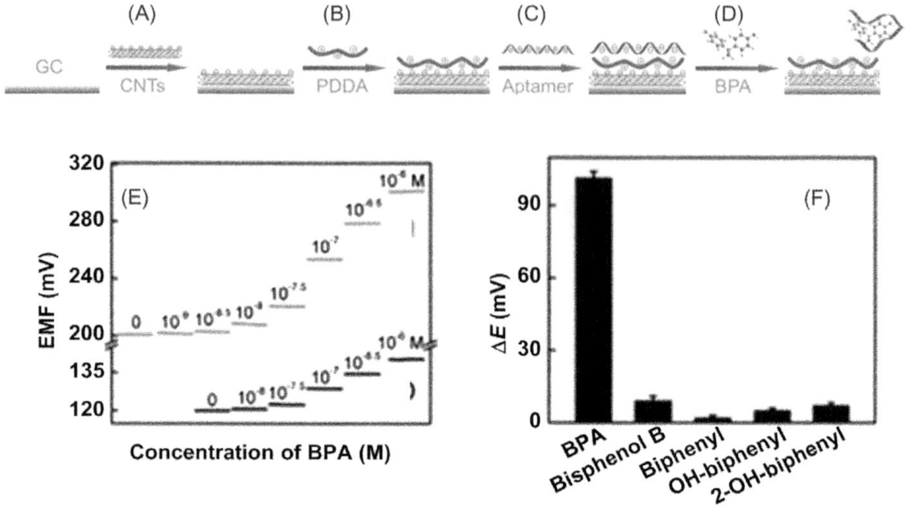

Figure 10.4 Schematic illustrations of GC/CNTs/PDDA/aptamer potentiometric sensor preparation (A–D) and potential responses to BPA (E) and other molecules (F) [17]. Source: Reproduced from E.G. Lv, J.W. Ding, W. Qin, Potentiometric aptasensing of small molecules based on surface charge change, Sens. Actuators B Chem. 259 (2018) 463–466 with permission from Elsevier (2018).

layer-by-layer technique, a potentiometric sensor was prepared for pollutant detection of bisphenol A (BPA) in water [17]. Fig. 10.4A–D showed modification process of electrode surface and the interaction with the targets. The CNTs was dropped on the polished GCEs (glass carbon) to obtain GC/CNTs electrode first. Poly(diallyldimethylammonium chloride) (PDDA) was adsorbed on the surface of the GC/CNTs electrode via the electrostatic interactions between PDDA and CNTs. In the following step, the aptamer was immobilized tightly on the electrode to prepare the GC/CNTs/PDDA/aptamer electrode (Fig. 10.4C). When the BPA is present in the sample, the aptamer detached from the surface via the conformation change (Fig. 10.4D). Consequently, the surface charge of the electrode changed, which was detected for sensing BPA in the water. A stable response to BPA was shown in the concentration range from 3.2×10^{-8} to 1.0×10^{-6} M with a detection limit of 1.0×10^{-8} M (Fig. 10.4E). The measurement results of other molecules with similar structure to BPA such as bisphenol B confirm the good selectivity toward BPA of this sensor (Fig. 10.4F).

10.3.2 Optical sensors

Optical biosensors are mainly based on the following sensing techniques: fluorescence sensors, surface plasmon resonance (SPR), infrared (IR) and Raman spectroscopy, colorimetric sensors, etc.

10.3.2.1 Fluorescence sensors

Fluorescence sensors involve specific fluorescence probes. By monitoring their fluorescence intensity changes after interacting with analytes, the concentration of analyte can be determined [18,19]. Apart from normal single fluorophore based sensors, fluorescence resonance energy transfer (FRET) technique was also used to sensitively detect chemicals. The acceptor (quencher species) could quench the fluorescence of the donor (fluorescent species) in the energy transfer process if they are within a certain distance. Once the analytes change the fluorescence intensity of either acceptor or donor during the FRET process, the analyte concentration can be determined. For example, a FRET biosensor was designed for bacteria detection recently [20], in which the AuNPs (acceptor) were modified with the bacteria targeting aptamers, while the corresponding complementary DNAs (cDNAs) were used to modify the upconversion nanoparticles (UCNPs, donor), as shown in Fig. 10.5A [20]. With the absence of bacteria, FRET phenomenon happened between AuNPs and UCNPs upon the light irradiation due to the DNA complexation reactions. However, the bacteria detach the AuNPs by binding with the targeting aptamers from the complex, consequently quench the FRET phenomenon. According the quenching effect, the bacteria can be detected quantitatively. The change of fluorescence intensity was monitored at 540 nm with the addition of different bacteria concentrations, as shown in Fig. 10.5B. The increase of fluorescence intensity (ΔF) exhibits linear response to the bacteria concentrations in a range of $5-10^6$ CFU/mL with the detection limit of 3 CFU/mL (Fig. 10.5C).

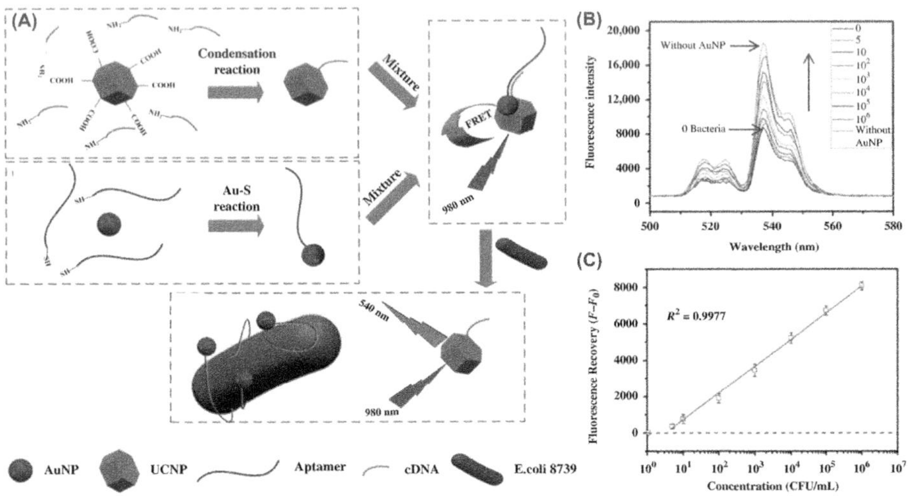

Figure 10.5 Schematic illustration of FRET-based biosensor (A). Fluorescence intensity results of target bacteria with different concentrations (B), and corresponding calibration curve (C) [20]. Source: Reproduced from B.R. Jin, S.R. Wang, M. Lin, Y. Jin, S.J. Zhang, X.Y. Cui, et al., Upconversion nanoparticles based FRET aptasensor for rapid and ultrasenstive bacteria detection, Biosens. Bioelectron. 90 (2017) 525−533 with permission from Elsevier (2017).

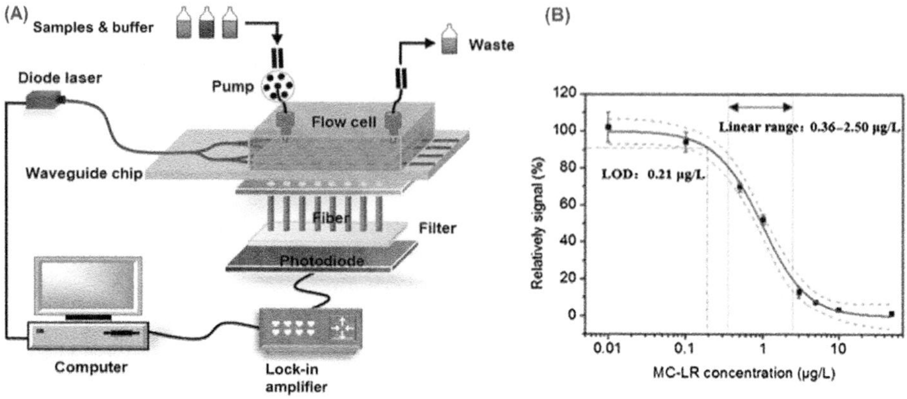

Figure 10.6 Schematic diagram (A) of the fiber-optic sensors and typical calibration curve for MC-LR detection (B) [23].
Source: Reproduced from L.H. Liu, X.H. Zhou, J.S. Wilkinson, P. Hua, B.D. Song, H.C. Shi, Integrated optical waveguide-based fluorescent immunosensor for fast and sensitive detection of microcystin-LR in lakes: optimization and analysis, Sci. Rep. 7 (2017) 3655 with permission from Springer Nature (2017) (open access article).

Another merit of fluorescence sensor is its flexible detection ways. Except for normal cuvette type measuring method, optical fibers provide a more flexible way for high throughput analysis. The fiber-optic sensors were developed for the real-time and on-site detection [21,22]. Another fluorescence immunosensor was developed by using an optical waveguide under the optimized geometry for microcystin-LR (MC-LR, a kind of cyanotoxins) detection in lake water [23]. The experimental setup was illustrated in Fig. 10.6A. The light from the diode laser illuminated samples in the 32 patches on the chip, which were collected by 32 polymer fibers underneath the functionalized waveguide chip. The filtered fluorescence light can be detected directly, which corresponded to the MC-LR concentration in samples directly. 0.36–2.50 μg/L MC-LR could be determined with linear range using this immunosensor (Fig. 10.6B).

10.3.2.2 Surface plasmon resonance sensors

SPR sensors depend on the propagation of surface waves along noble metals and refractive index changes resulting from the binding of analyte with the receptor immobilized at the sensing surface [24]. SPR technique is also suitable for the dynamic/kinetic measurements to investigate the binding constant of the analyte with the receptor. A wavelength modulation SPR biosensor was developed for human IgG detection [25], which involves silver nanocubes and carboxyl-functionalized graphene oxide (cGO). cGO was used to attach antihuman IgG on the surface of SPR chip. The concentration range from 0.075 to 40 μg/mL of human IgG was determined using this SPR biosensor. SPR sensors were also developed for pollutants detection [26–28]. The SPR biosensor based DNA hybridization was developed for the detection of *S. typhi* in water [29]. 5′-Thiolated single-stranded

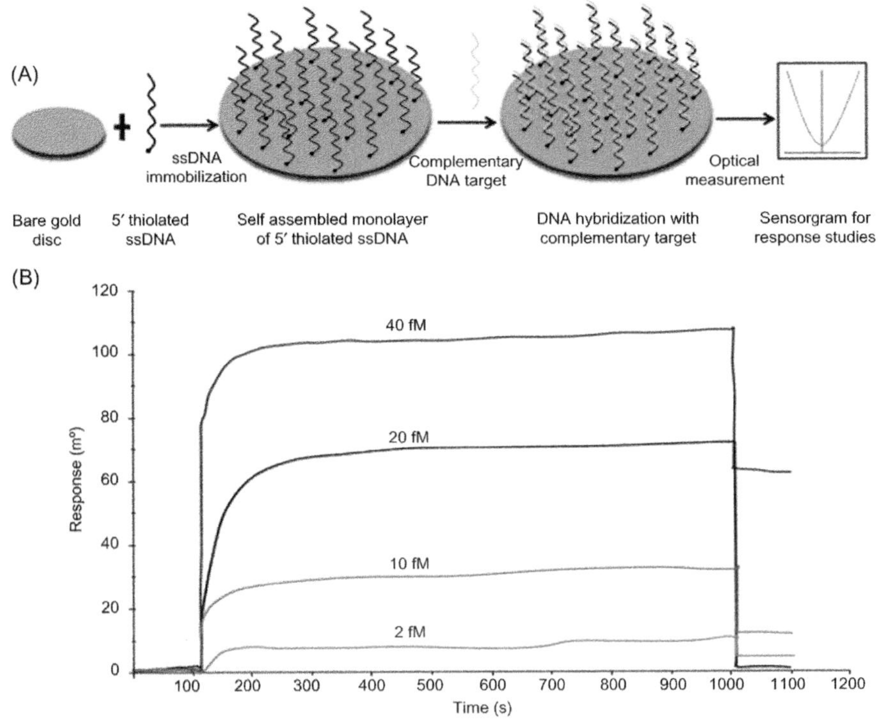

Figure 10.7 The schematic of the fabrication of SPR sensing interface (A) and the experimental data (B) [29].
Source: Reproduced from A. Singh, H.N. Verma, K. Arora, Surface plasmon resonance based label-free detection of *Salmonella* using DNA self assembly, Appl. Biochem. Biotechnol. 175 (2015) 1330–1343 with permission from Springer nature (2015).

DNA (ssDNA) monolayer was self-assembled on gold surface of SPR chip for capturing target DNAs to detect the cDNA extracted from *S. typhi*, as shown in Fig. 10.7A. *S. typhi* ssDNA immobilized gold disks can detect the complementary targets with the concentration from 2 to 40 fM (Fig. 10.7B). This sensor shows good selectivity toward *S. typhi* detection.

10.3.2.3 Infrared and Raman spectroscopy-based sensors

IR and Raman spectroscopy characterize the chemical groups of compounds based on the vibrational fingerprints. The analytes can be accurately monitored, determined, and characterized without destruction by using these vibrational spectroscopic techniques [30,31]. A combination of IR spectroscopy with cellular-based sensing was used to determine *poliovirus* (PV1) quantitatively [32]. The absorbency changes of kidney cells components after infecting by different concentration PV1 were monitored by IR spectroscopy with the detection range from 10^1 to 10^4 PFU/mL. Surface enhanced Raman spectroscopy (SERS) was developed to make up the deficiency of weak Raman signals and fabricate SERS biosensor platforms [33]. Both label-free and label-based strategies were designed to analyze the target analyte. A rapid, sensitive, and label-free SERS detection method for bacteria pathogens was reported [34]. The polyethylenimine (PEI)-modified Au-coated magnetic microspheres (Fe_3O_4@Au@PEI) with positive charges were able to catch the negatively charged bacteria. Those bacteria were fixed onto the substrate with a high density upon applying magnetic fields [34]. With the addition of concentrated Au@Ag nanoparticles, SERS sensing substrate for bacteria was established. The SERS signal from bacteria was enhanced by the metal particles on the microspheres, consequently used to detect bacteria. This is a typical label-free method to detect bacteria. Fig. 10.8A shows the underlining principle of this sensor.

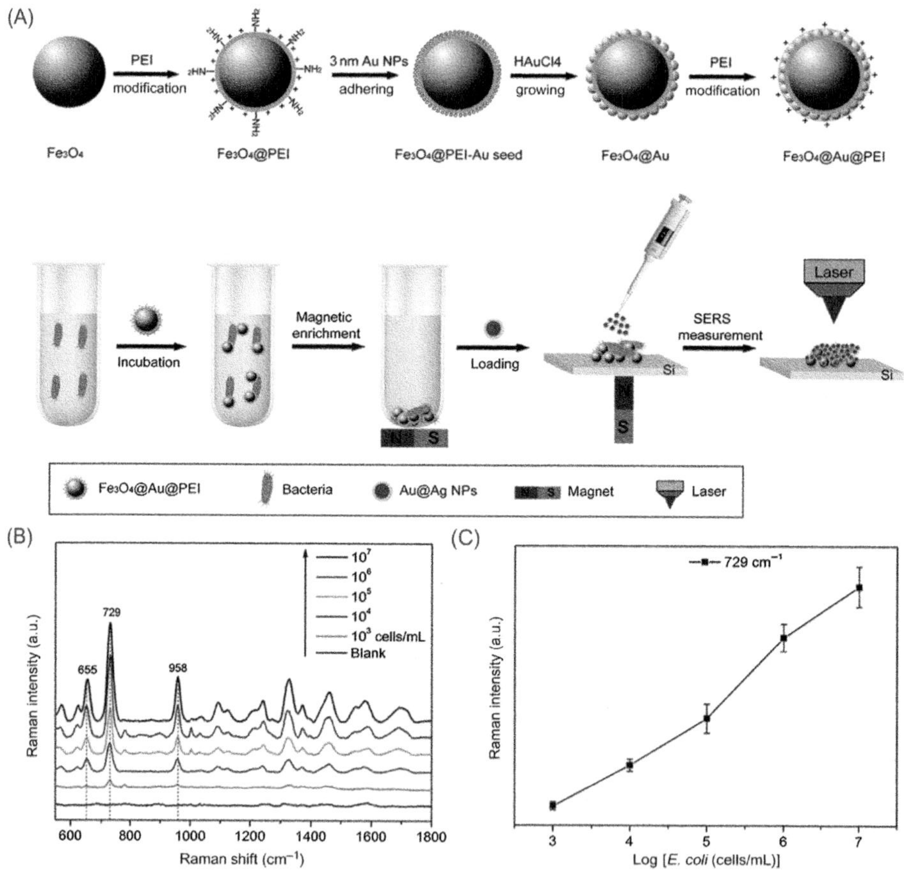

Figure 10.8 Schematic of the label-free SERS detection of bacteria (A). The SERS spectra (B) and intensity as a function of *Escherichia coli* concentration [34].
Source: Reproduced from C.W. Wang, J.F. Wang, M. Li, X.Y. Qu, K.H. Zhang, Z. Rong, et al., A rapid SERS method for label-free bacteria detection using polyethylenimine-modified Au-coated magnetic microspheres and Au@Ag nanoparticles, Analyst 141 (2016) 6226–6238 with permission from Royal Society of Chemistry (2016).

The intensity of the SERS signals of *E. coli* with different concentrations was shown in Fig. 10.8B. It obviously increased with *E. coli* concentration from 10^3 to 10^7 cells/mL. The detection of the *E. coli* from tap water and milk has a lower detection limit to be 10^3 cells/mL at 729 cm^{-1} of the strongest Raman peak (Fig. 10.8C) using this setup. It offered the significant advantages of short assay time, simple operating procedure, and higher sensitivity than previously reported methods of SERS-based bacteria detection.

10.3.2.4 Colorimetric sensors

Colorimetry provides a straightforward and convenient strategy for developing low-cost biosensors. The existence and concentration of sample are read-out according to the visual color changes. On account of the pronounced localized SPR of gold colloids within the visible spectrum, the distance among particles lead to the color change of the solution. Using this principle, the pathogen and bacteria can be determined [35,36]. A colorimetric sensor based on polyaniline nanoparticle (PAni NP) was developed to investigate the growth of bacteria by measuring their metabolic products [37]. The mechanism of this sensor was shown in Fig. 10.9A. It is well known that conducting polyaniline is sensitive to protons. In this case, the protonation of PAni NPs caused the color changing from blue to green, because the protons were the metabolic product of bacteria. Quantitative estimate was monitored based on the absorbance at 420 nm (Fig. 10.9B) and 600 nm (inset of Fig. 10.9B), respectively. A linear response is shown in the pH range from 4 to 8 at the wavelength of 420 nm, therefore 420 nm was chosen for detecting bacteria. The detection limit is 10^6 *E. coli*/mL within 120 min using this colorimetric sensor.

Electrochemical biosensors are the traditional and important sensors for detecting the pollutants such as heavy metal ions, pathogens, etc. Based on the redox

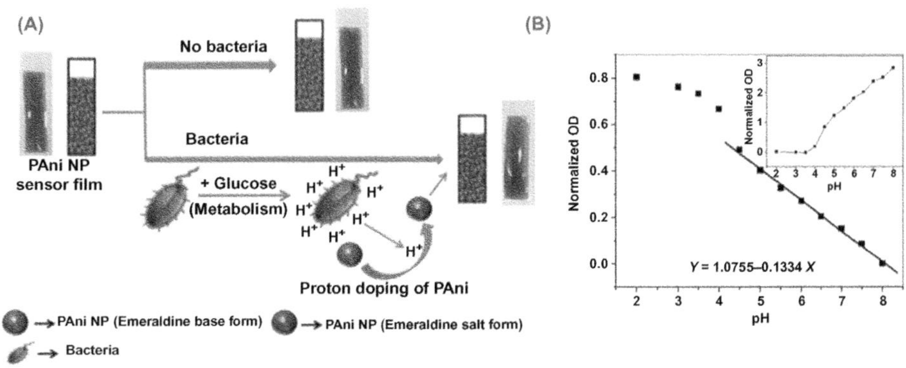

Figure 10.9 Schematic of the PAni NP colorimetric sensor for bacteria (A) and the normalized curves of experimental results (B) [37].
Source: Reproduced from B. Thakur, C.A. Amarnath, S.H. Mangoli, S.N. Sawant, Polyaniline nanoparticle based colorimetric sensor for monitoring bacterial growth, Sens. Actuators B Chem. 207 (2015) 262–268 with permission from Elsevier (2015).

reaction of metal ions, they can accurately measure the heavy metal ions in water, including lead, copper, mercury, silver, cadmium ions. Recently the introduction of nanomaterials to the sensing interface fabrication, their sensitivity and selectivity were greatly enhanced. The stability and reproducibility of some electrochemical biosensors still need to be improved. Optical sensors have also been explored to determine various pollutants in environment due to their property of sample analysis in real time and in situ. The optical sensors also detect samples in a noninvasive way, which make them suitable for biological samples. The smaller dynamic ranges and low analytical selectivity are the existing disadvantages of optical chemical sensors.

10.4 Conclusion

This chapter presented a brief review on chemical sensors for pollutant detection. Except for the definition and classification of chemical sensors, the fabrication and the application examples of chemical sensors on the detection of environmental pollutants especially heavy metal ions and the pathogens were described. With the improvement of technology and combination of nanomaterials, the characteristics of chemical sensors have made a great progress, including their sensitivity, selectivity, reproducibility, etc. Great efforts still need to be made to determine analytes in situ and in real time with minimal sample. Except for the aim of the improvement of sensitivity, and lower detection limit, the portable devices suitable for multianalytes determination with small sample volume is the future direction of chemical sensors for pollutant monitoring.

Acknowledgments

This work was supported by the National Natural Science Foundation of China (No. 21503072, 21773050), Program of Introduction Talents in University (No. XDB-2017-19), and Key Laboratory of Microsystems and Microstructures Manufacturing of Ministry of Education, Harbin Institute of Technology (No. 2017KM006).

References

[1] S. Pandey, S.B. Mishra, Chemical nanosensors for monitoring environmental pollution, Application of Nanotechnology in Water Research, John Wily & Sons, Weinheim, Germany, 2014.

[2] M. Zourob, S. Elwary, A. Turner, Principles of Bacterial Detection: Biosensors, Recognition Receptors and Microsystems, Springer, 2008.

[3] N. Kumar, Y. Hu, S. Singh, B. Mizaikoff, Emerging biosensor platforms for the assessment of water-borne pathogens, Analyst 143 (2018) 359–373.

[4] O.S. Wolfbeis, Chemical sensors-survey and trends, Fresenius J. Anal. Chem. 337 (1990) 522–527.
[5] A. Hulanicki, S. Glab, F. Ingman, Chemical sensors definitions and classification, Pure Appl. Chem. 63 (1991) 1247–1250.
[6] G.D. Francia, B. Alfano, V.L. Ferrara, Conductometric gas nanosensors, J. Sensors 2009 (2009) 659275.
[7] K.G. Reddy, G. Madhavi, B.E.K. Swamy, Mobilized lipase enzymatic biosensor for the determination of chlorfenvinphos and malathion in contaminated water samples: a voltammetric study, J. Mol. Liquids 198 (2014) 181–186.
[8] Y.C. Tsai, B.A. Coles, R.G. Compton, F. Marken, Microwave activation of electrochemical processes: square-wave voltammetric stripping detection of cadmium in the presence of the surfactant triton X, Electroanalysis 13 (2001) 639–645.
[9] S.H. Tan, S.P. Kounaves, Determination of selenium(IV) at a microfabricated gold ultramicroelectrode array using square wave anodic stripping voltammetry, Electroanalysis 10 (1998) 364–368.
[10] X.K. Tian, L. Liu, Y. Li, C. Yang, Z.X. Zhou, Y.L. Nie, et al., Nonenzymatic electrochemical sensor based on CuO-TiO_2 for sensitive and selective detection of methyl parathion pesticide in ground water, Sens. Actuators B Chem. 256 (2018) 135–142.
[11] L. Yu, Q. Zhang, B. Yang, Q. Xu, Q. Xu, X. Hu, Electrochemical sensor construction based on Nafion/calcium lignosulphonate functionalized porous graphene nanocomposite and its application for simultaneous detection of trace Pb^{2+} and Cd^{2+}, Sens. Actuators B Chem. 259 (2018) 540–551.
[12] M. Xu, R.H. Wang, Y.B. Li, An electrochemical biosensor for rapid detection of *E. coli* O157:H7 with highly efficient bifunctional glucose oxidase-polydopamine nanocomposites and Prussian blue modified screen-printed interdigitated electrodes, Analyst 141 (2016) 5441–5449.
[13] Q. Chen, J.H. Lin, C.Q. Gan, Y.H. Wang, D. Wang, Y.H. Xiong, et al., A sensitive impedance biosensor based on immunomagnetic separation and urease catalysis for rapid detection of *Listeria monocytogenes* using an immobilization-free interdigitated array microelectrode, Biosens. Bioelectron. 74 (2015) 504–511.
[14] Y. Wan, D. Zhang, B.R. Hou, Selective and specific detection of sulfate-reducing bacteria using potentiometric stripping analysis, Talanta 82 (2010) 1608–1611.
[15] G.A. Zelada-Guillen, J.L. Sebastian-Avila, P. Blondeau, J. Riu, F.X. Rius, Label-free detection of *Staphylococcus aureus* in skin using real-time potentiometric biosensors based on carbon nanotubes and aptamers, Biosens. Bioelectron. 31 (2012) 226–232.
[16] Y.X. Huang, X.C. Dong, Y.X. Liu, L.J. Li, P. Chen, Graphene-based biosensors for detection of bacteria and their metabolic activities, J. Mater. Chem. 21 (2011) 12358–12362.
[17] E.G. Lv, J.W. Ding, W. Qin, Potentiometric aptasensing of small molecules based on surface charge change, Sens. Actuators B Chem. 259 (2018) 463–466.
[18] U. Dogan, E. Kasap, D. Cetin, Z. Suludere, I.H. Boyaci, C. Turkyilmaz, et al., Rapid detection of bacteria based on homogenous immunoassay using chitosan modified quantum dots, Sens. Actuators B Chem. 233 (2016) 369–378.
[19] L. Wang, S.H. Ma, X.J. Wang, D.Q. Liu, S.Q. Liu, X.J. Han, Electrochemiluminescent TiO_2/CdS nanocomposites for efficient immunosensing of HepG2 cells, J. Mater. Chem. B 1 (2013) 5021–5027.
[20] B.R. Jin, S.R. Wang, M. Lin, Y. Jin, S.J. Zhang, X.Y. Cui, et al., Upconversion nanoparticles based FRET aptasensor for rapid and ultrasenstive bacteria detection, Biosens. Bioelectron. 90 (2017) 525–533.

[21] X.D. Wang, O.S. Wolfbeis, Fiber-optic chemical sensors and biosensors (2013–2015), Anal. Chem. 88 (2016) 203–227.
[22] A. Leung, P.M. Shankar, R. Mutharasan, A review of fiber-optic biosensors, Sens. Actuators B Chem. 125 (2007) 688–703.
[23] L.H. Liu, X.H. Zhou, J.S. Wilkinson, P. Hua, B.D. Song, H.C. Shi, Integrated optical waveguide-based fluorescent immunosensor for fast and sensitive detection of microcystin-LR in lakes: optimization and analysis, Sci. Rep. 7 (2017) 3655.
[24] S.G. Patching, Surface plasmon resonance spectroscopy for characterisation of membrane protein-ligand interactions and its potential for drug discovery, Biochim. Biophys. Acta 2014 (1838) 43–55.
[25] Q. Wu, D.Q. Song, D. Zhang, Y. Sun, An enhanced SPR immunosensing platform for human IgG based on the use of silver nanocubes and carboxy-functionalized graphene oxide, Microchim. Acta 183 (2016) 2177–2184.
[26] P. Su, Z.J. He, L.Q. Wu, L. Li, K.L. Zheng, Y. Yang, SI-traceable calibration-free analysis for the active concentration of G2-EPSPS protein using surface plasmon resonance, Talanta 178 (2018) 78–84.
[27] B.J. Yakes, E. Papafragkou, S.M. Conrad, J.D. Neill, J.F. Ridpath, W. Burkhardt, et al., Surface plasmon resonance biosensor for detection of feline calicivirus, a surrogate for norovirus, Int. J. Food Microbiol. 162 (2013) 152–158.
[28] Y. Wang, W. Knoll, J. Dostalek, Bacterial pathogen surface plasmon resonance biosensor advanced by long range surface plasmons and magnetic nanoparticle assays, Anal. Chem. 84 (2012) 8345–8350.
[29] A. Singh, H.N. Verma, K. Arora, Surface plasmon resonance based label-free detection of *Salmonella* using DNA self assembly, Appl. Biochem. Biotechnol. 175 (2015) 1330–1343.
[30] A. Alvarez-Ordonez, D.J.M. Mouwen, M. Lopez, M. Prieto, Fourier transform infrared spectroscopy as a tool to characterize molecular composition and stress response in foodborne pathogenic bacteria, J. Microbiol. Methods 84 (2011) 369–378.
[31] X.N. Lu, H.M. Al-Qadiri, M.S. Lin, B.A. Rasco, Application of mid-infrared and Raman spectroscopy to the study of bacteria, Food Bioprocess Technol. 4 (2011) 919–935.
[32] F.T. Lee-Montiel, K.A. Reynolds, M.R. Riley, Detection and quantification of poliovirus infection using FTIR spectroscopy and cell culture, J. Biol. Eng. 5 (2011) 16.
[33] J.H. Granger, N.E. Schlotter, A.C. Crawford, M.D. Porter, Prospects for point-of-care pathogen diagnostics using surface-enhanced Raman scattering (SERS), Chem. Soc. Rev. 45 (2016) 3865–3882.
[34] C.W. Wang, J.F. Wang, M. Li, X.Y. Qu, K.H. Zhang, Z. Rong, et al., A rapid SERS method for label-free bacteria detection using polyethylenimine-modified Au-coated magnetic microspheres and Au@Ag nanoparticles, Analyst 141 (2016) 6226–6238.
[35] P.C. Ray, S.A. Khan, A.K. Singh, D. Senapati, Z. Fan, Nanomaterials for targeted detection and photothermal killing of bacteria, Chem. Soc. Rev. 41 (2012) 3193–3209.
[36] M.S. Verma, J.L. Rogowski, L. Jones, F.X. Gu, Colorimetric biosensing of pathogens using gold nanoparticles, Biotechnol. Adv. 33 (2015) 666–680.
[37] B. Thakur, C.A. Amarnath, S.H. Mangoli, S.N. Sawant, Polyaniline nanoparticle based colorimetric sensor for monitoring bacterial growth, Sens. Actuators B Chem. 207 (2015) 262–268.

Part II

Flexible, Wearable, and Mobile Sensors and Related Technologies

Smart clothing with wearable bioelectrodes "hitoe"

11

Hiroshi Nakashima and Shingo Tsukada
NTT Basic Research Laboratories, Nippon Telegraph and Telephone Corporation, Tokyo, Japan

11.1 Introduction

Wearable devices that can monitor the wearer's mental and physical wellbeing are a useful way of helping people live healthy lives and detect illnesses at an early stage. With the arrival of the Internet of Things (IoT) technology, all sorts of things are being connected to the Internet. As part of this trend, progress is being made in the creation of mechanisms for collecting information from wearable devices and using artificial intelligence (AI) to analyze this information. Wearable devices in the form of eyewear, wristwatches, and wrist bands are already available, but recent developments include items that can be worn every day, like shirts, and highly sensitive, multifunctional, and lightweight flexible devices made using thin films and elastomers.

Since wearable devices can be worn on a daily basis, they are very useful for healthcare and lifestyle monitoring applications such as logging changes in heart rate and electrocardiogram, and for gathering data while people are sleeping or exercising. In other words, they allow lifestyle habits to be visualized by collecting accurate data over long periods of time. There is a particular need for wearable devices in the fields of medical care, rehabilitation, and nursing. In such field, it is important to obtain routine measurements of heart rate and cardiac potential to anticipate and/or prevent problems such as arrhythmia in heart patients. Even for patients who have undergone heart surgery can be rehabilitated at home if wearable devices can be used to continuously monitor their heart rates and/or electrocardiograms with medical-quality signals. On the other hand, in the working generation (i.e., people in their 30s and 40s), people start to develop diseases caused by lifestyle habits. Mental and physical health management could play an important role in addressing stress-related issues at home or in the workplace. In addition, people who take part in sports and other fitness activities are increasingly turning to heart rate monitoring to improve their performance.

When electrocardiogram measurements are made using conventional medical electrodes, the electrodes have to be held in contact with the skin by an adhesive electrolyte paste. This is uncomfortable for the user and can cause problems such as skin rashes, making it unsuitable for long-term use. To measure the amount of exertion during exercises such as running, it is instead possible to use synthetic fibers

Chemical, Gas, and Biosensors for the Internet of Things and Related Applications.
DOI: https://doi.org/10.1016/B978-0-12-815409-0.00011-5

(usually metalized with silver) as electrodes. However, these metalized electrodes are water repellent and poor biocompatible and they not only carry the risk of provoking metal allergies, but are also uncomfortable due to the tough metalized fibers coming into contact with the wearer's skin. Since the signals can be very noisy, the electrodes have to be pressed hard against the skin by using a rubber belt or the like. This is somewhat unsuitable for long-term use, since it can cause problems such as rashes from the wearer's sweat at the locations of the electrodes or belt. The routine monitoring of biomedical signals is useful for detecting stress and heart disease before subjective symptoms appear. However, due to the constraints of biomedical electrodes, it is technically difficult to subject people to routine long-term mental and physical monitoring. To address this situation, we require a tool that is comfortable to wear and can record stable biological signals for extended periods of time without having to use electrolyte pastes.

Against this background, we developed a functional material called "hitoe" by coating a nanofiber material with a conductive polymer, resulting in a soft material with excellent flexibility and breathability (Fig. 11.1). This material is also hydrophilic, which means it is able to draw sweat and humidity away from the skin, and it is able to obtain stable electrocardiograms rivaling those obtained using medical electrodes without the need for electrolyte paste. By developing hitoe clothing with embedded bioelectrodes, it is possible to obtain heart rate and electrocardiogram information simply by wearing these items. This should make it possible to perform routine monitoring of biosignals with a much lower burden on the patient or user.

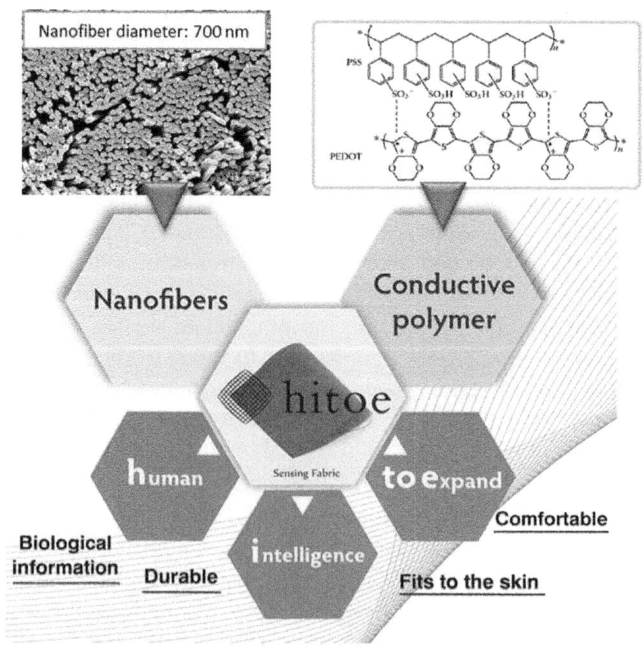

Figure 11.1 Functional material "hitoe".

11.2 Functional material "hitoe"

11.2.1 Composite material of a conductive polymer and fibers

To produce body-friendly bioelectrodes, we used a biocompatible and conductive polymer called PEDOT-PSS [poly(3,4-ethylenedioxythiophene)-polystyrenesulfonate]. The most common application of PEDOT-PSS is as a substitute for indium tin oxide in the field of organic electronics, where it is widely used in industry for transparent electrodes on touch panels and flexible liquid crystal displays. Since PEDOT-PSS has highly hydrophilic and biocompatible qualities, it is also a useful material for bioelectrodes. However, these applications are limited because it is a water-soluble polymer that can become brittle in humid environments, and thus has poor water resistance and processability. In our laboratory, we have deposited PEDOT-PSS to the surface of metal electrodes to form microelectrodes that can measure the action potential of nerve cells, and electrodes that can be implanted into the cerebral cortex [1]. Even fragile nerve cells in the hippocampus and cerebral cortex are able to form networks on the electrodes, allowing the action potential to be recorded stably. However, due to the hydrophilic nature of PEDOT-PSS, the mechanical strength of these electrodes is reduced in wet environments where cells grow. To impart a sufficient level of water resistance and prevent moisture from reducing the mechanical strength, we fabricated a composite material consisting of a fiber base material such as silk with a coating of PEDOT-PSS chemically fixed to its surface. Silk was used for surgical sutures due to its excellent biocompatibility and hydrophilicity, so it is used as the base material for bioelectrodes. In the scanning electron microscope image of a silk/PEDOT-PSS composite material, we can observe the surface of the silk is coated with a uniform film of PEDOT-PSS (Fig. 11.2). It is now possible to apply a coating of PEDOT-PSS not only to the surface of silk fibers but also to the fibers of fabrics with a large surface area, including synthetic fibers such as polyester and nylon. As shown in Fig. 11.3, although ordinary metal-coated fabrics are hydrophobic and water-repellent, fibers coated with PEDOT-PSS are highly hydrophilic and readily absorb water (Fig. 11.3). This composite material retains the conductivity, hydrophilicity, and biocompatibility of PEDOT-PSS, while combining these properties with the strength and processability of the base fiber. By devising a method for fixing the PEDOT-PSS, we were able to prevent it from eluting into aqueous solutions or peeling from the fiber surfaces.

Using the bioelectrodes made from this composite material, we measured electrocardiograms from live animals (rats), and showed clear cardiographic waveforms comparable to the results obtained with medical electrodes without the need for electrolytic pastes or conductive gels (Fig. 11.4A and B). For comparison, electrocardiograms obtained with the silver-plated nylon electrodes (Fig. 11.4C) were often unstable due to the occurrence of baseline fluctuations caused by respiratory motion, or noise originating from changes in the contact resistance at the surface of the body. With bioelectrodes made from conductive polymer–fiber composite material, the hydrophilic properties of PEDOT-PSS make the material softer and better able to absorb sweat and moisture from the skin, allowing stable low-noise measurements to be obtained while remaining adequately in contact with the skin [2].

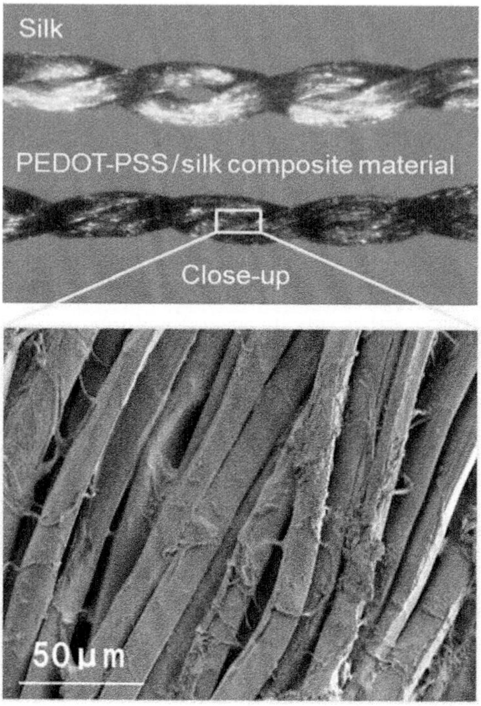

Figure 11.2 PEDOT-PSS/silk composite material.

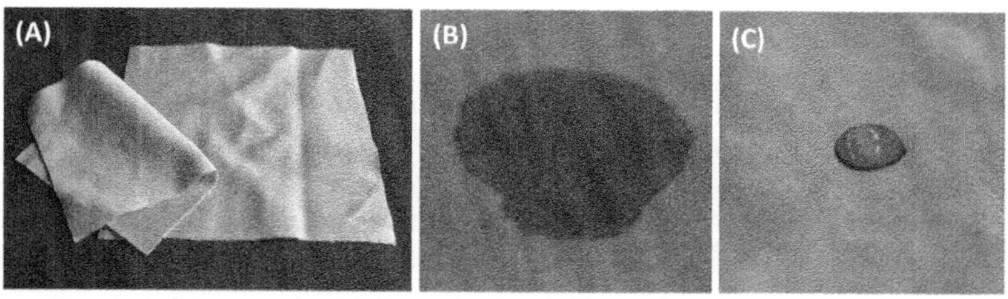

Figure 11.3 PEDOT-PSS coated fabric and metal deposited fabric.

11.2.2 The development of hitoe smart clothing

The characteristics of the fibrous conductive material led to the development of hitoe smart clothing that allows people to measure their heart rate and cardiographic waveforms simply by wearing an item of clothing. In biosignal measurements using smart clothing, problems can arise due to the lower adhesion (cling pressure) of the electrodes, and dry skin can lead to noisier signals and even broken connections. We therefore decided to use a state-of-the-art nanofiber material developed by Toray Industries Inc. as a functional base material that remains in contact with the

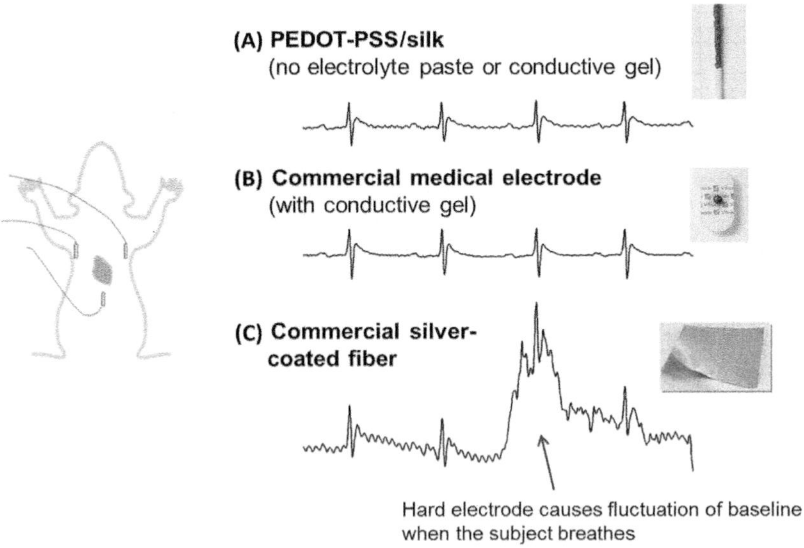

Figure 11.4 Electrocardiogram measurement with PEDOT-PSS/silk composite electrode.

skin and retains humidity. Single filaments of this nanofiber material have a uniform diameter of about 700 nm (Fig. 11.1) and the textile fabric has countless gaps formed from ultrafine fibers. PEDOT-PSS penetrates deeply into the gaps between these fibers, forming a continuous layer of stable and durable conductive resin. This is how we developed the hitoe functional conductive fiber material. In addition, since the nanofibers have a large contact area with the skin and a large frictional resistance, hitoe works as high sensitive bioelectrode and maintains good skin adhesion.

We have also made improvements to smart clothing incorporating hitoe, and we have adopted an optimal interface configuration that can efficiently sense biosignals. This material integrates advanced technologies to ensure that shirts make good contact with the body so that electrodes can be suitably placed for measurements, control the pressure applied to the electrodes without making the shirt feel too tight, and include a structure that prevents short circuits due to perspiration, rain or the like. When wearing hitoe smart clothing, the fabric electrodes come into contact with the wearer's skin, allowing them to obtain stable cardiographic waveform measurements. This biosignal data is sent wirelessly (via Bluetooth) from a transmitter incorporated into the item of clothing to a smartphone device (Fig. 11.5). It has been confirmed that R-waves can be measured stably by continuous monitoring of cardiographic waveforms over long periods of time. In hitoe smart clothing, the fabric's breathability, softness, and stretchability result in garments that do not feel uncomfortable, yet are able to obtain biosignals for prolonged periods without having to use electrolyte pastes or conductive gels. Through a partnership between GOLDWIN, Inc., and NTT docomo, Inc., hitoe smart clothing for

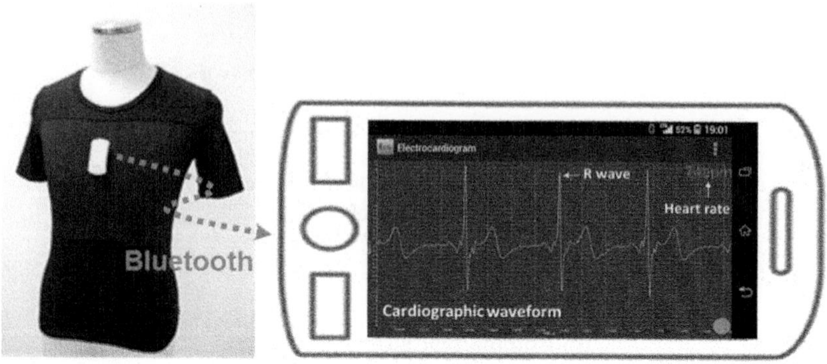

Figure 11.5 "hitoe" smart clothing and example of electrocardiographic waveform displayed on smartphone.

use in sports is now commercially available in the C3fit IN-pulse sportswear, which measures the wearer's heart rate during physical exertion.

11.3 Application examples

In addition to the field of daily personal healthcare, the applications of hitoe smart clothing include a variety of use cases for which there is a growing need. In particular, it is expected to find applications in the fields of medicine/rehabilitation, sports, and worker health/safety management.

11.3.1 Medicine/rehabilitation

The wearable devices could possibly be used a lot more for the early detection of diseases and the improvement of lifestyle habits. To promote the use of hitoe in medical applications, we improved its electrodes to bring them in line with medical standards, and we completed the registration of hitoe products for electrocardiogram (hitoe medical electrodes and hitoe medical leads) as general medical appliances at the Pharmaceuticals and Medical Devices Agency (PMDA) (Fig. 11.6) [3]. The clothing for medical application was designed so that multiple hitoe electrodes can be arranged to comply with the requirements of electrocardiograms in medical examinations. In medical applications, hitoe would make it possible to collect electrocardiogram data for even longer periods than a conventional 24-hour Holter monitor. Such long-term data will enable the early detection of more potentially fatal disorders such as arrhythmia and atrial fibrillation (which can lead to strokes). We are currently preparing for clinical trials of hitoe medical wear at medical institutions.

Furthermore, in the field of rehabilitation, we are collaborating with the rehabilitation institute to verify the effectiveness of IoT monitoring system of a rehabilitation patient. By monitoring a patient's heart rate during rehabilitation, it is possible

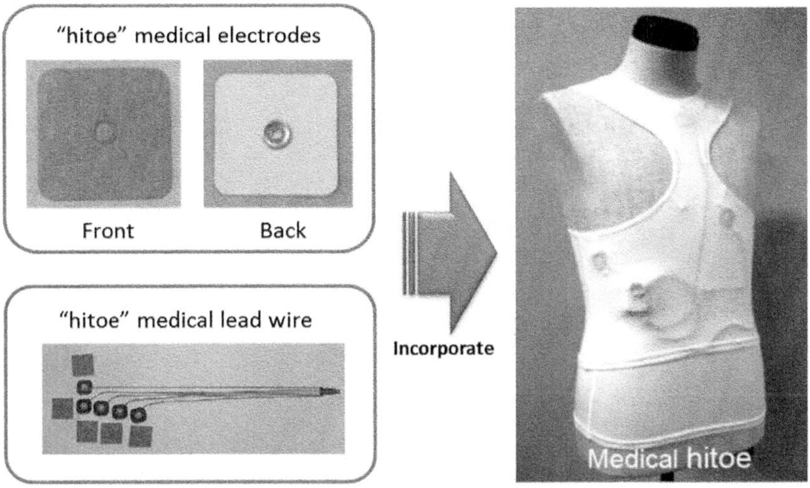

Figure 11.6 "hitoe" electrode and wear for medical use.

to gather and display data on the patient's level of activity (e.g., exercising, resting, standing, or lying down) in real time, allowing healthcare providers to ascertain whether or not the patient has an appropriate daily exercise load. The accumulated data can be used by doctors in planning highly effective rehabilitation. We aim to conduct trials to examine how the use of hitoe smart clothing affects the planning and implementation of rehabilitation programs and the recovery of patients, and to acquire knowledge for use in future rehabilitation support services.

11.3.2 Sports

11.3.2.1 Heart rate measurement

A major advantage of hitoe in sports applications is its ability to continue performing accurate biosignal measurements from people even when they are sweating from prolonged activity, or covered with water. Training based on heart rate variations provides a simple way of visualizing the level of exertion, and is therefore thought to be useful for the efficient management of exercise regimes in endurance sports, or for ascertaining the effectiveness of dietary changes. hitoe can accurately measure heart rate without causing discomfort, even when training for long periods of time. The obtained data will support individual optimal training.

As an example of a field test in sports, we obtained heart rate data from badminton players. Badminton is a game that requires a combination of muscular strength and fast movement, involving intense bursts of activity that cause instantaneous fatigue (high lactose concentration in the blood, low oxygen uptake, and high heart rate). We performed tests in which a badminton expert and a beginner wore hitoe smart clothing while playing the game. In both players, we observed high heart rates in the region of 150–170 bpm (Fig. 11.7A). But interestingly, the heart rates

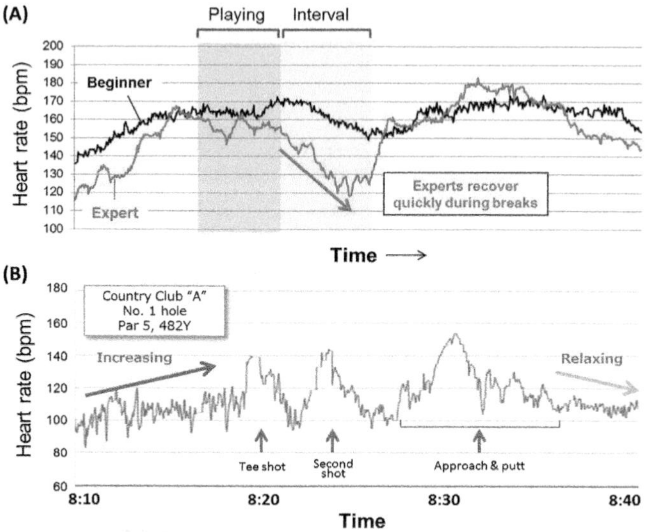

Figure 11.7 Change in heart rate while playing (A) badminton and (B) golf.

of the expert recovered faster during the rest intervals, and there was also a difference in the degree of fatigue recovery between the two. This sort of data should be useful for ascertaining the condition of each player according to his or her skill level. We also confirmed that hitoe smart clothing is able to obtain stable heart rate measurements even in sports like badminton that involve vigorous activity.

In the field test of golf, we observed an increase in the heart rate of golfers during each actual shot (Fig. 11.7B). The heart rate of the test subjects reached 140–160 bpm during each shot. The player's mental state, such as an excitement when teeing off, and a release of tension after completing a hole is speculated by the heart rate data. The correlation between a player's mental condition and physical performance is considered to vary widely according to the characteristics of each individual. Thus, it will be recommended to assess the individual biosignals at a deeper level to improve a player's performance and lead better training.

11.3.2.2 Surface electromyography measurements

In sport science, many studies have been made to evaluate the psychological and physical characteristics of athletes, and these findings have provided useful guidelines for the improvement of training and coordination. However, with the growing demand for more accurate measurements, these evaluations are becoming increasingly dependent on limited laboratory methods, which are disconnected from the actual sports environment. Hardware limitations have been a major obstacle to performing measurements in real sport environments. For example, when electromyography (EMG) signals are recorded to evaluate muscle activity and muscle fatigue, this is usually done using wet electrodes that are connected to leads. This method not only takes a long time to set up, but also significantly limits the movements of players. In recent years, dry wireless electrodes have started to appear. These allow measurements to be made with less of a burden on the individual being tested, but still have numerous issues such as vulnerability to sweat and a level of discomfort that makes them difficult to ignore during actual game play. It is therefore thought that wearable hitoe bioelectrodes integrated with clothing and supporters as shown in Fig. 11.8A will be able to acquire stable biosignals such as surface EMG signals with little interference from the players or their movements.

We performed an experiment in which test subjects were asked to perform squat exercises while wearing leggings fitted with hitoe bioelectrodes. The surface EMG signals were measured from four hitoe electrodes placed at the left and right sides of the quadriceps and gluteus. The test subjects performed Hindu squat exercises without weights 100 times. Fig. 11.8B shows the EMG waveforms obtained from the muscles on the left side during 10 squat motions. In the quadriceps, the variation of muscle activity with the squat exercises can be clearly seen, and although the amplitude is smaller for the gluteus, it is still possible to see periodic activity. Even after 100 repetitions, it was possible to obtain stable EMG signals with a baseline that remained almost unchanged. A similar tendency was also confirmed in the muscles on the right side.

Furthermore, as an example of motion measurement, we measured EMG signals from golf players while striking the ball. Professional and amateur golfers wore supporters fitted with hitoe bioelectrodes on both arms and both legs so that EMG signals could be obtained from a total of eight muscles (the flexor carpi radialis, extensor carpi radialis, tibialis anterior, and gastrocnemius muscles on the left and

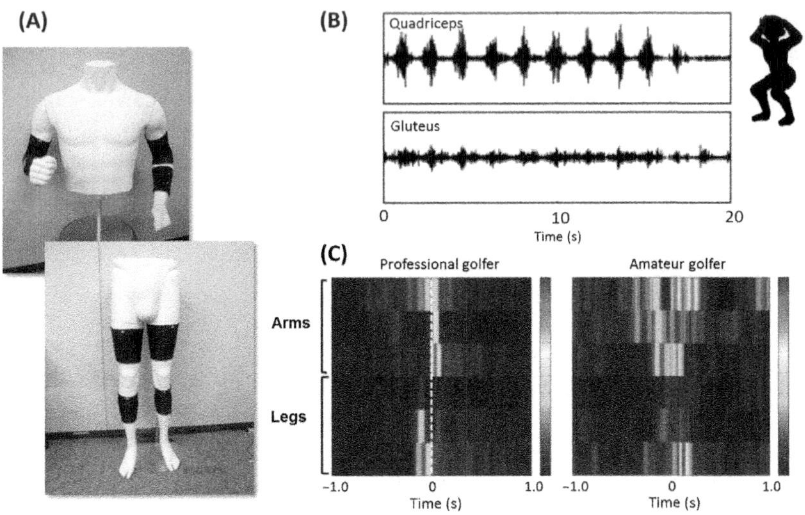

Figure 11.8 Surface electromyography (EMG) measurements with "hitoe" electrodes: (A) Supporters with "hitoe", (B) EMG signals during squat motions and (C) EMG signals at impact of golf.

right sides). The measured signals were routed via leads to a compact module with an integrated biosignal amplifier and transmitter, and were transmitted wirelessly to a PC terminal at a rate of 1 kHz. Fig. 11.8C is a heat map showing the muscle activity occurring during driver shots in both groups of test subjects, where the relative level of muscle activity is represented using different colors. Time 0 corresponds to the instant at which the ball is struck. The professional golfers exhibited concentrated activity in both the arms and legs around the time of impact, whereas the amateur golfers exhibited a wider range of activation timing. In other words, these findings suggest that a lot of the muscle activity in amateur golfers serves no useful purpose and is not connected to efficient force transmission or golfing performance.

11.3.3 Worker health/safety management

People who work at construction sites and other outdoor locations during recent hot summers have been subjected to much greater physical burdens. It is therefore important to have a monitoring system so that the physical condition of these workers can be managed. We have started a hitoe worker protection service that takes advantage of the ease of use of hitoe smart clothing to implement health management and safety protection measures for each individual worker. Toray Industries Inc. and NTT Group have performed verification trials at multiple workplaces in the transportation and manufacturing sectors. To manage the safety of individual workers, especially those that have to work at night or outside in hot weather, we developed hitoe smart clothing with an improved level of comfort. We also built a system that collects and analyzes heart rate data from workers, and provides efficient feedback to the managers (Fig. 11.9).

In this service, we collect data including heart rates and acceleration (the acceleration data obtained from hitoe transmitter) from workers wearing hitoe smart clothing. The information is sent to a smartphone via hitoe transmitter and then sent to a cloud-based safety management system from the smartphone together with

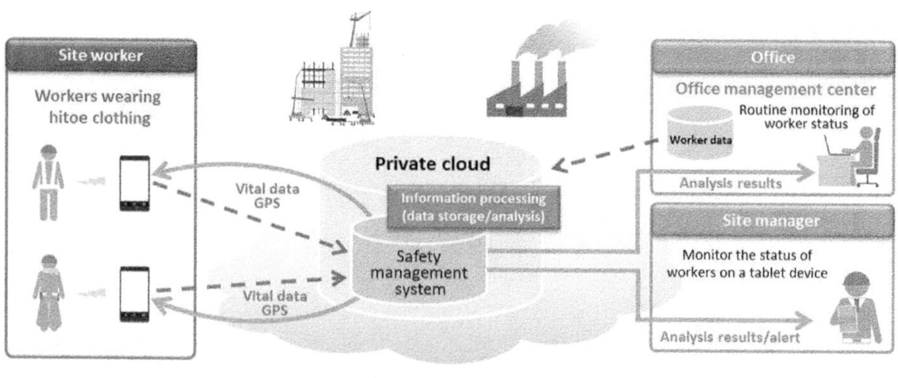

Figure 11.9 Health and safety monitoring system for workers.

GPS information. From this information, we provide visual feedback by analyzing and estimating seven types of information: (1) heart rate, (2) degree of heat exposure, (3) work intensity (physical exertion based on increase of heart rate), (4) psychological stability (whether or not the worker is relaxed), (5) occurrence of falls (based on posture and inclination), (6) energy consumption, and (7) position information. This system enables workers, site managers, and administrative centers to check the situation in real time on various types of terminal including smartphones, tablets, and PCs. It is also possible to monitor and display the health and safety status of multiple workers at the same time. When an abnormal situation arises, an alert notification function reports this occurrence so that countermeasures can be implemented promptly in case of an emergency. It is expected that safety management systems that use IoT technology to combine cloud systems with vital data from hitoe smart clothing will be widely used as tools for protecting the safety and security of workers.

11.4 State estimation based on heart rate variability and other data

The biological information obtained from hitoe smart clothing can include not only heart rate and cardiographic waveform information, but also acceleration sensor information from wireless communication transducers. Based on this acceleration information, it is possible to estimate the wearer's actions such as posture and gait. It is also possible to estimate the wearer's breathing patterns from the electrocardiogram waveform [4]. Here, we introduce a method for estimating states such as posture/gait, breathing, and sleeping that are calculated as secondary information from the cardiographic waveforms and accelerometer data measured directly by smart clothing.

11.4.1 Estimating posture information from accelerometer data

The wearer's posture is estimated by performing calculations based on the inclination of the accelerometer relative to the direction of gravitational acceleration to determine if the upper body is oriented upright or lying down, for example. Similarly, using gait information obtained from the acceleration data, it is possible to derive information such as the total number of steps taken, the walking pace, the stride length, the speed of movement, and distance traveled, from which it is possible to recognize if the subject is at rest, walking, or running [5]. By obtaining this sort of gait information together with heart rate measurements, it should be easier to interpret changes in the wearer's heart rate. That is, the state information inferred from the accelerometer data can play an important role in the interpretation of changes in the life log data. The state estimation calculations are performed in a terminal device such as a smartphone or cloud server. We have also made user-friendly application developer kits available so that users can easily implement various kinds of monitoring based on diverse biological information.

11.4.2 Estimating respiratory activity from electrocardiogram data

It is also possible to use hitoe smart clothing to estimate respiratory activity based on electrocardiogram measurements. People's chests expand and contract as they breathe, and it is known that this chest movement causes fluctuations in the R-wave and S-wave voltage differences (RS amplitude) in cardiographic waveforms [6]. It is therefore possible to estimate the wearer's breathing activity from this relationship between chest movements and cardiographic waveforms. Since hitoe smart clothing makes it possible to obtain cardiographic waveform information on a routine basis, it will make it easy for people to visualize their state of respiratory activity in various scenes. For example, using cardiographic waveform data from hitoe smart clothing after compensation processing to obtain accurate RS amplitude measurements, we conducted an experiment to verify the correlation between the measured RS amplitude fluctuations and the actual respiratory air flow (Fig. 11.10). When we compared the measured RS amplitude fluctuations with the air flow patterns measured with a temperature sensor inside a face mask, we confirmed that similar patterns were obtained for the two data. In practice, if it is possible to estimate a person's breathing patterns simply by having them wear an item of clothing, then this will be much less restrictive than having them wear a face mask. If a person's state of breathing during daily life can be obtained easily, then this information will be very useful not only for monitoring respiratory diseases, but also for the development of tools that can be used in activities such as sports and yoga.

11.4.3 Estimating sleep states

Quality of sleep is thought to have a large influence on both physical and mental wellbeing. Poor sleep quality can trigger various diseases including depression [7]. In addition to its health effects, sleep impairment can also reduce a person's

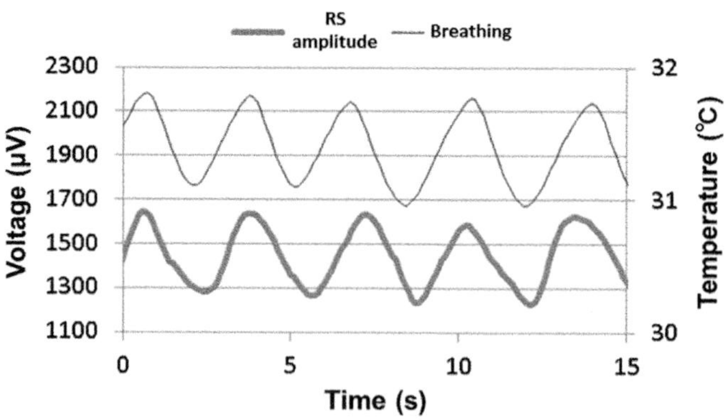

Figure 11.10 Correlation between RS amplitude change calculated from electrocardiogram waveform and respiratory air flow measurement.

concentration, memory, reasoning, and other mental faculties, and can lead directly to a decline in job performance. If people wear hitoe smart clothing while sleeping, they can review their daily sleep patterns by visualizing their sleep states based on their heart rate variability. By using frequency-domain indicators to analyze heart rate variability during sleep, we can analyze the high-frequency (HF) and low-frequency (LF) spectral components to determine the balance of the sympathetic and parasympathetic nervous systems as HF and LF indicators [8,9]. In other words, the state of sleep can be estimated from heart rate variability by using the relationship between sleep states (awake/REM sleep/non-REM sleep) and the activity level of autonomic nerves [10,11]. Since hitoe smart clothing is suitable for long-term monitoring while sleeping, it is therefore expected that the use of hitoe smart clothing to analyze sleep patterns will make it possible to ascertain how a patient's sleeping patterns differ from those of a person in good health.

11.5 Conclusion

Recent IoT information terminals include various wearable devices such as watches and eyewear in addition to smartphones and smart pads. In the sports and healthcare sectors, we can also see a growing market for wearable devices such as wrist bands and chest belts, and these products refine the output of biosignal measurements. Although wearable devices are highly promising, there are also many hardware and software issues that need to be resolved before they can be launched on the market. In particular, there is an urgent need to reduce the size and weight of these devices, improve their comfort and usability, and increase their battery life.

The hitoe smart clothing introduced here is a new concept of wearable device that enables highly accurate measurements of heart rate and electrocardiogram waveforms simply when someone is wearing it. It takes a slightly different approach compared with conventional electronic devices, and the most important aspect of its design is that it blends in with people's daily life activities without causing any discomfort. When hitoe smart clothing is used to obtain biosignal measurements, it is possible to monitor these signals for long periods of time without placing a burden on the wearer because it uses a flexible and breathable electrode material that requires no electrolyte paste. In the near future, we hope that hitoe smart clothing will be used to support the provision of medical and nursing care to elderly people in their own homes. In cooperation with hospitals and universities, we have started looking into expanding the range of applications of this technology to fields such as home medical care and telemedicine as medical ICT devices. On the other hand, the ability to monitor heart rate variability during exercise can play an effective role in helping people to manage their own exercise regimes and manage the risk of heat stroke and other such hazards. Heart rate variability is closely related to autonomic nerve function and stress and is expected to be applicable for quantitative evaluation of both physical and mental factors including sleeping time. We hope to realize a wide range of services to create new added value by combining hitoe smart clothing with various wearable devices, mobile terminals, and other forms of ICT.

References

[1] T. Nyberg, A. Shimada, K. Torimitsu, Ion conducting polymer microelectrodes for interfacing with neural networks, J. Neurosci. Methods 160 (1) (2007) 16–25.
[2] S. Tsukada, H. Nakashima, K. Torimitsu, Conductive polymer combined silk fiber bundle for bioelectrical signal recording, PLoS One 7 (4) (2012) e33689.
[3] hitoe medical electrode: medical device report number 13B1X00015000031; hitoe medical lead: medical device report number 13B1X00015000032.
[4] E.J. Bowers, A. Murray, P. Langley, Respiratory rate derived from principal component analysis of single lead electrocardiogram, Comput. Cardiol. 35 (2008) 437–440.
[5] T. Ogasawara, Y. Itoh, K. Kuwabara, R. Kasahara, Gait analysis using a wearable t-shirt type sensor, NTT Techn. Rev. 14 (4) (2016) 1–7.
[6] H. Fujisawa, T. Uozumi, K. Ono, Estimation of respiratory rate by analysis of electrocardiogram R wave amplitude variation, Med. Electron. Bioeng. 36 (4) (1998) 337–342.
[7] Y. Doi, Frequency of sleep disorders in Japan and their health effects, J. Natl. Inst. Public Health 61 (1) (2012) 3–10.
[8] K. Kameyama, T. Suzuki, M. Yukitani, A sleep judgment and sleep monitoring system for restful sleep, Toshiba Rev. 61 (10) (2006) 41–44 (in Japanese).
[9] T. Hori, Fundamentals of sleep—including development, aging and gender differences, Nippon Rinsho 66 (2) (2008) 27–33 (in Japanese).
[10] T. Takeda, O. Mizuno, T. Tanaka, Time-dependent sleep stag transition model based on heart rate variability, Proc. 37th Annual Intl. Conf. IEEE Eng. Med. Biol. Soc. (2015) 2343–2346.
[11] T. Takeda, T. Watanabe, K. Yoshida, O. Mizuno, Time-dependent sleep stage transition model using heart rate variability, Database Soc. Japan 14-J (16) (2016) (in Japanese).

Cavitas bio/chemical sensors for Internet of Things in healthcare

12

Kohji Mitsubayashi, Koji Toma and Takahiro Arakawa
Department of Biomedical Devices and Instrumentation, Institute of Biomaterials and Bioengineering, Tokyo Medical and Dental University, Tokyo, Japan

12.1 Introduction

A measurement of biophysical quantities of human body has been investigated for the medical and health care fields. So far, many wearable sensors have been developed and commercialized in the world because of their perspectives for human monitoring of relevant parameters in sports, health care, and medical applications [1,2]. The majority of existing wearable devices focus on measurement and evaluation of physical parameters (activities). Utilization of biophysical information with the wearable devices is expected to provide proactive management of health that could improve public health and reduce medical expenditure. However, an investigation of noninvasive bio/chemical sensing has been delayed. However, it is required for collecting the biological and chemical information in our daily life. The noninvasive techniques of bio/chemical measurement would also affect the control of the physical condition in patients suffering from lifestyle related diseases and health conditions. For example, self-monitoring of blood glucose is most commonly performed with finger-prick testing using a blood sugar meter for diabetic patients. However, the compliance often impaired due to the unpleasant, painful, carries a risk of infection and may induce anxiety or fear by blood sampling and insulin injection. An invasive continuous glucose monitoring can overcome some limitation, enable short-term fluctuations to be monitored, and demonstrate immediate effects of dietary and therapeutic interventions [3].

The bio/chemical samples excreted from human body are not only urine and feces, but include also sweat, exhaled air (breath), saliva, nasal secretion (rhinorrhea), skin gas, tears, etc. Though urine and feces have been applied in the medical field, these samples do not provide temporal information for daily health care. Human secretions would provide bio/chemical information about the health and well-being of an individual. Extracting this information would be the goal of developing noninvasive approaches for diagnosis, medical screening, and health care monitoring. In these aspects, the wearable sensors also have tremendous potential for evaluation of bio/chemical markers relevant to disease and/or metabolic condition [4,5].

Since the 1990s, authors have paid attention to human body cavities [6,7]. Our body has many cavities such as cavitas oris (oral cavity), cavitas pharyngis, saccus

Chemical, Gas, and Biosensors for the Internet of Things and Related Applications.
DOI: https://doi.org/10.1016/B978-0-12-815409-0.00012-7

lacrimalis (lacrimal sac), cavitas nasi, cavitas abdominalis, cavitas infraglotticum, cavitas larynges, etc. *Cavitas* word is the etymological origin of the word *cavity* in Latin. Authors have developed many sensors to apply to the human cavities (cavitas oris, cavitas pharyngis, conjunctival sac) for noninvasive monitoring of bio/chemical information in the permanent body fluid in the human cavities. Then authors have called them "Cavitas sensors." "Cavitas" is a new category of self-attachable medical sensors between "Implantable" and "Wearable" (Fig. 12.1) [8]. While the implantable one is the medical device applied via the medical surgery by medical doctors (nondetachable by subject-self), the wearable one is an attachable device applied by subject-self but no enough to collect the fruitful bio/chemical information. Recently, some cavitas sensors have been investigated in the world. Author's belief that new consciousness of daily medicine (health care, presymptomatic, and preventive medicine) with cavitas and wearable sensors and Internet of Things (IoT) techniques is necessary to improve the quality of life (QOL) in view of the aging of society and the rapid changes in living environments.

This chapter provides an update of various cavitas sensors and techniques along with their advantages, challenges, and future potential. In particular, authors focused on soft contact lens (SCL) type sensors for tear chemicals and transcutaneous gas at eyelid conjunctiva, and mouthguard type sensor for saliva analysis.

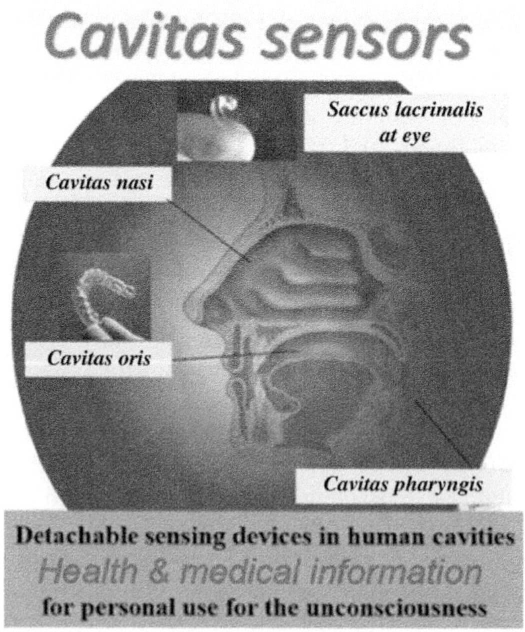

Figure 12.1 Concept of cavitas sensors for noninvasive bio/chemical monitoring at body cavities.

12.2 Soft contact lens type bio/chemical sensors

12.2.1 *Tear fluid in conjunctiva sac*

A tear film forms an interface between the air and ocular tissues [9]. The lacrimal secretary system has two components: main lacrimal gland at the large orbital and smaller palpebral portions (about 95% of the aqueous component of tears), and accessory lacrimal glands of Krause and Wolfring located in the conjunctival stroma [10]. The important functions of tears are lubrication of the eyelids, formation of a smooth and even layer over an otherwise irregular corneal surface, and provision of antibacterial systems for the ocular surface and nutrients for the corneal epithelium [10]. Average specific gravity of 1.01 for tears has been reported, and the pH generally is around 7.4, but pH values from 5.2 to 8.3 have been observed because alkaline tears are shed after corneal injuries [11]. The quantity of electrolyte in tears is composed chiefly of Na^+ and Cl^-, though considerable K^+ also may be present (Na^+ 120–165 mmol/L, K^+ 20–42 mmol/L, and Cl^- 118–135 mmol/L) [12]. The principal tear proteins are lysozyme, lactoferrin, and tear-specific prealbumin. Comparison with serum levels indicates similar concentrations for Na^+, Cl^-, HCO^{3-} (20–42 mmol/L), Mg^{2+} (0.5–0.9 mmol/L), and urea. Other electrolytes are present at markedly different levels: K^+ and lactate (2–5 mmol/L) are higher, and Ca^{2+} (0.4–1.1 mmol/L), glucose (0.1–0.60 mmol/L), and protein (5–9 mg/mL) are lower in tears than in serum [13].

In general, tear fluid for bio/chemical analysis is collected by either filter paper or glass capillary pipettes. The tear samples are pooled and then analyzed by a variety of analytical techniques. However precise estimations of the chemical composition, or the physical properties of tear fluid are subject to large error owing to a number of factors such as the small sample size (approximately 7 μL) [14]; evaporation during collection; wide variability between individuals; diurnal variations [15]; and the method of collection [12]. There are inherent flaws that make the collection of perfectly reliable and reproducible samples for analysis very difficult. A continuous measurement at lacrimal sac would be one of approaches to eliminate the collection error. From these points of view, the eye site with tear fluid in conjunctiva sac is important for noninvasive monitoring of physiological bio/chemicals and blood gases related some diseases and/or metabolic condition. The eye-cavity sensor such as a contact lens would be useful for be applied directly at lacrimal sac for monitoring the tear chemicals and conjunctiva volatiles without the several noted problems.

12.2.2 *Flexible conductivity sensor for tear flow function*

An electrical conductivity in the liquid phase (even body fluids) would be an indirect function of electrolyte activity, or osmolality [16]. The fluid-specific conductivity is the sum of the contributions from all charged species. The specific conductivity will vary nearly linearly with concentration at very low concentrations [17]. The conductivity measurement would not distinguish among the types of ions present in the fluids. But comparison between the conductivity values for several physical conditions and monitoring of fluctuations in fluid conductivity are possible [16], because the relative proportions of each ion are known for biologic fluids such as sweat, tears, saliva and airway fluid in the body cavities [18,19]. Microfabrication technology could produce planar integrated electrode arrays with small areas on the order of a few square millimeters [20], decreasing the sample volume required to evaluate the conductivity of body fluid. Polyimide has been used as a substrate to achieve flexibility, but toxicity problems remain because of the use of substances such as chromium and nitric acid in the preposition and etching stages of the microfabrication techniques for obtaining the desired electrode patterns [16].

A safety and flexible sensor with a hydrophilic polytetrafluoroethylene membrane placed between two gold deposited layers was evaluated for use as the electrical conductimetric sensor in biologic fluids (i.e., tear fluid in lacrimal sac) [6]. The conductivity was measured using the device at frequencies ranging from 100 Hz to 100 kHz, and the device was calibrated at 100 kHz against sodium chloride solutions over the range of 0.1–50.0 g/L, which include physiologic ion concentrations. As an animal experiment, the flexible conductimetric sensor can be placed directly onto the surface of the rabbit eye like contact lens to monitor the electrical conductivity of tear fluids. A tear flow function with a mean turnover rate could be elicited by eye-drops of various concentration NaCl solutions and distilled water.

As in vivo physiological application, the flexible conductimetric sensor (3 mm wide) was placed inside the human subject's temporal lower cul-de-sac (similar with Schirmer test strip), and was used to evaluate the electrolyte concentration and the turnover rate in tear for normal healthy volunteers (aged from 20 to 85 years) and patients suffering from keratoconjunctivitis sicca (KCS) (aged from 50 to 69 years) in a normal light indoor environment [7,21] (Fig. 12.2). The tear electrolyte concentration for normal eyes was calculated from tear electrical conductivity to give a mean value of 297 mEq/L (SD: 30 mEq/L; $n = 33$) which was consistent with previously reported values. The mean concentration of tear electrolyte for KCS eyes 325 mEq/L (SD: 41 mEq/L; $n = 29$) was higher than that for normal

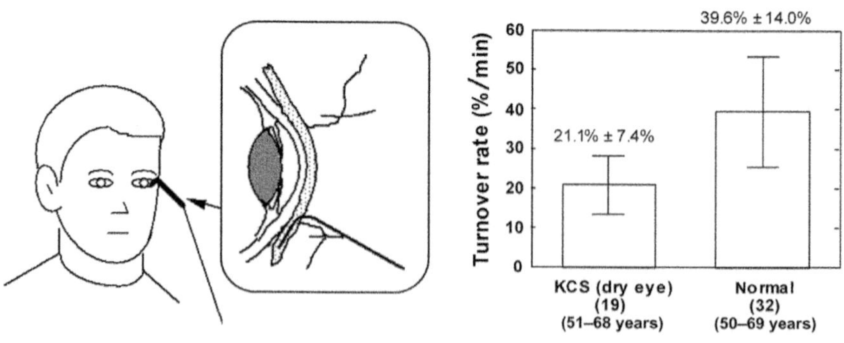

Figure 12.2 Schematic diagram of tear conductivity measurement at human conjunctiva (left). Comparison of tear turnover rates between KCS patient and normal eyes (right).

eyes. The tear turnover rate was calculated by a single exponential equation of tear conductivity change, following the eye-drop applications of 40.0 g/L NaCl solution. The mean turnover rate was 40.4% per minute (SD: 14.8% per minute; $n = 86$), being in agreement with previously reported values. The mean tear turnover rate for KCS eyes (22.1% per minute, SD: 7.4% per minute; $n = 19$) is much lower than that for control normal eyes (39.6% per minute, SD: 14.0% per minute; $n = 32$, 50–69 years). The high electrolyte concentration and low turnover rates of tears for KCS eyes were considered to be related to the lower rate of tear secretion from the lacrimal gland. The flexible sensor could be used for estimating the tear dynamics with static and dynamic sensing techniques.

12.2.3 Soft contact lens type biosensors using biocompatible polymers

Glucose monitoring does not measure blood sugar directly, but relies instead on measurement of the glucose levels in other biological fluids in the human body [22]. Relationships between general physical conditions and constituents of body fluids such as tears, mucus, sweat, and saliva in human cavities were reported [23–27]. Especially, a correlation between glucose concentration in tears and blood glucose was reported [28,29]. The tear glucose level changes with a few minutes' time delay in comparison with the blood sugar level [29].

In 1995, a flexible glucose sensor was developed by immobilized glucose oxidase (GOD) within a gold-coated, hydrophilic polytetrafluoroethylene membrane without the use of harmful substances in a simple fabrication process [30]. The biosensor was calibrated against glucose solutions from 6.7 to 662 mg/L including tear sugar level. From the medical and physiological points of view, flexible and biocompatible polymers should be used for all materials in a glucose sensor to apply in the lacrimal sac. In particular, phospholipid polymer, so-called MPC polymer, was used for the sensing region. Hydrophilic 2-methacryloyloxyethyl phosphorylcholine (MPC) polymer has molecular configuration, which is similar to a cell membrane. Such a configuration was carried out by the techniques of polymer chemistry [31,32]. Utilizing this sort of a polymer as a contacting part to measuring site, biocompatible sensor can be achieved. In this study, a novel biocompatible polymer PMEH was copolymerized with MPC and 2-ethylhexylmethacrylate.

A SCL biosensor for in situ monitoring of tear glucose was fabricated using the biocompatible PMEH polymer and another hydrophobic polydimethylsiloxane (PDMS) polymer (Fig. 12.3) [33,34]. PDMS is also one of biocompatible polymers used for an artificial lung. The SCL biosensor was designed for in situ monitoring of tear glucose level. PDMS was used as the body material of the SCL (base curve radius: 8.6). GOD was immobilized by entrapping with the biocompatible PMEH polymer. The signal of the SCL sensors was tentatively read using a wired electrical sheet in the preclinical experiment. The SCL sensor showed a good flexibility and soft characteristics as a commercial available one. This device kept the electrical performance without electrical breaking after mechanical bending.

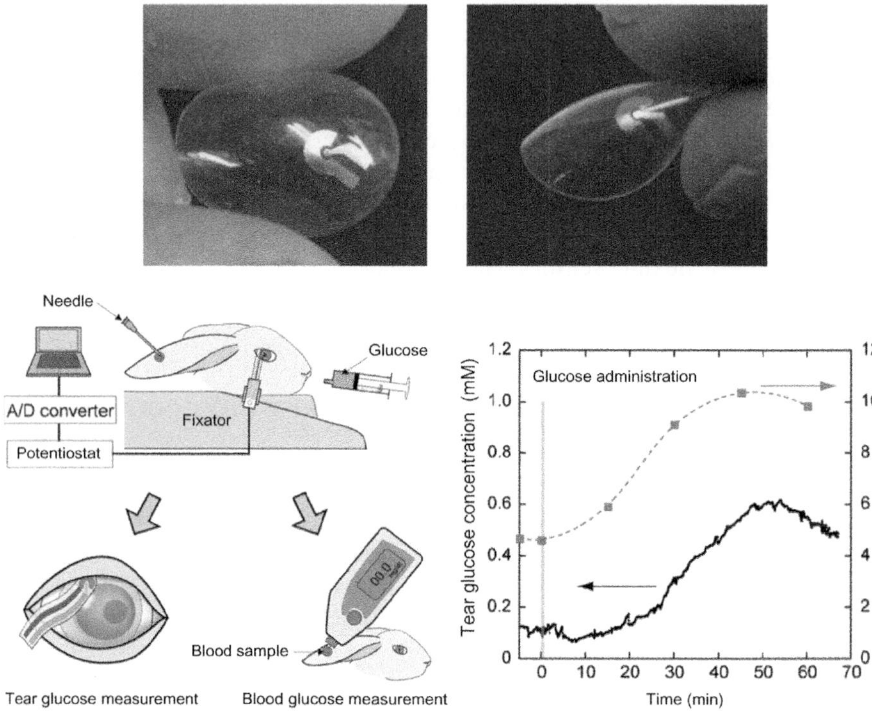

Figure 12.3 Photographs of SCL biosensor showing the flexibility not only lens but also coated electrodes (upper photos). Comparison experiments between tear glucose by SCL biosensor and blood one by a commercially available test-kit (left-lower) after glucose oral-administration. Temporal changes both in tear and blood glucose (right-lower).

The amperometric SCL biosensor showed a good relationship between the output current and glucose concentration in a range of 0.03–5.0 mmol/L, with a correlation coefficient of 0.999. Also, the SCL biosensor was applied to the rabbit eye for the purpose of tear glucose monitoring. The SCL sensor succeeded to monitor tear glucose concentration, which is approximately one-tenth that of blood glucose. The basal tear glucose was estimated to 0.13 mmol/L. Also, the change in tear glucose induced by the change in blood sugar level was assessed by the oral glucose tolerance test. As a result, tear glucose level increased with a delay of 8 minutes from blood sugar level (Fig. 12.3). The result demonstrated a meaning relationship between the tear sugar level and blood one with a few minutes of delay. The result showed that the SCL biosensor would be expected to provide further detailed information about the relationship between dynamics of blood glucose and tear one.

Another ocular biomonitoring with the SCL biosensor was also carried out. The SCL biosensor worn on the rabbit eye and the glucose levels in tear fluids were monitored in situ. The SCL biosensor was confirmed to be useful both in the static and the dynamic state. The tear turnover rate was assessed using the SCL biosensor. The tear turnover was evaluated using SCL biosensor by an instillation of different

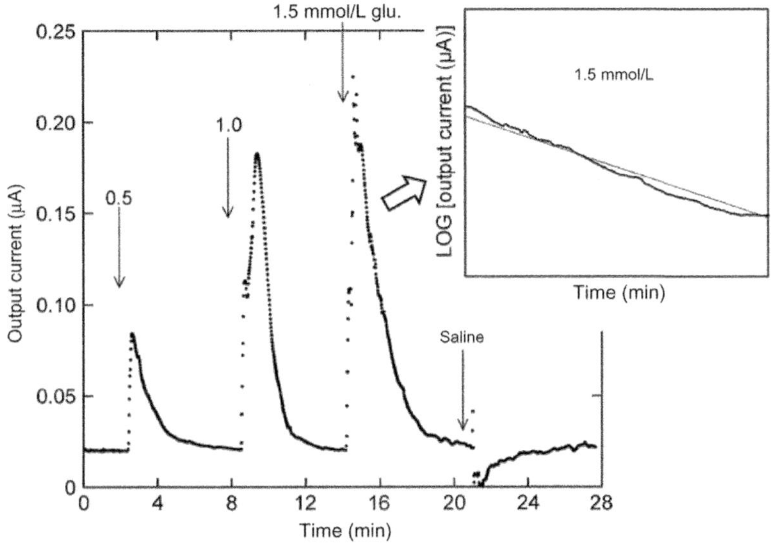

Figure 12.4 Temporal change in tear glucose level due to ocular instillation of standard glucose (0.5, 1.0, and 1.5 mmol/L glucose, respectively). The inverse peak indicates dilution of tear glucose by instillation of physiological sodium chloride solution.

glucose solutions: 0.5, 1.0, and 1.5 mmol/L glucose, respectively. The peak current depended on the glucose concentration by the ocular instillation (Fig. 12.4). The inverse peak indicates dilution of tear glucose by instillation of physiological sodium chloride solution [35]. The tear turnover was estimated using semi−log regression curve. The rabbit's tear turnover rate due to secretion of fresh tears was calculated as 29.6% ± 8.42% per minute, which is slightly slower than the human quoted rate of 40% per minute. The result indicated that the SCL biosensor would be also useful for advanced biomonitoring on eye. Additionally, since PDMS is a flexible and workable polymer, the PDMS based biosensor could be optimized to any surface of the human body. It is expected to realize "ubiquitous biomonitoring" by further studies such as communication techniques.

12.2.4 Transcutaneous gas sensor at eyelid conjunctiva

Transcutaneous oxygen sensor has been developed and been in use for monitoring arterial oxygen pressure ($tcPO_2$) in premature infants to prevent retinopathy of prematurity at neonatal intensive care unit as a noninvasive blood gas assessment [36,37]. The commercially available $tcPO_2$ sensors with a rigid cylindrical cell and a heating unit for oxygen diffusion are fixed to the infant skin with adhesive plaster, thus resulting in common skin rashes and general discomforts on the infants. A novel oxygen sensor with good flexibility and wearability, such as a clinical wet-pack, has been required for transcutaneous monitoring in comfort not only for infants but also for adult humans.

Eyelid conjunctiva has high gas penetration and supplies oxygen molecule to the cornea tissue [38,39]. Conjunctival oxygen monitoring [40–42] was considered as a possible application for the new oxygen sensor, which would obviate the need of heating. A conjunctival oxygen monitor using 10 newborn subjects was reported as a pilot report in 2002 [43]. The correlation coefficient between conjunctival oxygen tension and pulse oximetry was significant ($P < .001$). Although the sensor unit was placed on the eye, the size of the sensor was too large for general application, especially on the infants.

A thinner and flexible oxygen sensor was fabricated using functional polymers and microelectrical mechanical system (MEMS) techniques to monitor transcutaneous oxygen tension from eyelid conjunctiva [44–47]. The wearable oxygen sensor with membrane structure was constructed by pouching KCl electrolyte solution by nonpermeable membrane and gas-permeable membrane in which Pt and Ag/AgCl electrodes were formed by sputtering deposition with photolithography. The wearable oxygen sensor (width: 3 mm, thickness: 84 μm) was used as an amperometric sensing device with a fixed potential of −550 mV versus Ag/AgCl. The calibration range of the wearable oxygen sensor was from 0.01 to 8.0 mg/L of DO (dissolved oxygen) in the liquid phase. The sensor was also evaluated in the gas phase by purging with 10% oxygen gas and the response time to reach 90% of the steady output after purging was approximately 45 seconds. The sensor outputs and responses were stable during repeated measurements (CV: 3.66%). As the physiological application, the wearable sensor was placed into the lacrimal sac to attach to the eyelid conjunctiva of a Japanese white rabbit without any thermoregulation. Then, the rabbit inhaled standard air (oxygen: 20.9%) and high concentration oxygen (60% and 90%) (Fig. 12.5). As the result of the rabbit experiment, the sensor output increased and decreased synchronously with high concentration oxygen and

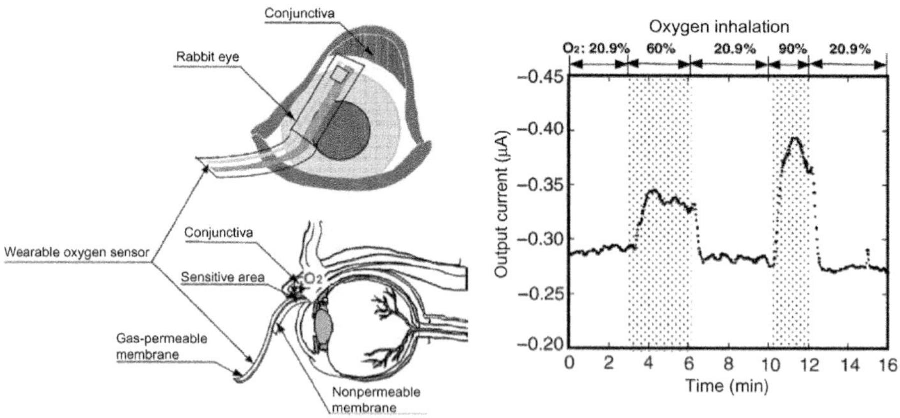

Figure 12.5 Illustration of affixing the wearable oxygen sensor at eyelid conjunctiva (left upper figure) and cross section of the sensor attachment (left-lower). Typical responses of flexible sensor for transcutaneous oxygen monitoring (inhaled O_2 conc.: 20.9%, 60%, 90%) at the rabbit conjunctiva (right).

standard air inhaling, respectively. The thinner wearable oxygen sensor succeeded to monitor the transcutaneous oxygen level at the eyelid conjunctiva as a noninvasive blood gas measurement. The eyelid conjunctiva in lacrimal sac would be a suitable site for noninvasive biomonitoring of blood gases without thermoregulation.

12.3 Mouthguard type biosensor for saliva biomonitoring

12.3.1 Salivary fluids in oral cavity

Saliva is secreted mainly by the parotid, the submaxillary, and sublingual glands (Fig. 12.6). Though parotid saliva is nonviscous, sublingual and submaxillary saliva is viscous because of its mucoprotein content [48]. The saliva fluid contains many kinds of components such as amino acids, ions, proteins, sugar (glucose), nucleotides, microorganism, etc. In general, saliva is also a complex biofluid comprising numerous constituents permeating from blood via transcellular or paracellular paths. Several researchers have developed portable in vitro salivary diagnostic tools [49]. Saliva is a noninvasive diagnostic fluid providing an alternative to direct blood analysis via the permeation of blood constituents without any skin piercing for blood sampling. For example, a saliva glucose measurement would be useful to estimate the blood sugar level for diabetic patients.

12.3.2 Wireless mouthguard sensor for salivary glucose

The saliva glucose concentrations range approximately from 20 to 200 μmol/L in normal and diabetic patients, closely follow circadian blood sugar fluctuations [50], and offer promising opportunities for noninvasive assessment of blood sugar level [51]. Saliva and blood glucose levels correlate reasonably in a sample of individuals [52–55]. In particular, a much stronger correlation is observed within the same

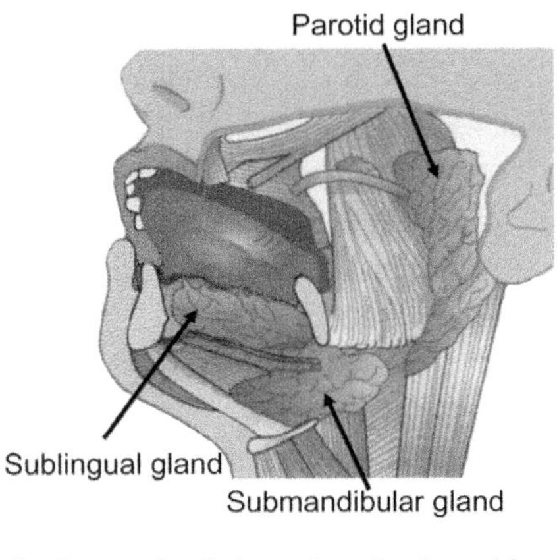

Figure 12.6 Illustration of salivary glands in oral cavity (parotid, submaxillary, sublingual glands).

individual, enabling blood glucose concentrations to be estimated from saliva glucose measurements [56]. To continuously measure the glucose concentration in salivary fluid, a mouthguard biosensor has been developed using an enzyme electrode [8].

A salivary glucose sensor was developed by incorporating Pt and Ag/AgCl electrodes on a mouthguard support as dental hard material with a GOD membrane. Fig. 12.7 is a schematic image of the glucose sensor on the polyethylene terephthalate glycol (PETG) mouthguard support. At first, Pt and Ag electrodes were deposited on the PETG through a sputtering process with thinner Ti stencils mask. The base sensor consisted of a 0.20 mm^2 Pt working electrode and a 4.0 mm^2 Ag/AgCl reference/counter electrode, both insulated with PDMS on a 0.5-mm-thick PETG layer. GOD (30 units) was immobilized to the sensing region of the Pt working electrode. To optimize enzyme immobilization, 2.0 mL of 1.0 wt.% PMEH solution was spread over the Pt sensing region and then PMEH was overcoated again.

The telemetric mouthguard biosensor seamlessly was integrated with the salivary glucose sensor and a wireless measurement system. One watch battery (1.5 V button battery) was utilized as a power source between the mouthguard materials inside of the cheek side (Fig. 12.8). When investigating in vitro performance, the mouthguard biosensor exhibits a robust relationship between output current and glucose concentration in artificial saliva. Artificial saliva containing various proteins, was prepared from disodium hydrogen phosphate, anhydrous calcium chloride, potassium chloride, sodium chloride, urea, and type II mucin from porcine stomachs according to a protocol reported by Fusayama et al. [57]. In artificial saliva, the mouthguard biosensor is capable of highly sensitive detection over a range of 5–1000 μmol/L of glucose, which encompasses the range of glucose concentrations found in human saliva (Fig. 12.9).

The ability of the sensor and wireless communication module to monitor saliva glucose in a phantom jaw imitating the structure of the human oral cavity was

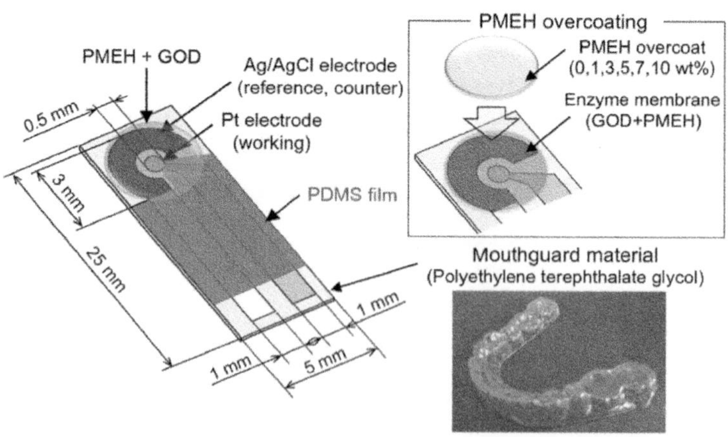

Figure 12.7 Schematic image of GOD enzyme electrode on the polyethylene terephthalate glycol (PETG) mouthguard support. Pt and Ag electrodes were formed on the PETG through a sputtering deposition.

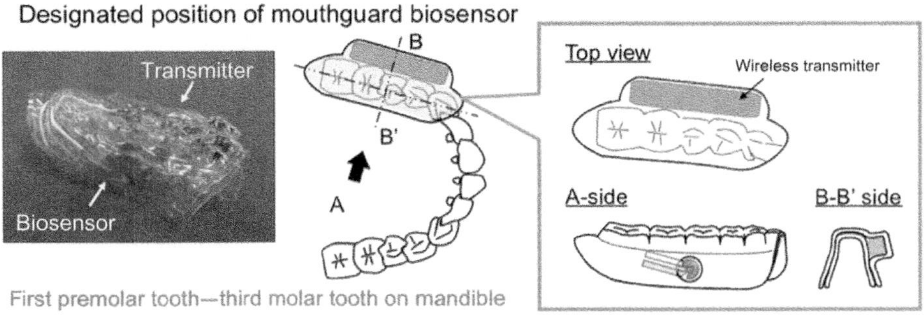

Figure 12.8 Schematic diagram of mouthguard biosensor to custom-fit the patient's dentition. The device consisted of a GOD electrode and wireless transmitter incorporating a potentiostat for stable glucose measurement. The wireless transmitter was neatly encased in PETG.

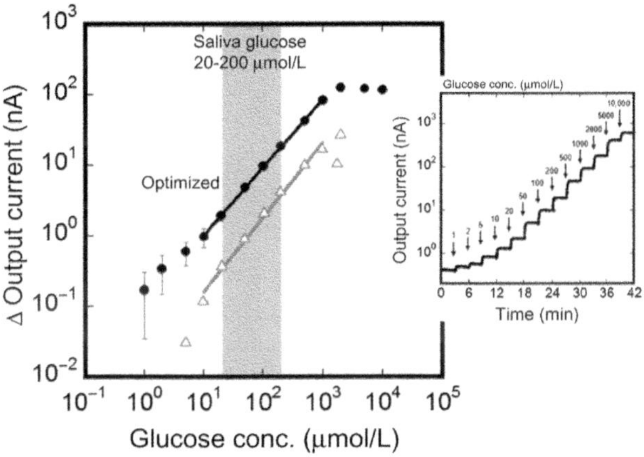

Figure 12.9 Calibration curves of the mouthguard biosensor. Inset: Typical response to glucose solution. The calibration range was 10–1000 μmol/L, which encompassed the physiological saliva glucose range in humans (20–200 μmol/L).

demonstrated. Stable and long-term real-time monitoring (exceeding 5 hours) with the wireless system was achieved. The mouthguard biosensor would be useful as a novel method for real-time saliva glucose monitoring. In addition, selectivity of the mouthguard sensor was evaluated by comparing output current in response to 100 μmol/L glucose, galactose, fructose, mannitol, sorbitol, and xylitol solutions. The mouthguard biosensor was shown to be highly selective for glucose based on the substrate specificity of GOD. Fructose, mannitol, sorbitol, and xylitol including some foods were not detected, producing a negligible output current less than 0.05% of the magnitude of the output current produced by glucose. Galactose was detected to a minimal extent, producing an output current at 0.265% of the magnitude of the output current produced by glucose.

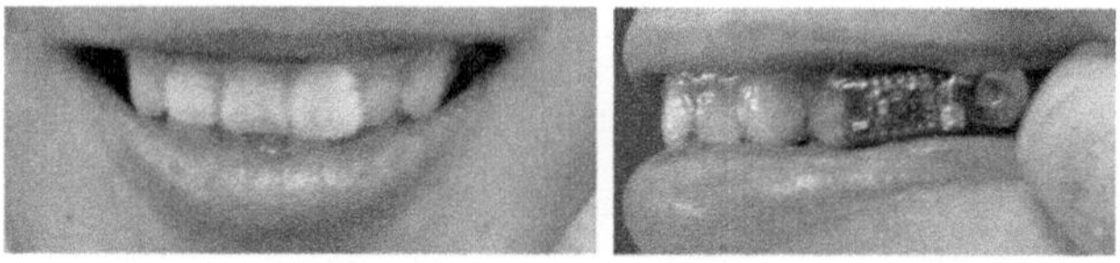

Figure 12.10 Front and side view photographs of telemetric IoT mouthguard sensor for monitoring salivary glucose. The invisible mouthguard biosensor to custom-fit the patient's dentition. The device consisted of a glucose sensor and wireless BLE transmitter incorporating a potentiostat for stable glucose measurement.

A novel IoT mouthguard biosensor using thinner and transparent dental materials integrated with a GOD electrode and a small Bluetooth low energy (BLE 4.0) wireless module for real-time monitoring of saliva glucose has also been developed (Fig. 12.10). Stable and long-term monitoring (24 hours) using telemetry system has been established. The IoT mouthguard biosensor would be useful as a real-time, noninvasive method for the subject person as a novel health care management in the dental field. And the invisible "Smart mouthguard gear" would be a suitable device not only for patients but also healthy subjects in their daily life.

12.4 Conclusion

Human secretions such as tear, saliva, sweat, body gas, etc., will provide important bio/chemical information about the health and well-being of an individual. Extracting this information would be the goal of developing noninvasive techniques for future diagnosis. Additionally, the body cavities would be a suitable site for monitoring the body bio/chemicals related human diseases and metabolic conditions. "Cavitas sensors" such as the SCL type and mouthguard type sensors are a novel category of self-attachable devices in daily life. In the near future, many types of cavitas medical sensor cooperating with IoT techniques would be developed and commercialized for managing the aging society in the world. The fabrication of the cavitas sensor is required to use safe, nontoxic, nonharmful chemicals and techniques and to apply the right materials in human body sites (not only soft and flexible to skin and mucosa, but also rigid and hard to tooth and bone). The IoT daily medicine (health care, presymptomatic, and preventive medicine) with cavitas and wearable sensors would be necessary to improve the QOL in view of the aging of society and the rapid changes in living environments.

Acknowledgments

This work is partly supported by Japan Society for the Promotion of Science (JSPS) Grants-in-Aid for Scientific Research System, by Japan Science and Technology Agency (JST) and by MEXT (Ministry of Education, Culture, Sports, Science and Technology) Special Funds for Education and Research "Advanced Research Program in Sensing Biology."

References

[1] E. Ghafar-Zadeh, Wireless integrated biosensors for point-of-care diagnostic applications, Sensors 15 (2015) 3236–3261.

[2] S.R. Corrie, J.W. Coffey, J. Islam, K.A. Markey, M.A.F. Kendall, Blood, sweat, and tears: developing clinically relevant protein biosensors for integrated body fluid analysis, Analyst 140 (2015) 4350–4364.

[3] B.W. Bode, T. Battelino, Continuous glucose monitoring, Int. J. Clin. Practice 64 (166) (2010) 11–15.

[4] G. Matzeu, L. Florea, D. Diamond, Advances in wearable chemical sensor design for monitoring biological fluids, Sens. Actuat. B: Chem. 211 (2015) 403–418.

[5] A.J. Bandodkar, W. Jia, C. Yardımcı, X. Wang, J. Ramirez, J. Wang, Tattoo-based noninvasive glucose monitoring: a proof-of-concept study, Anal. Chem. 87 (2015) 394–398.

[6] K. Mitsubayashi, K. Yokoyama, T. Takeuchi, E. Tamiya, I. Karube, Flexible conductometric sensor, Anal. Chem. 65 (1993) 3586–3590.

[7] K. Mitsubayashi, K. Ogasawara, K. Yokoyama, T. Takeuchi, T. Tsuru, I. Karube, Measurement of tear electrolyte concentration and turnover rate using a flexible conductimetric sensor, Technol. Health Care 3 (1995) 117–121.

[8] T. Arakawa, Y. Kuroki, H. Nitta, P. Chouhan, K. Toma, S. Sawada, et al., Mouthguard biosensor with telemetry system for monitoring of saliva glucose: a novel cavitas sensor, Biosens. Bioelectron. 84 (2016) 106–111.

[9] S.Y. Borelho, Tears and the lacrimal gland, Sci. Am. 211 (1964) 78–86.

[10] E.R. Berman, Biochemistry of the Eye, Plenum Press, New York and London, 1991, pp. 63–88.

[11] A. White, Principles of Biochemistry, McGraw-Hill Kogakusha, Ltd, Tokyo, 1973, pp. 905–927.

[12] N.J. Haeringen, Clinical biochemistry of tears, Surv. Ophthalmol. 26 (1981) 84–96.

[13] R.M. Stanifer, Tear film, in: R.E. Anderson (Ed.), Biochemistry of the Eye, Mannuals Program, American Academy of Ophthalmology, San Francisco, 1983, pp. 7–22.

[14] S. Mishima, A. Gasset, S.D. Klyce Jr, Determination of tear volume and tear flow, Invest. Opthalmol. Vis. Sci. 5 (1966) 264–276.

[15] R.J. Fullard, L.G. Carney, Diurnal variation in human tear enzymes, Exp. Eye Res. 38 (1984) 15–26.

[16] J.M. Fouke, A.D. Wolin, K.G. Saunders, M.R. Neuman, E.R. McFadden Jr., Sensor for measuring surface fluid conductivity in vivo, IEEE Trans. Biomed. Eng. 35 (1988) 877–881.

[17] D.F. Evans, M.A. Matesich, The measurement and interpretation of electrolytic conductance, in: E. Yeager, A.J. Salkind (Eds.), Techniques of Electrochemistry, Wiley, New York, 1973, pp. 1–59.

[18] H. Shwachman, Electrical conductivity of sweat: a simple diagnostic test in children, Pediatrics 32 (1963) 85.

[19] T. McKendrick, Sweat sodium levels in normal subjects, in fibrocystic patients and their relatives, and in chronic bronchitic patients, Lancet 1 (1962) 183.

[20] N.F. Sheppard Jr., R.C. Tucker, C. Wu, Electrical conductivity measurements using microfabricated interdigitated electrodes, Anal. Chem. 65 (1993) 1199–1202.

[21] K. Ogasawara, K. Mitsubayashi, T. Tsuru, I. Karube, Electrical conductivity of tear fluid in healthy persons and keratoconjunctivitis sicca patients measured by a flexible conductimetric sensor, Graefes Arch. Clin. Exp. Ophthalmol. 234 (1996) 542–546.

[22] G.S. Wilson, R. Gifford, Biosensors for real-time in vivo measurements, Biosens. Bioelectron. 20 (2005) 2388–2403.
[23] K.M. Daum, R.M. Hill, Human tear glucose, Invest. Ophthalmol. Vis. Sci. 22 (1982) 509–515.
[24] S.F.P. Man, G.K. Adams, D.F. Proctor, Effects of temperature, relative humidity, and mode of breathing on canine airway secretions, J. Appl. Physiol.: Respir. Environ. Exerc. Physiol. 46 (1979) 205–211.
[25] K. Mitsubayashi, M. Suzuki, E. Tamiya, I. Karube, Analysis of metabolites in sweat as a measure of physical condition, Anal. Chim. Acta 289 (1994) 27–34.
[26] A. Romano, F. Rolant, A non-invasive method of blood glucose evaluation by tear glucose measurement, for the detection and control of diabetic states, Metab. Pediatr. Syst. Ophthalmol. 11 (1988) 78–85.
[27] T.R. Stolwijk, J.A. van Best, H.H.P.J. Lemkes, R.J.W. de Keizer, J.A. Oosterhuis, Determination of basal tear turnover in insulin-dependent diabetes mellitus patients by fluorophotometry, Int. Ophthalmol. 15 (1991) 377–381.
[28] D.K. Sen, G.S. Sarin, Tear glucose levels in normal people and in diabetic patients, Br. J. Ophthalmol. 64 (1980) 693–699.
[29] W.F. March, F.E. Smith, P. Herbrechtsmeier, A. Domschke, C. Morris, Clinical trial of a non-invasive ocular glucose sensor, Diabetes 50 (Suppl) (2001) A125.
[30] K. Mitsubayashi, J.M. Dicks, K. Yokoyama, T. Tamiya, I. Karube, A flexible biosensor for glucose, Electroanalysis 7 (1) (1995) 83–87.
[31] K. Ishihara, T. Ueda, N. Nakabayashi, Preparation of phospholipid polylners and their properties as polymer hydrogel membranes, Polym. J. 22 (1990) 355–360.
[32] K. Ishihara, H. Nomura, T. Mihara, K. Kurita, Y. Iwasaki, N. Nakabayashi, Why do phospholipid polymers reduce protein adsorption? J. Biomed. Mater. Res. 39 (2) (1998) 323–330.
[33] M.X. Chu, H. Kudo, T. Shirai, K. Miyajima, H. Saito, N. Morimoto, et al., A soft and flexible biosensor using a phospholipid polymer for continuous glucose monitoring, Biomed. Microdev. 11 (2009) 837–842.
[34] M.X. Chu, K. Miyajima, D. Takahashi, T.O. Arakawa, K. Sano, S. Sawada, et al., Soft contact lens biosensor for in situ monitoring of tear glucose as non-invasive blood sugar assessment, Talanta 83 (2011) 960–965.
[35] M.X. Chu, T. Shirai, D. Takahashi, T. Arakawa, H. Kudo, K. Sano, et al., Biomedical soft contact-lens sensor for in situ ocular biomonitoring of tear contents, Biomed. Microdev. 13 (2011) 603–611.
[36] R. Huch, A. Huch, D.W. Lubbers, Transcutaneous measurement of blood Po2 (tcPo2) — Method and application in perinatal medicine, J. Perinat. Med. 1 (1973) 183–191.
[37] D.W. Lübbers, A. Huch, R. Huch, Unblutige, kontinuierliche Erfassung des Säure-Basen-Status beim Menschen durch transcutane Messung des arteriellen PO2 und PCO_2 sowie der Hämoglobinkonzentration Bei Körpertemperatur, Klin. Wschr. 51 (1973) 411–420.
[38] W.C. Shoemaker, P.M. Lawner, Method for continuous conjunctival oxygen monitoring during carotid artery surgery, Crit. Care Med. 11 (1983) 946–947.
[39] B.A. Weissman, I. Fatt, J. Rasson, Diffusion of oxygen in human corneas in vivo, Invest. Ophthalmol. Vis. Sci. 20 (1981) 123–125.
[40] S. Podolsky, J. Wertheimer, S. Harding, The relationship of conjunctival and arterial blood gas oxygen measurements, Resuscitation 18 (1989) 31–36.

[41] H. Haljamae, I. Frid, J. Holm, S. Holm, Continuous conjunctival oxygen tension (Pcjo2) monitoring for assessment of cerebral oxygenation and metabolism during carotid artery surgery, Acta Anaesthesiol. Scand. 33 (1989) 610–616.
[42] M. Brown, J.S. Vender, Noninvasive oxygen monitoring, Crit. Care Clin. 4 (1988) 493–509.
[43] S.J. Isenberg, D. Neumann, S. Fink, R. Rich, Continuous oxygen monitoring of the conjunctiva in neonates, J. Perinat. 22 (2002) 46–49.
[44] K. Mitsubayashi, T. Endo, S. Sasaki, I. Karube, Optical-transparent oxygen sensor with a thinner membrane structure, Anal. Sci. 17 (Suppl) (2000). i773/i775.
[45] K. Mitsubayashi, Y. Wakabayashi, D. Murotomi, T. Yamada, T. Kawase, S. Iwagaki, et al., Wearable and flexible oxygen sensor for transcutaneous oxygen monitoring, Sens. Actuat. B Chem. 95 (2003) 373–377.
[46] S. Iguchi, K. Mitsubayashi, T. Uehara, M. Ogawa, A wearable oxygen sensor for transcutaneous blood gas monitoring at the conjunctiva, Sensors Actuat. B 108 (2005) 733–737.
[47] H. Kudo, S. Iguchi, T. Yamada, T. Kawase, H. Saito, K. Otsuka, et al., A flexible transcutaneous oxygen sensor using polymer membranes, Biomed. Microdev. 9 (2007) 1–6.
[48] A. White, P. Handler, E.L. Smith, Principles of Biochemistry, McGraw-Hill Kogakusha, Ltd, Tokyo, 1973, pp. 905–927.
[49] A. Vasudeva, A. Kaushik, Y. Tomizawa, N. Norena, S. Bhansali, An LTCC-based microfluidic system for label-free, electrochemical detection of cortisol, Sens. Actuat. B 182 (2013) 139–146.
[50] M. Yamaguchi, M. Mitsumori, Y. Kano, Noninvasively measuring blood glucose using saliva, IEEE Eng. Med. Biol. Mag. 17 (3) (1998) 59–63.
[51] P. Mascarenhas, B. Fatela, I. Barahona, Effect of diabetes mellitus type 2 on salivary glucose–a systematic review and meta-analysis of observational studies, PLoS One 9 (7) (2014) e101706.
[52] A. Sener, C. Jurysta, N. Bulur, B. Oguzhan, I. Satman, T.M. Yilmaz, et al., Salivary glucose concentration and excretion in normal and diabetic subjects, J. Biomed. Biotechnol. (2009) 430426.
[53] P. Abikshyeet, V. Ramesh, N. Oza, Glucose estimation in the salivary secretion of diabetes mellitus patients, Targets Ther. 5 (2012) 149–154.
[54] C. Liu, Y. Sheng, Y. Sun, J. Feng, S. Wang, J. Zhang, et al., A glucose oxidase-coupled DNAzyme sensor for glucose detection in tears and saliva, Biosens. Bioelectron. 70 (15) (2015) 455–461.
[55] A. Soni, S.K. Jha, A paper strip based non-invasive glucose biosensor for salivary analysis, Biosens. Bioelectron. 67 (2015) 763–768.
[56] W. Zhang, Y. Du, M.L. Wang, On-chip highly sensitive saliva glucose sensing using multilayer films composed of single-walled carbon nanotubes, gold nanoparticles, and glucose oxidase, Sens. Bio-Sens. Res. 4 (2015) 96–102.
[57] T. Fusayama, T. Katayori, S. Nomoto, Corrosion of gold and amalgam placed in contact with each other, J. Dent. Res. 42 (1963) 1183–1197.

Point of care testing apparatus for immunosensing

13

Tomoyuki Yasukawa, Fumio Mizutani and Masato Suzuki
Graduate School of Material Science, University of Hyogo, Kamigori, Japan

13.1 Introduction

On-demand medical testing is often required at medical sites. Point-of-care testing (POCT), or bedside testing is available to obtain valuable information for diagnostics, therapeutics, nursing care, disease prevention, and promotion of healthy lifestyles at a patient's bedside, and can lead to an improvement of the quality of life [1]. POCT is a system of tests that physicians can move at a moment's notice to wherever the POCT apparatuses are needed. Test results can be obtained within tens of minutes by using POCT apparatuses. Thus, POCT is necessary for today's medical personnel, especially in diagnostic departments that require rapid test results, such as the intensive care unit, critical care, and in ambulances.

Instruments that test blood glucose levels are a typical example of POCTs. A drop of blood is obtained from the finger of a patient and the amount of glucose present is detected by glucose sensors. POCT apparatuses that can measure the blood glucose level in situ immediately after blood is drawn from human bodies are very useful because the glycolysis reaction proceeds even after blood has been sampled. Instruments for the self-monitoring of blood glucose (SMBG) have recently made surprising progress. However, the SMBG instruments are not included as a POCT apparatus because the test is performed by individual patients at their homes without the presence of medical workers.

The principle of blood-glucose determination is identical for both POCT apparatuses and SMBG instruments. The enzymes that catalyze the oxidation of glucose, such as oxidase and dehydrogenase, have been applied to develop sensing systems for glucose monitoring. Fig. 13.1 shows the reaction steps involved in the detection of glucose. When glucose in a blood sample is oxidized by the reaction with glucose dehydrogenase, the oxidized form of the mediator contained in a glucose-sensing chip is reduced. The reduced form of the mediator is then oxidized at an electrode on the chip to produce a current signal. Thus, the concentration of glucose can be indirectly determined from the current produced by the oxidation of the reduced form of the mediator. Glucose dehydrogenase acts as an element for both the recognition of glucose and the conversion of signal molecules (electrochemically active molecules) from the oxidized form to the reduced form of the mediator. Moreover, the oxidized form of the mediator that's produced by the electrochemical

Chemical, Gas, and Biosensors for the Internet of Things and Related Applications.
DOI: https://doi.org/10.1016/B978-0-12-815409-0.00014-0

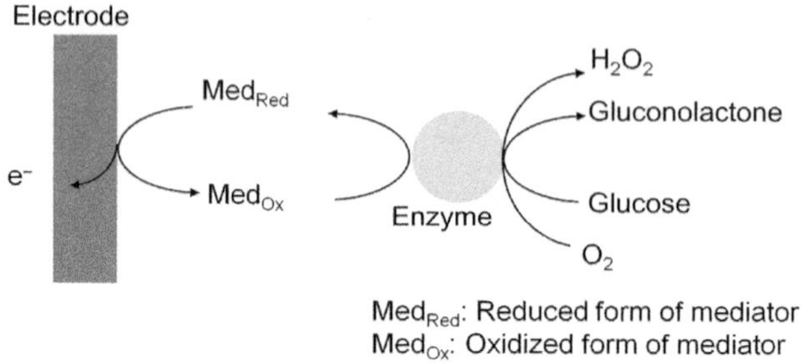

Figure 13.1 Reaction steps to detect glucose.

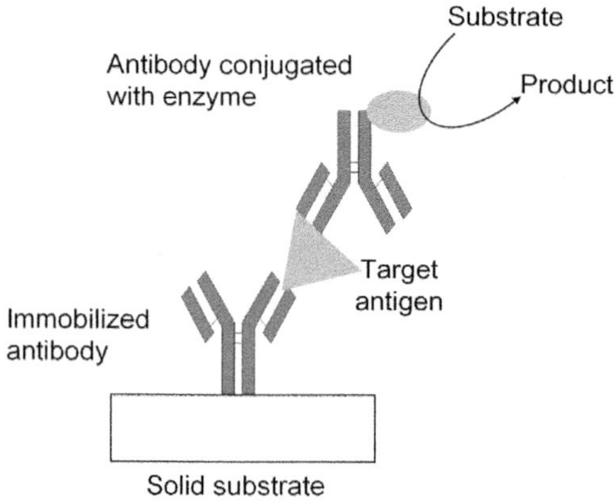

Figure 13.2 Schematic illustration of the immunocomplex created on the solid substrate for ELISA.

reaction can be reused as a substrate for the enzyme reaction. The electrochemical signal is also amplified by redox cycling between the enzyme reaction and electrochemical reaction.

The tests run by POCT apparatuses and their reagents play an important role in the early diagnostics of infectious diseases. A large part of the testing for infectious diseases is performed by immunochromatography assays (ICA), using antigen–antibody reactions. For ICA, antibodies are used to recognize the target antigen, and their low dissociation constants (10^{-10}–10^{-12} mol/L) mean that once both antibody and antigen have reacted, they are unlikely to separate. This is the main reason why antibodies are ideal for selectively targeting a selected antigen. However, antibodies do not have the ability to convert electrochemically active molecules from reduced forms to oxidized forms, and vice versa. Therefore, if we want to include the ability to determine the presence of antigens via electrical current, we have to incorporate a system to transduce the chemical behavior to electrical signal. An antibody conjugated with an enzyme has been frequently used to recognize the presence of an antigen and give an appropriate signal (enzyme linked immunosorbent assay, ELISA). Fig. 13.2 shows the schematic illustration of the immunocomplex created on the solid substrate for ELISA. The target molecules are trapped by the antibody immobilized on the substrate and then labeled with enzyme via an immunoreaction of the antibody conjugated with enzyme. After separating the unreacted antibody conjugated with enzyme, the product generated by the enzyme reaction can be detected by absorbance, fluorescence, and electrochemistry-based measurements.

Separation systems of bound and unbound elements for signal conversion must be included. The development of the mechanisms for sensing using antigen–antibody reactions is difficult compared with that of using enzymes. However, we can obtain a predominantly large variety of antibodies compared with that of enzymes because of progress made in hybridoma technology. Thus, currently novel detection strategies by using an antibody have been actively studied to develop rapid, simple, and quantitative mechanisms for POCT apparatuses. In this chapter, we describe the principles behind ICA and the present state of its use in POCT apparatuses that apply an antigen–antibody reaction. We also describe the development of a quantitative ICA device that incorporates an electrochemical detection system and a quick, simple, and quantitative immunosensing method using particle manipulation techniques based on dielectrophoresis.

13.2 Immunochromatography assay

ICA was developed by combining the principles of ELISA with chromatography; usually, a lateral flow platform is used [2]. ICA is a well-established technique and is widely employed in kits for clinical diagnoses and food inspections. Results of tests performed by the ICAs in these kits are rapidly obtained without any special apparatus. The success of the home pregnancy test by the ICA led to the increase of motivation to use the ICAs as the POCT for other fields. Today, ICA is widely applied to clinical diagnoses at a patient's bedside and early diagnoses of the initial stages of infectious diseases.

Fig. 13.3A shows a schematic illustration of ICA kits used to detect an antigen with the formation of a sandwich type immunocomplex, which consists of a nitrocellulose membrane (lateral flow strip), sample pad, conjugate pad, and absorbent pad. The antibody, which can be specifically reacted with an epitope of a target antigen with high molecular weight, such as a protein or a virus, is immobilized on the nitrocellulose membrane to form the A line. The antibody that can be reacted with the labeled antibody is also immobilized on the lower section of the nitrocellulose membrane to form the B line. The sample pad and conjugate pad are arranged at the end of the nitrocellulose membrane, while the absorbent pad is at the other end of the membrane. The antibody labeled with a reporter, such as gold nanoparticles or colored latex particles, is included in the conjugate pad [3]. When a solution containing the target antigen is dropped on the sample pad the solution passes through the conjugate pad and migrates toward the other end of the membrane by capillary force. At the conjugate pad, the epitope of the target antigen reacts with

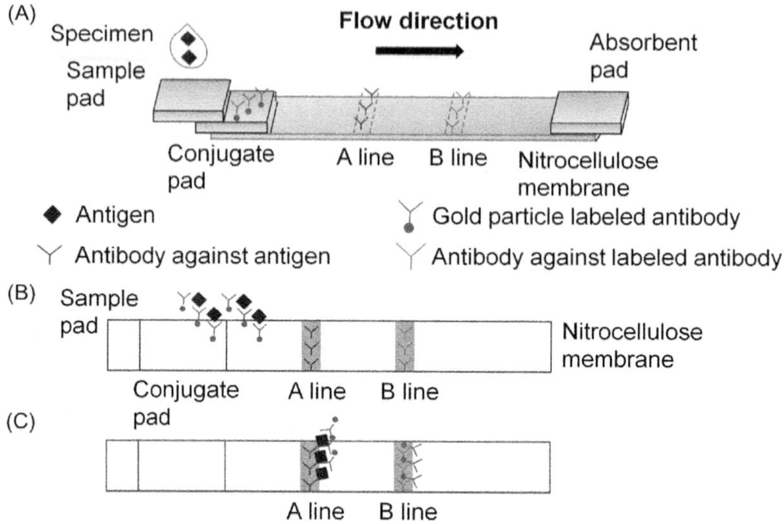

Figure 13.3 Schematic illustrations of (A) ICA kits used to detect an antigen with the formation of a sandwich type of immunocomplex. (B) Formation of an immunocomplex of the target antigen in a specimen with the labeled antibody incorporated into the conjugate pad. (C) Formation of a sandwich type of immunocomplex at the A line and the immunocomplex between the immobilized antibody and labeled antibody.

the labeled antibody to form an immunocomplex of antigen and labeled antibody (Fig. 13.3B). When the immunocomplex arrives at the A line where the antibody for the target antigen is immobilized, the immunocomplex is captured by the immobilized antibody via the other epitope of the target antigen, resulting in the formation of the sandwich type immunocomplex of antibody-antigen—labeled-antibody at the A line (Fig. 13.3C). The gold nanoparticles used as a label are integrated into the A line, resulting in the visualization (reddish purple color) of the A line. The intensity of the A line increases with the increasing concentration of the target antigen because the number of captured immunocomplex of antigen and labeled antibody at the A line increases along with the increasing the concentration. The immunocomplex of antigen and labeled antibody unreacted with the antibody immobilized on the A line pass through the A line and arrive at the B line. Here they are integrated into the B line owing to the presence of an antibody against labeled-antibody immobilized on the B line (Fig. 13.3C). Finally, the absorbent pad absorbs the solution. In the presence of antigen, the A line is colored, while the B line should be colored regardless of the presence of antigen. A colorless B line suggests a failure of the test.

Kits based on ICA to detect target antibodies are also available. Fig. 13.4 shows a schematic illustration of the ICA kit used to detect an antibody. For these kits, antigen is immobilized on the A line of the membrane and labeled antibody, which can recognize the target antibody, is placed onto the conjugate pad (Fig. 13.4A). When the solution containing the target antibody is applied onto the sample pad,

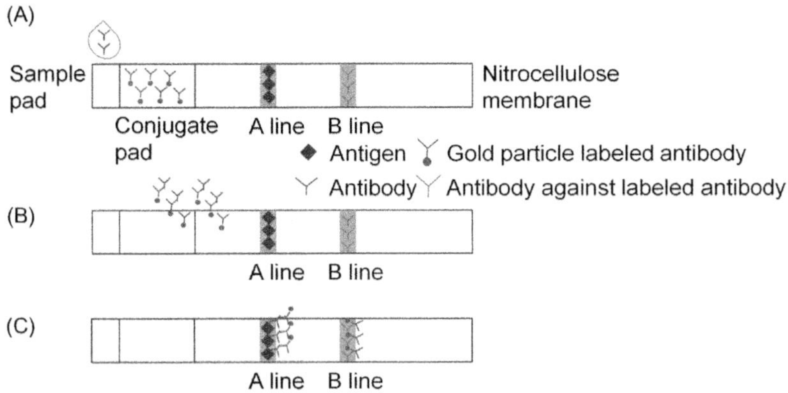

Figure 13.4 Schematic illustrations of (A) ICA kits to detect an antibody. (B) Formation of an immunocomplex of a target antibody in a specimen and labeled antibody incorporated into the conjugate pad. (C) Formation of the immunocomplex of antigen–target antibody–labeled antibody at the A line.

the immunocomplex of target antibody and labeled antibody that is produced at the conjugate pad is transferred and captured by the antigen immobilized on the A line to form the immunocomplex of antigen–antibody labeled antibody (Fig. 13.4B and C). The antibody immobilized on the B line recognizes the labeled antibody, and can therefore be used to evaluate the success of the test.

13.3 Immunochromatography assay for infectious diseases

ICA is a single-step test without any special equipment. It can rapidly produce results, generally within 15–30 minutes. This single-step rapid test kit has been developed for the diagnosis of infectious diseases to allow medical professionals to make rapid decisions about a patient's healthcare. There are two types of detection methods used to identify infections. The first is to directly detect the pathogenic microbes that are the cause of infection, and the second is to detect the antibody induced by the infection. The direct detection system of the pathogenic microbes has been adopted for test kits for respiratory viral infections such as influenza virus, respiratory syncytial virus, and adenovirus, as well as viral infections of the digestive system, such as norovirus and rotavirus, and bacterial infections such as the pathogenic *Escherichia coli*, *Streptococcus pneumoniae*, and *Streptococcus agalactiae*. There are not many ICA kits available for the detection of antibodies induced by the infection compared with those for the direct detection of microbes. This may be due to the difference of the application; the detection of pathogenic microbes is used to determine the presence of infections, but, for example, the test for the antibody found in a hepatitis B infection is used to determine the potential for healing from hepatitis B and for screening patients to see if they have been vaccinated against hepatitis B. For pathogen detections, pituita, phlegm, and liquids from wiping the pharynx are used as specimens, while blood serum and plasma are used for the detection of antibodies. The kits with two or more bands of immobilized antibodies are commercially available to differentiate between multiple pathogenic microbes. The use of antibodies against influenza A and B, for example, allows differential diagnosis.

13.4 Reliability of the examination kits

ICA is a simple and rapid test with high sensitivity (the population of true positive results in positive patients) and high specificity (the population of true negative results in negative patients). In these tests, positive or negative results are judged by a visual colorimetric measurement using the naked eye. Thus, the problems for false positive and false negative are inevitable owing to this qualitative or semiquantitative decision. Especially, decrease of the visible sensitivity of aged judges causes an oversight of weak coloring of the test line. To try and prevent these errors from occurring, portable strip readers, for example, BD Berita plus analyzer, Spotchem IL, have been developed to recode the intensity of the test line for a more quantitative judgment [4,5]. However, the limit of detection (LOD) obtained by ICA is larger than that of other detection methods (gene analysis method by PCR). Quantitative determination strategies with lower LOD are desired for accurate POCT use.

13.5 Signal amplification

Silver amplification systems were employed as a method to improve the LOD for ICA [6–11]. The solution containing silver ions and reducing agent, that is, hydroquinone and ascorbic acid, was dropped on the test lines after target antigen was captured on the test lines and labeled with gold nanoparticles. Silver ions were catalytically reduced to metallic silver at the surface of gold nanoparticles captured on the test lines by immunoreactions, resulting in the enlargement of gold nanoparticles by the silver deposition (Fig. 13.5). The coloring can be clearly observed due to the increased particle size, even when a small amount of gold nanoparticles are trapped at the line for antigen detection. This amplification procedure has been applied to immunohistochemistry [12] and ELISA [13]. However, sequential two-step procedures for trapping the antigen and silver deposition are required, while substantial nanoparticles remain in the strips. Therefore, washing steps are required to suppress the background signals obtained from the strips, excepting the lines for target detection. The addition of silver ion solution and the washing solution complicates the ICA process. The nonspecific adsorption of gold nanoparticles with antibody at the line for antigen detection frequently gives rise to a false positive,

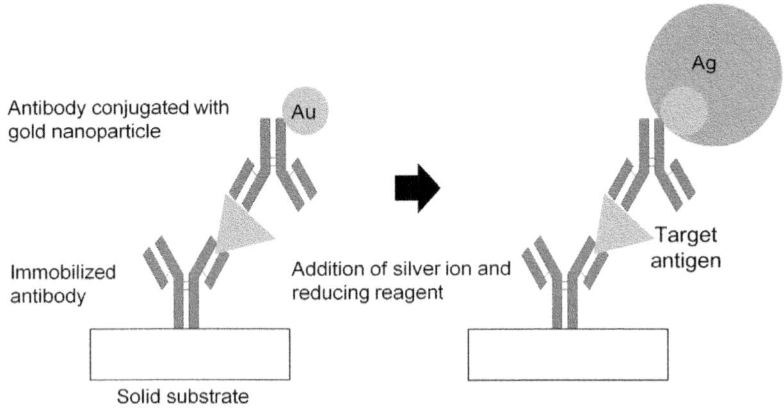

Figure 13.5 Principle of the diagnostic test for a specific antigen by silver amplification systems.

because the silver amplification of nonspecific nanoparticles increases the possibility of signal observation. Even with these extra steps and the possibility of false positives, ICA systems incorporating solution containers for washing and amplification have been developed and commercialized for rapid and sensitive diagnosis, such as IMMUNO AG1 and IMMUNO AG2 from FUJIFILM Co., Japan. Both the washing solutions and the solutions for silver deposition were sequentially and automatically applied to the test line to increase the visibility of the test.

Improvement of the labeled materials has been energetically studied to lower the LOD of ICA. Quantum dots [10,11,14], europium chelate-doped nanoparticles [15,16]), and fluorescent silica nanoparticles [17] have been developed to enhance the signal from labels accumulated on the test line. Nanoparticles as reporters for ICA were introduced in a recent review [18]. The remarkable improvement of the detectability achieved by the streptavidin–biotin binding has been attributed to the increase of captured nanoparticles per each immunorecognition event at the test line. This method achieved single-step signal amplification without any further posttreatments.

13.6 Quantitative ICA by electrochemical detection systems

Electrochemical detection systems were adopted to develop a quantitative ICA [19]. An electroactive species encapsulated into a liposome [20]) and a bismuth ion trapped with a chelator [21] were used as signal sources for ICA. Metallic nanoparticles and enzymes were also used as labels to enhance the electrochemical signals of ICA. Metallic ions generated by dissolving the labeled nanoparticles produce a large reduction current, hence the signal amplification [22–24]). Enzymes also produce large numbers of product through the catalytic reaction of enzymes to concentrate signal molecules [25]. However, additional complex steps were required to obtain the electrochemical signal, such as cutting the strips, transferring them into an electrochemical cell, and adding a strong acidic solution to dissolve the captured nanoparticles or the solution containing enzyme substrate.

Recently, electrochemical detection of electrochemically active species flowing in the membrane strips was achieved by attaching a microelectrode directly to the strips, which were then applied to the ICA [26,27]. A dual electrochemical sensor based on a test strip assay has also been studied for the determination of albumin and creatinine [28]. Fig. 13.6 shows the principle of electrochemical detection based on an immunochromatographic device. Each nitrocellulose membrane immobilized antibody and enzyme was prepared on the left side and right side of the fluidic devices, respectively (Fig. 13.6A). The solution containing albumin and creatinine was applied to the membrane from the center of the device and allowed to flow equally into both immobilization regions. Hydrogen peroxide generated by the three enzymes was oxidized at the working electrode 2 (W2) downstream from the enzyme immobilization region to detect the creatinine concentration (Fig. 13.6B and C). Oxidation current of the mediator generated by the reaction of glucose oxidase (GOx), which was used as a label for immunoreactions, was detected at the working electrode 1 (W1) and used to detect the albumin concentration (Fig. 13.6D and E). The present electrochemical immunochromatographic assay allowed for the quantitative determination of albumin and creatinine in a test sample. Ideally, however, substrate for the labeled enzyme in the ICA device should be incorporated to develop the one-step ICA. The achievement of one-step assay strategies and improvement of the sensitivity could lead to a becoming standard for quantitative ICA due to the ease of miniaturization of electrochemical system, such as the portable glucose meter.

13.7 Rapid and Quantitative ICA based on dielectrophoresis

Lastly, the quick and simple immunoassay procedure based on a particle manipulation technique by dielectrophoresis (DEP) will be introduced. Particles placed in a spatially heterogeneous electric field experience DEP force by a polarization effect induced in the particles. DEP is attractive for the manipulation of micro- and nanoobjects, including biological living cells and bacteria [29], and has been used in a wide range of applications, such as separation and sorting [30–32], trapping [33,34], and patterning [35–37].

By using dielectrophoretic particle manipulation, we can determine the concentration of the target molecule by only switching on/off the voltage, after injecting the mixture of the solution containing the target molecule and particle suspension [38,39]. Fig. 13.7 shows the cross-sectional view of the DEP device with an interdigitated array (IDA) electrode and the fundamentals of immunosensing based on

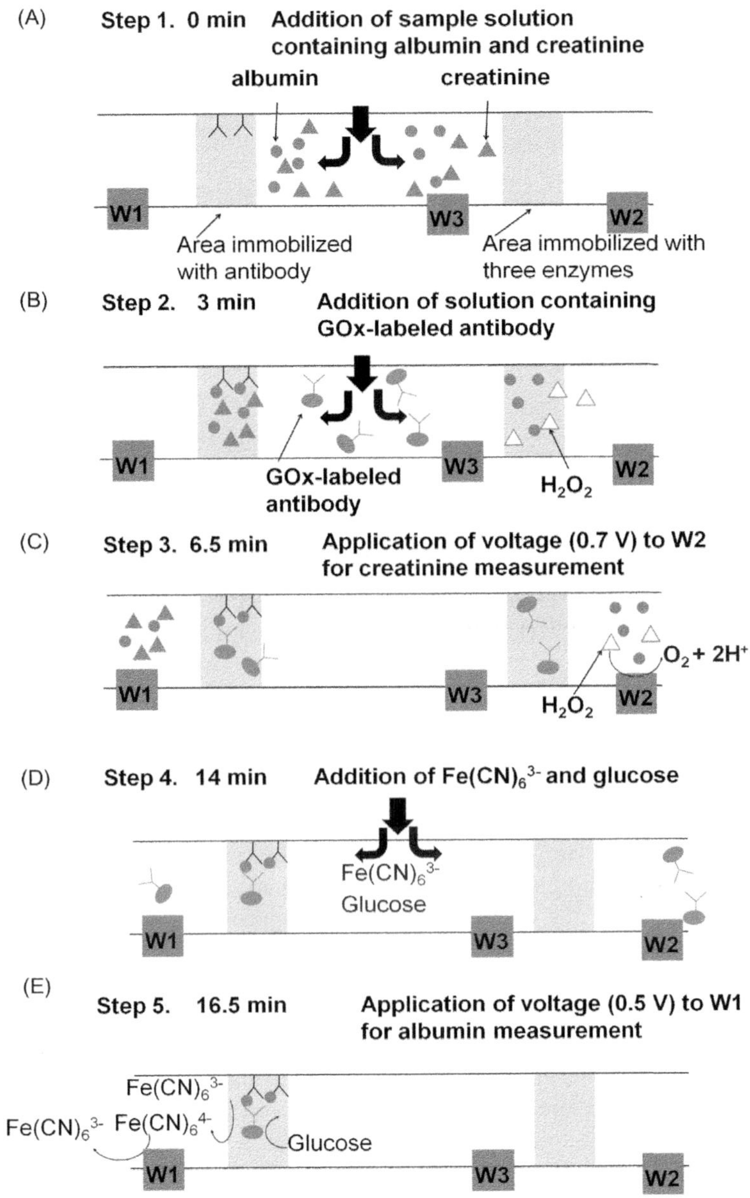

Figure 13.6 Principle of the electrochemical detections of creatinine and albumin with the dual electrochemical sensor. (A) The solution containing different concentrations of albumin and creatinine was added to the window positioned in the center of the test strip. (B) Hydrogen peroxide was generated by the enzyme reactions of creatinine, albumin was trapped at the antibody immobilization area, and a GOx-labeled antibody was introduced. (C) The generated hydrogen peroxide was oxidized at W2 and the captured albumin was labeled with a GOx-labeled antibody. (D) Substrates for the enzyme reaction, $Fe(CN)_6^{4-}$, and glucose were introduced. (E) $Fe(CN)_6^{4-}$ produced by the enzyme reaction of GOx was detected with W1. Source: Reprinted with permission from T. Yasukawa, Y. Kiba, F. Mizutani, A dual electrochemical sensor based on a test-strip assay for the quantitative determination of albumin and creatinine, Anal. Sci. 31 (2015) 583−589 [28], Copyright (2018) Japan Society for Analytical Chemistry.

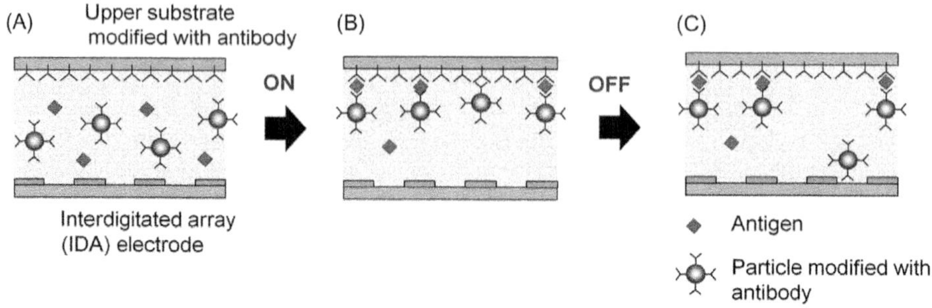

Figure 13.7 Cross-sectional view of the DEP device with an interdigitated array electrode and the principle of immunosensing based on DEP.

DEP. Both particles and the upper substrate were modified with an antibody against the target molecule. The particle suspension mixed with the target molecule is then that introduced into the DEP device. Before AC voltage is applied, particles are randomly dispersed (Fig. 13.7A). When an AC voltage with a frequency for a negative DEP (n-DEP) is applied to the IDA electrode to direct the particles to the region with the lower electric field, the microparticles are forced to move toward the upper substrate, thereby arranging in a line pattern along the band electrodes of the IDA within several seconds (Fig. 13.7B). The restrictive enforced contact between the antibody on the particles and the antibody immobilized on the upper substrate, along with the convection flow by joule heating at the electrodes, accelerate the reaction between the target molecules with the antibodies to form antibody–target–antibody complexes. The formation of immunocomplexes leads to the irreversible trapping of the microparticles on the upper substrate even after the AC voltage is turned off (Fig. 13.7C). In contrast, unreacted microparticles are redispersed from the upper substrate after switching off the AC voltage. Thus, unreacted microparticles are automatically separated from the upper substrate. The number of particles captured on the upper substrate increases with the increasing the concentration of the target molecules and can be easily calculated using optical or fluorescent microscopy. Therefore, the present procedure does not require the separation for stepwise immunoreaction and the removal of unreacted microparticles by washing. Furthermore, the simultaneous determination of multiple target molecules was achieved by preparing different types of antibodies on the upper substrate [39]. The incorporation of the microparticles modified with antibody would allow for the development of a rapid and single-step platform for immunosensing.

13.8 Conclusion

The principle of ICA and the present state of ICA technology as POCT apparatuses were introduced in this chapter. When a solution containing a target molecule is added to the ICA device, the capillary force causes the solution to migrate automatically and react with the antibodies that have been modified with gold nanoparticles. The gold nanoparticles that react with the target molecule become trapped at the test line. Within 15 minutes, the user of the POCT can judge if the target molecule is present or not, while unreacted nanoparticles are automatically separated from the test line by capillary flow. Because of ICA's simplicity and its ability to give rapid results, ICA has become an incredible tool for immunoassays. However, the ICA apparatus does not have the ability to quantify target molecules. It also has a high LOD and lacks sensitivity. A future trend for the practical use of ICA is the development of strategies that provide an increased sensitivity (lower LOD) and an ability for an ICA to give quantitative information. The incorporation of miniaturized electrochemical detection systems to ICA will allow for the production of affordable and simple POCT apparatuses for rapid and highly sensitive testing, as well as provide a platform for quantitative determination of target molecules. Novel determination strategies based on recognition events by immunoreaction should be developed in the research field of microfluidics while working jointly with an external force, such as DEP.

References

[1] J.K. Gerald, Goals, guidelines and principles for point-ofcare testing, in: J.K. Gerald (Ed.), Principles & Practice of Point-of-Care Testing, Lippincott Williams & Wilkins, New York, 2001, pp. 1–12.

[2] E. Eltzov, S. Guttel, A. Low Yuen Kei, P.D. Sinawang, R.E. Ionescu, R.S. Marks, Lateral flow immunoassays–from paper strip to smartphone technology, Electroanalysis 27 (2015) 2116–2130.

[3] M. Beggs, M. Novotny, S. Sampedro, A selfperforming chromatographic immunoassay for the qualitative determination of human chorionic gonadotrophin (HCG) in urine and sera, Clin. Chem. 36 (1990) 1084–1085.

[4] X. Mao, Y. Ma, A. Zhang, L. Zhang, L. Zeng, G. Liu, Disposable nucleic acid biosensors based on gold nanoparticle probes and lateral flow strip, Anal. Chem. 81 (2009) 1660–1668.

[5] A.K. Ellerbee, S.T. Phillips, A.C. Siegel, K.A. Mirica, A.W. Martinez, P. Striehl, et al., Quantifying colorimetric assays in paper-based microfluidic devices by measuring the transmission of light through paper, Anal. Chem. 81 (2009) 8447–8452.

[6] D.J. Chiao, R.H. Shyu, C.S. Hua, H.Y. Chiang, S.S. Tang, Colloidal gold-based immunochromatographic assay for detection of botulinum neurotoxin type B, J. Chromatogr. B 809 (2004) 37–41.

[7] K. Mitamura, H. Shimizu, M. Yamazaki, M. Ichikawa, K. Nagai, K. Katada, et al., Clinical evaluation of highly sensitive silver amplificationimmunochromatography systems for rapid diagnosis of influenza, J. Virol. Methods 194 (2013) 123–128.

[8] I.H. Cho, S.M. Seo, E.H. Paek, S.H. Paek, Immunogold-silver staining-ona-chip biosensor based on cross-flow chromatography, J. Chromatogr. B Anal. Technol. Biomed. Life Sci. 878 (2010) 271–277.

[9] A. Wada, Y. Sakoda, T. Oyamada, H. Kida, Development of a highly sensitive immunochromatographic detection kit for H5 influenza virus hemagglutinin using silver amplification, J. Virol. Methods 178 (2011) 82–86.

[10] Q. Yang, X. Gong, T. Song, J. Yang, S. Zhu, Y. Li, et al., Quantum dot-based immunochromatography test strip for rapid, quantitative and sensitive detection of alpha fetoprotein, Biosens. Bioelectron. 30 (2011) 145–150.
[11] W. Yang, X.B. Li, G.W. Liu, B.B. Zhang, Y. Zhang, T. Kong, et al., Colloidal gold probe-based silver enhancement immunochromatographic assay for the rapid detection of abrin-a, Bioasens. Bioelectron. 26 (2011) 3710–3713.
[12] C.S. Holgate, P. Jackson, P.N. Cowen, C.C. Bird, Immunogold–silver staining: new method of immunostaining with enhanced sensitivity, J. Histochem. Cytochem. 31 (1983) 938–944.
[13] C.-H. Yeh, C.-Y. Hung, T.C. Chang, H.-P. Lin, Y.-C. Lin, An immunoassay using antibody–gold nanoparticle conjugate, silver enhancement and flatbed scanner, Microfluid Nanofluid 6 (2009) 85–91.
[14] Z. Zou, D. Du, J. Wang, J.N. Smith, C. Timchalk, Y. Li, et al., Quantum dot-based immunochromatographic fluorescent biosensor for biomonitoring trichloropyridinol, a biomarker of exposure to chlorpyrifos, Anal. Chem. 82 (2010) 5125–5133.
[15] X. Xia, Y. Xu, X. Zhao, Q. Li, Lateral flow immunoassay using europium chelate-loaded silica nanoparticles as labels, Clin. Chem. 55 (2009) 179–182.
[16] Y. Xu, Q. Li, Multiple fluorescent labeling of silica nanoparticles with lanthanide chelates for highly sensitive time-resolved immunofluorometric assays, Clin. Chem. 53 (2007) 1503–1510.
[17] X. Huang, Z.P. Aguilar, H. Li, W. Lai, H. Wei, H. Xu, et al., Fluorescent $Ru(phen)_3{}^{2+}$-doped silica nanoparticles-based ICTS sensor for quantitative detection of enrofloxacin residues in chicken meat, Anal. Chem. 85 (2013) 5120–5128.
[18] X. Huang, Z.P. Aguilar, H. Xu, W. Lai, Y. Xiong, Membrane-based lateral flow immunochromatographic strip with nanoparticles as reporters for detection: a review, Biosens. Bioelectron. 75 (2015) 166–180.
[19] B. Liu, D. Du, X. Hua, X.–Y. Yu, Y. Lin, Paper-based electrochemical biosensors: from test strips to paper-based microfluidics, Electroanalysis. 26 (2014) 1214–1223.
[20] K.S. Lee, T. Kim, M. Shin, W. Lee, J. Park, Disposable liposome immunosensor for theophylline combining an immunochromatographic membrane and thick-film electrode, Anal. Chim. Acta 380 (2005) 17–26.
[21] F. Lu, K.H. Wang, H.Y. Lin, Rapid, quantitative and sensitive immunochromatographic assay based on stripping voltammetric detection of a metal ion label, Analyst 130 (2005) 1513–1517.
[22] G. Liu, Y.–Y. Lin, J. Wang, H. Wu, C.M. Wai, Y. Lin, Disposable electrochemical immunosensor diagnosis device based on nanoparticle probe and immunochromatographic strip, Anal. Chem. 79 (2007) 7644–7653.
[23] X. Mao, M. Baloda, A.S. Gurung, Y. Lin, G. Liu, Multiplex electrochemical immunoassay using gold nanoparticle probes and immunochromatographic strips, Electrochem. Commun. 10 (2008) 1636–1640.
[24] H. Nian, J. Wang, H. Wu, J.–G. Lo, K.H. Chiu, J.G. Pounds, et al., Electrochemical immunoassay of cotinine in serum based on nanoparticle probe and immunochromatographic strip, Anal. Chim. Acta 713 (2012) 50–55.
[25] K. Inoue, P. Ferrante, Y. Hirano, T. Yasukawa, H. Shiku, T. Matsue, A competitive immunochromatographic assay for testosterone based on electrochemical detection, Talanta 73 (2007) 886–892.
[26] Y. Kiba, Y. Otani, T. Yasukawa, F. Mizutani, Electrochemical detection of redox species flowing in a nitrocellulose membrane and application to quantitative immunochromatography, Electrochim. Acta 81 (2012) 14–19.

[27] K. Tominaga, S. Arimoto, K. Shimono, T. Yoshioka, F. Mizutani, T. Yasukawa, Quantitative and single-step enzyme immunosensing based on an electrochemical detection coupled with lateral-flow system, Anal. Sci. 33 (2017) 531–536.
[28] T. Yasukawa, Y. Kiba, F. Mizutani, A dual electrochemical sensor based on a test-strip assay for the quantitative determination of albumin and creatinine, Anal. Sci. 31 (2015) 583–589.
[29] H. Morgan, N.G. Green, AC Electrokinetics: Colloids and Nanoparticles, Research Studies Press, Baldock, Hertfordshire, England, 2003.
[30] T. Yasukawa, M. Suzuki, T. Sekiya, H. Shiku, M. Matsue, Flow sandwich-type immunoassay in microfluidic devices based on negative dielectrophoresis, Biosens. Bioelectron. 22 (2007) 2730–2736.
[31] M.D. Vahey, J. Voldman, An equilibrium method for continuous-flow cell sorting using dielectrophoresis, Anal. Chem. 80 (2008) 3135–3143.
[32] J. Voldman, Electrical force for microscale cell manpulation, Annu. Rev. Biomed. Eng. 8 (2006) 425–454.
[33] M.P. Hughes, H. Morgan, F.J. Rixon, J.P.H. Burt, R. Pethig, Manipulation of herpes simplex virus type 1 by dielectrophoresis, Biochim. Biophys. Acta, Gen. Subj. 1425 (1998) 119–126.
[34] F. Grom, J. Kentsch, T. Muller, T. Schnelle, M. Stelzle, Accumulation and trapping of hepatitis A virus particles by electrohydrodynamic flow and dielectrophoresis, Electrophoresis 27 (2006) 1386–1393.
[35] H.J. Lee, T. Yasukawa, M. Suzuki, Y. Taki, A. Tanaka, M. Kameyama, et al., Rapid fabrication of nanoparticles array on polycarbonate membrane based on positive dielectrophoresis, Sens. Actuat. B: Chem. 131 (2) (2008) 424–431.
[36] M. Suzuki, T. Yasukawa, Y. Mase, D. Oyamatsu, H. Shiku, T. Matsue, Dielectrophoretic micropatterning with microparticle monolayers covalently linked to glass surfaces, Langmuir 20 (25) (2004) 11005–11011.
[37] M. Suzuki, T. Yasukawa, H. Shiku, T. Matsue, Negative dielectrophoretic patterning with colloidal particles and encapsulation into a hydrogel, Langmuir 23 (7) (2007) 4088–4094.
[38] H.J. Lee, T. Yasukawa, H. Shiku, T. Matsue, Rapid and separation-free sandwich immunosensing based on accumulation of microbeads by negative dielectrophoresis, Biosens. Bioelectron. 24 (2008) 1000–1005.
[39] H.J. Lee, S.H. Lee, T. Yasukawa, J. Ramón-Azcón, F. Mizutani, K. Ino, et al., Rapid and simple immunosensing system for simultaneous detection of tumor markers based on negative-dielectrophoretic manipulation of microparticles, Talanta 81 (2010) 657–663.

IoT sensors for smart livestock management

14

Wataru Iwasaki, Nobutomo Morita and Maria Portia Briones Nagata
Advanced Manufacturing Research Institute, National Institute of Advanced Industrial Science and Technology, Tokyo, Japan

14.1 Introduction

IoT technology has been attracting attention in the field of agricultural technology. It is expected that IoT technology can make a breakthrough in livestock management by connecting biological information of livestock and environmental information obtained by IoT sensors to farmers whose farm is located in remote location via cloud (Fig. 14.1). A large amount of the information obtained with IoT sensors is sent to and stored in a data server at a cloud database. Some of these data are analyzed, diagnosed, and changed into simple and valuable information by artificial intelligence (AI) computing. Rearing conditions of livestock are automatically or manually controlled based on the information. For example, if biological sensors detect increase of body temperature, AI turns on the air conditioner or fan automatically. If AI diagnoses possibility of disease based on total information such as increase of body temperature and decrease of activity, AI tells it to farmers, and they can take care of the livestock quickly. This helps to improve production efficiency and reduce physical labor and labor cost. Harrop estimated that the market for wearable sensors for animals will increase more than 2.5 times from 2017 to 2027 [1]. Over the last century, a large amount of studies related to relationships among environmental and biological information and growth of plants and animals have been conducted. Moreover, knowledge and experience have been utilized not only for research but also for real production. However, for environmental and agroindustrial applications, information obtained from the environment, crops, and livestock has not been fully utilized because of difficulty in monitoring various environmental and biological information. Therefore, a large amount of research findings on biological and environmental condition obtained so far had not been effectively utilized in the actual production of crops and livestock. All that changed dramatically with the development of microelectromechanical systems (MEMS) technology. MEMS technology enables fabrication of various sensors at small size, low power consumption, and cost-effective means. Actually, wearable sensors for humans, for example, smart watches, which can measure various biological information such as activity, heart rate, and so on, have already been commercialized.

Chemical, Gas, and Biosensors for the Internet of Things and Related Applications.
DOI: https://doi.org/10.1016/B978-0-12-815409-0.00015-2

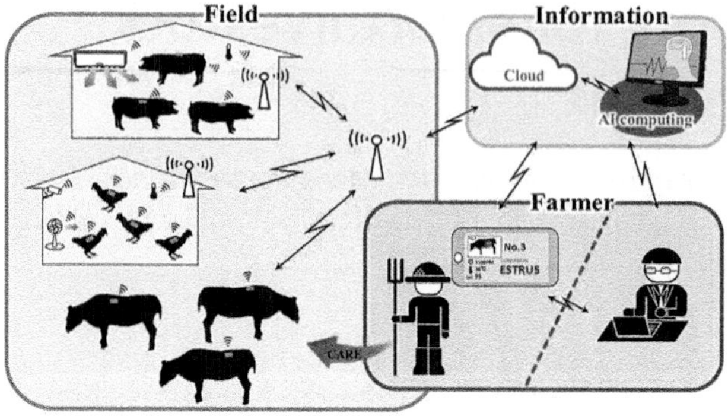

Figure 14.1 Schematic of a smart livestock monitoring system based on IoT technology.

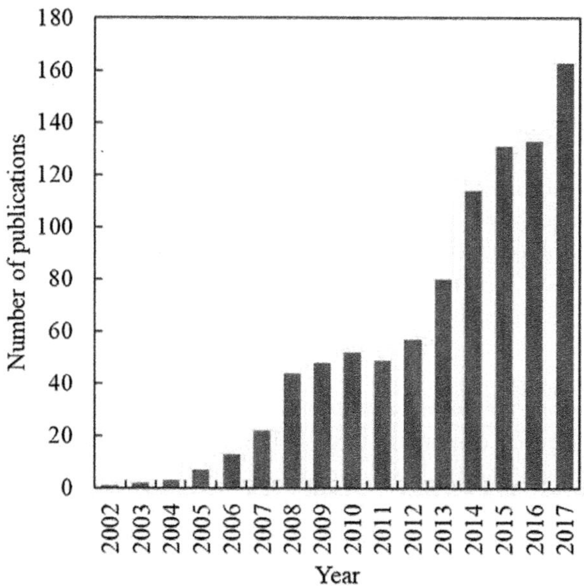

Figure 14.2 The number of publications about wireless sensor for agriculture based on Web of Science offered by Clarivate Analytics Co., Ltd.

Recently, application of these technologies has been tried in agricultural fields, especially in livestock. In fact, research on wireless sensors for agriculture has increased for the last decade (Fig. 14.2). However, there are some differences in the requirements for wearable sensors between human and livestock as mentioned below because unlike humans, animals cannot wear the sensors by themselves.

14.2 Measurement site and fixing method

The location of sensor attachment on the animal body is very important to obtain valuable biological information. It should be considered that the data obtained from the measurement site can measure correct biological information and that the obtained information can be used for diagnostics of health conditions. Almost all wearable sensors measure biological data such as body temperature, circulation, sweat, and so on, from the sensor surface. Therefore, the sensors must remain in contact with the skin. However, there are limitations imposed by dense hair layers of the animal. In particular, thick hair poses not only constraints in terms of sensor's contact with the body but it significantly affects the temperature on the outer surface of the body by forming a layer of air, thus making it difficult to make an accurate measurement of the skin temperature. In the case of optical measurement, light is absorbed and scattered by hair, therefore preventing a correct measurement. Even when dense hair layers are shaved for the purpose of sensor attachment, hair grows soon and creates a gap between the skin and sensor. In addition to where sensors should be mounted on the body, it is also important to consider how to mount sensors on the body. Of course, unlike humans, animals don't wear the sensors by themselves. In addition, animals dislike being attached with something that they are not used to on their body. Therefore, if sensors are attached on the body, animals would try to extricate themselves from the material by shaking their body or hooking the sensors to something. If large and heavy sensors are fixed by wrapping with a magic tape band or a rubber band to a body part that moves wildly such as the tail, wing, head and so on, the sensors can easily fall off the body. Tight wrapping will prevent blood circulation. Fixing with an adhesive bond is also not a good method for long-term monitoring, because the adhesive bond is weak to water and degrades by sweat and rain. In addition, the sensors get easily broken when they fall off the ground, fence, or when they hit a wall. Therefore, it is better to attach a sensor to a suitable site where it is protected by body parts such as armpit, tail base, or internal body. Therefore, we should consider using efficient mounting methods and suitable mounting sites.

14.3 Size and weight

Size requirement of sensors depends on the target animal and the site where sensors are mounted, for example, in the case of sensors that are fitted on the outside of the body (external sensors), or sensors implanted to the animal (internal sensors). The weight of the sensor is also an important consideration to decrease the physiological and psychological load of the animal. Bulky and heavy wearable sensors make the wearer uncomfortable and this has been proven even in humans. In the case of humans, the weight of wearable devices such as cell phone, smart watch, and smart glasses are about 150 g, 50 g, and 70∼700 g, respectively. On the other hand, in small animals like chickens, the sensor weight should be less than 50 g because the body weight is only about only 2–3 kg.

14.4 Power consumption

One of the main essentials for sensors used in IoT is that they must be energy-efficient. The lifetime of wearable sensors should be extended because long term monitoring is required for IoT in livestock and frequent replacement or recharging batteries of worn out batteries is not practical. Ideally, sensors with no batteries or parts that wear out, or batteries that can last until the life of the animal or even longer are desired. For example, the lifetime of broilers and layer chickens are from several weeks to a few months and a few years, respectively. In the case of beef cattle and dairy cows, it is about 30 months and about 5 years, respectively. It is required for IoT sensors intended for livestock to overcome the challenge of battery depletion, develop better power management, and improve the battery lifespan for a long-term use when using a small battery. The battery capacity depends on its volume and weight. For example, one of the representative coin cell batteries (CR2032) with 20 mm in diameter, 3.2 mm in height, and 2.9 g in weight has 220 mAh, which is less than one-tenth of the battery capacity of the latest cell phone. Thus, power consumption of IoT sensors should be reduced considering the limitations posed by the volume and weight of sensor battery.

14.5 Frequency bands of radio wave

One of the challenges in the development of IoT sensors for livestock is the choice of frequency bands to be used for the IoT system. There exists a wide range of choices of frequency bands for carrying information and there is no definite designated frequency for the IoT system because radio characteristics differ for each frequency band, and available frequency band is regulated by radio laws, which differ among various places in the world. Therefore, general features of representative frequency bands as shown in Fig. 14.3 are introduced. Firstly, 315,

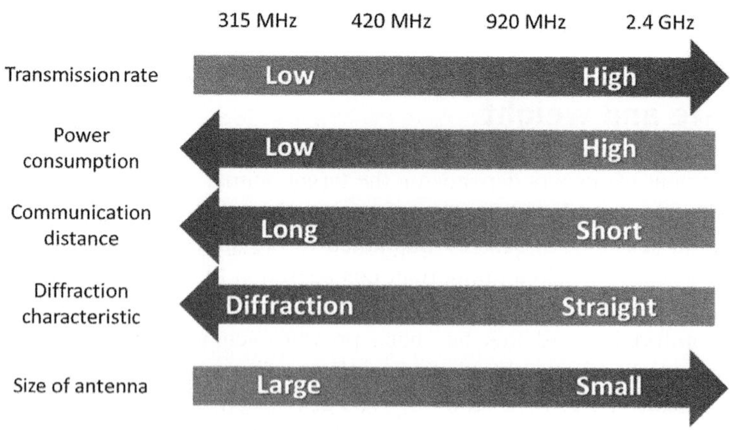

Figure 14.3 Features of each frequency band of radio wave.

420, 920 MHz, and 2.4 GHz bands are considered candidates for an IoT system. Transmission rate becomes faster with increase in frequency. It is difficult to transfer continuous data with less than 920 MHz band. Therefore, if we want to transfer continuous data, we must choose 2.4 GHz band. In addition, using higher frequency bands has an advantage in terms of miniaturization of wearable sensor because the size of antenna can be reduced with increase in frequency. On the other hand, 2.4 GHz frequency band has disadvantages in terms of power consumption, communication distance, and diffraction characteristic. Lower frequency bands can reduce power consumption due to its lower volume of data and intermittent operation. Moreover, radio wave with longer wavelength has smaller attenuation than shorter wavelength because radio wave attenuation is inversely proportional to the square of its wavelength. In addition, shorter wavelength of radio wave has larger diffraction angle. These features enable better coverage performance. Therefore, if we don't need continuous transmission, lower frequency bands have some advantages against 2.4 GHz band. Thus, we should use the optimal radio frequency band for such situations because of the trade-offs among transmission rate, distance, and power consumption.

Thus, there are many requirements for IoT sensors for livestock that are different with human. Although various developments and trials of IoT sensors have continued for a few decades, these challenges have not yet been completely overcome up to now. Some prototypes and commercialized sensors and their application to chicken and cattle are introduced in the next subsection.

14.6 Applications of wearable biosensors for livestock

14.6.1 Chickens

Monitoring the health information of animals is an important component of the application of information and communication technology in agriculture, aiming for smart and effective farming and preventing or detecting diseases in farm animals. For example, there is an urgent need to find effective methods for the early detection of avian influenza, which is of international importance [2–9]. Health monitoring of chickens has been conducted for early detection and prevention of a pandemic avian influenza [10–13]. In animal production, monitoring of body temperature is most important for health monitoring. Suzuki et al. measured body temperature of chicken using wireless thermometer (AirSence, Hitachi, Tokyo, Japan) and found that chickens infected with DkYK10 avian influenza virus had a fever. The fever appeared at 30 hours after infection and persisted for about 2 days. The peak of the fever was more than 2.0°C. However, they also found that chickens infected with other strong virus, such as CkYM7, did not show any changes of body temperature before dying. Therefore, it is not sufficient to measure body temperature for early detection and prevention of avian influenza. Activity is also one of the most common barometers for measuring health condition in addition to body temperature. Okada et al. developed a wireless sensor system that can measure temperature and acceleration of chickens [12]. The sensor has 24 mm diameter and 9 mm length, 2.4 GHz wireless transmission, and its weight is only 5.2 g. It can work for more than 2 weeks with 140 mAh coin cell battery (CR1632). To achieve long-term monitoring with a small and low capacity battery, intermittent operation was adopted for reducing power consumption. The 3-axis acceleration was measured 20 times with sampling frequency of about 120 Hz and wirelessly transmitted to a receiver with 2.4 GHz band in every 20 seconds intermittent operation. They found that the activity calculated from 3-axis acceleration decreased prior to several hours of death due to debilitation even for strong virus CkYM7, which did not cause fever, and 3-axis acceleration was a good indicator than body temperature for detection of avian influenza infection. However, it is difficult to obtain a large signal from the chicken motion with low frequency using small volume sensor. This is because large signal of movement can be detected when the cantilever, which is one of the standard systems of acceleration sensor, resonates, and large volume is required for resonance at low frequency. Nogami et al. developed 3.25×2 mm small cantilever with low and wide resonance frequency from 5 to 15 Hz by adapting multimodal S-shaped cantilever design [11]. Nishihara et al. conducted another approach by blood flow monitoring [10]. They considered the possibility of avian influenza detection from blood flow because birds infected with avian influenza show swelling and congestion in their feet [13], and it is believed that these symptoms might have been reflected by blood flow. Blood flow can be measured noninvasively by laser Doppler flowmetry [14]. Fig. 14.4 shows a

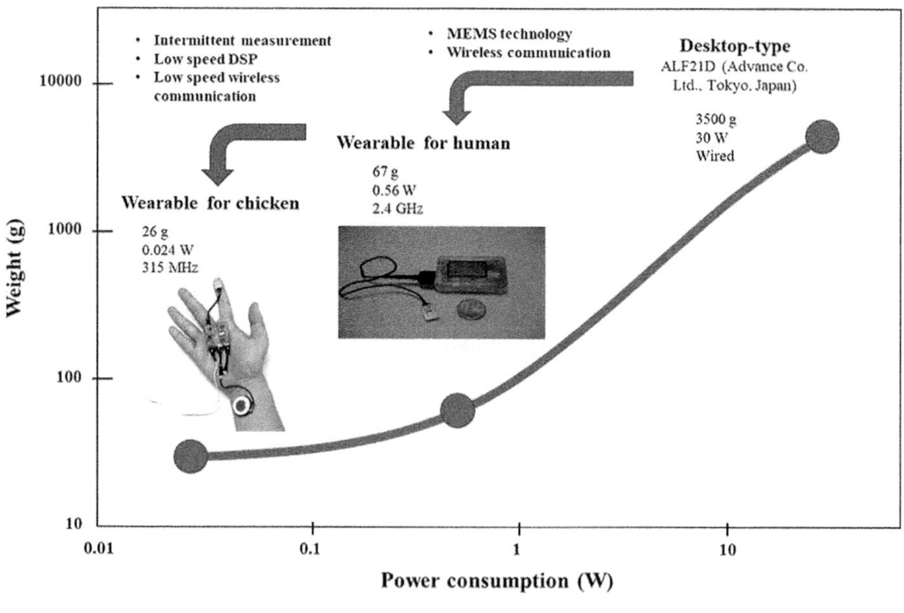

Figure 14.4 Miniaturization and lowering power consumption of laser Doppler blood flowmeter.

progress of miniaturization and reduction of power consumption of laser Doppler flowmeter (LDF). Firstly, Higurashi et al. developed a microintegrated LDF (MEMS-LDF) [15]. They miniaturized the MEMS-LDF by applying MEMS technology and using bare chip of optical components such as laser diode and photodetector. In addition, MEMS-LDF adapted wireless communication, resulting in wearable sensors for humans. However, when chickens wore the MEMS-LDF, which weighs 67 g, chickens lost their balance and were unable to stand. Therefore, Nishihara et al. reduced the weight further by discarding the display, a plastic case, and some other parts, resulting in sensor probe size of $19 \times 13 \times 7$ mm, electric circuit box size of $38 \times 38 \times 11$ mm, and 18 g in weight, and Li-ion rechargeable coin cell battery with a size of $\varphi 30 \times 4.8$ mm, and 9.3 g in weight, with a total weight reduced to 27.3 g [10]. When chickens wore the newly developed MEMS-LDF for chickens, they were able to move freely. Nishihara et al. also reduced the power consumption by adapting low speed digital signal processor instead of CPU, low speed wireless transmission at 315 MHz, and a 5 minutes intermitted operation, resulting in operation lasting for more than 5 days. Finally, its weight and power consumption were reduced to 2 digits and 3 digits lower than those of commercialized desktop-type LDF (LPF21D, Advance Co. Ltd., Tokyo, Japan), respectively.

Thus, for these devices, volume, weight, and power consumption were reduced by using low power consumption chip, low-frequency wireless bands, and adapting intermittent operation.

14.6.2 Cattle

14.6.2.1 Automated milking system

Various IoT sensors and technology have been developed for reproduction of cattle and dairy farming. The development concerning individual management of cows started in the 1980s [16]. Firstly, the automated milking system was developed and commercialized in 1992. De Koning reported that the automated milking system can manage larger herds with less physical labor and labor cost, and the system was installed in over 8000 farms at the end of 2009 [17]. Early detection of mastitis is important because contaminated milk from affected cows renders it unsuitable for human consumption thus resulting in wasting of milk. Therefore, farmers should detect bacterial contamination of milk from mastitis infected cows so contaminated milk could be discarded and prevented from being mixed with good quality milk. Mastitis can be detected by various methods such as electrical conductivity, electrical resistance, milk color, and in-line monitoring for automatic milking system [18–20]. Herd Navigator (DeLaval Ltd., Cardiff, UK) can measure lactate dehydrogenase (LDH) in the milk online by enzymatic reaction [21]. LDH is highly correlated to somatic cell counts (SCC), which is closely linked to the presence of mastitis. These measurement technologies have fewer challenges for the wearable sensors mentioned above because these sensors are built in large milking systems.

14.6.2.2 Importance of wearable sensors

Wearable sensors for cattle were developed in the 1990s for health monitoring and estrus detection. Improvement of conception rate of cattle is one of the most important challenges for stable and highly efficient management. The difference of conception rate between 45% and 60% with the same production level results in the difference in overall income by 10% [22]. Since the timing of artificial insemination is strongly correlated with conception rate [23,24], it is necessary to precisely detect estrus of cattle to obtain high conception rate. Behaviors of mounting on other cows (mounting) and allowing to be mounted without escaping (standing) are good indicators of estrus [25,26]. However, farmers miss the mounting behavior a lot because cows show the behavioral signs only a few hours, and such estrous behaviors are weakened by the influence of environmental conditions [27]. Especially, an increased temperature in summer season causes the weakened estrous behaviors by disturbing ovarian function, resulting in inappropriate timing of artificial insemination. Moreover, the increase of maternal body temperature causes embryonic death [28]. Therefore, the conception rate in artificial insemination is further decreased in summer, which is a serious problem in calf production with artificial insemination. Monitoring of rectal temperature is a standard method to detect estrus and maintain health. However, manual work such as collecting body temperature and measurement of rectal temperature of a large number of cattle is extremely taxing to farmers. Therefore, recently, automatic health monitoring system for cows using wireless sensors has been developed and commercialized. The representative wireless sensors are listed in Table 14.1 and their advantages and disadvantages are introduced in the subsequent subsection.

14.6.2.3 Pedometers

Firstly, a representative commercialized wearable wireless sensor system is pedometer based on a 3-axis accelerometer such as CowAlert powered by IceQube (IceRobotics Ltd., Edinburgh, Scotland, UK), GYHUO SaaS (Fujitsu Kyushu Systems Ltd., Fukuoka, Japan). They can detect estrus by monitoring increasing activity and can notify the farmer as soon as it happens. To optimize the use of pedometers as a tool for estrus detection, it is important to choose an optimal threshold of increase in number of steps because the detection rate of estrous changes from 51% to 87% according to the threshold [37]. They sometimes miss and mistakenly detect estrus because of weakened estrous behavior and influence by other surrounding cattle, respectively. Therefore, farmers may risk missing estrus by using only a pedometer.

14.6.2.4 Ruminal sensors

Rumen monitoring is one of the methods to monitor deep body temperature. Some rumen sensors were developed and it has been reported that ruminal temperature has a potential to predict estrus and parturition [33,38]. However, these sensors are at risk of being broken and excreted. In addition, it consumes high power for wireless transmission due to low transparency of rumen fluid and thick body to radio waves. Nogami et al. also reported rumen temperature rapidly decreases just after

Table 14.1 IoT sensors for cattle monitoring

Sensor	Manufacture	Parameter	Site	Size (mm)	Weight (g)	Radio frequency bands	Lifetime of battery	Communication distance	References
Bio-Thermo microchip	Digital Angel Corp.	Temperature	Implant	Dia. 2 × 12	N/A	134.2 kHz	N/A[a]	N/A	[29]
Custom made	A.M. Lefcourt	Temperature	Implant	Dia. 35 × 100	N/A	150 MHz	≧174 days with 3 min intervals	N/A	[30,31]
Vel'Phone	Medria	Temperature	Vagina	N/A	N/A	N/A	≧6 years days with 3 min intervals	≧200 m	
Mobile Gyuonkei	Remote Inc.	Temperature	Vagina	Dia. 20 × 115 without stopper	N/A	315 MHz	5 years with 5 min intervals	7.5 m	
Custom made	Andersson et al.	Temperature Conductivity	Vagina	Dia. 20 × 120 or 160 without stopper	N/A	429 MHz and 920 MHz	5 years with 30 min intervals	≧100 m	[32]
Bolus	Smart Stock	Temperature	Rumen	Dia 31.7 × 82.5	114	N/A	N/A	90 m	
Custom made	Nogami et al.	Temperature Accelerometer	Rumen	Dia. 30 × 70	94	315 MHz	≧600 days with 5 s intervals	≧20 m	[33]
Custom made	Sato et al.	pH	Rumen	Dia. 30 × 140	184	429 MHz	3 months with 10 min intervals	≧32 m	[34]
Custom made	Miura et al.	Temperature	Tail base	25 × 25 × 9.6	10.7	920 MHz	≧6 months with 2 min intervals	N/A	[35]
SensOor	Agis Automatisering BV	Activity, Temperature	Ear	60 × 50 × 22	32	N/A	N/A	N/A	[36]
Cow Alert IceQube	IceRobotics Ltd.	Activity	Leg	55 × 55 × 26	72	N/A	≧5 years	N/A	
GYUHO SaaS	Fujitsu Kyushu Systems	Activity	Leg	N/A	N/A	N/A	≧5 years with 1 h intervals	150 m	

[a]Batteryless system based on radio frequency identifier (RF-ID).

drinking water [33]. Therefore, rumen is suitable for monitoring meals, though it is not suitable for monitoring core body temperature. Rumen monitoring has been originally conducted for monitoring of dysfunctional rumen system as represented by subacute ruminal acidosis (SARA) by measuring the ruminal pH. SARA is an acute disease causing other diseases such as ruminitis, abomasal displacement, ruminal tympany, fertility, and so on [39]. Some studies reported that 11%–18% of early lactation, and 19%–26% of midlactation had SARA [40,41]. Sato et al. developed a wireless ruminal pH sensor that was 30 mm in diameter, 140 mm length, and 184 g in weight, and demonstrated continuous measurement of ruminal pH that markedly decreased after eating a SARA-inducing diet [34]. The continuous monitoring of ruminal pH is effective in the detection of SARA. However, pH sensor needs internal liquid in the reference electrode, and it limits the life of the sensor. To resolve this challenge, a pH sensor combining a metal-oxide-semiconductor field-effect transistor (MOSFET) with sensing electrode made of indium tin oxide (ITO) film that doesn't need the internal liquid for the reference electrode, was developed and has demonstrated long-term monitoring of ruminal pH [42].

14.6.2.5 Vaginal sensors

There has been growing interest in measurement of core-body temperature in cattle using internal probes. Monitoring of vaginal temperature allows for the continuous capture of cow's core temperature. Vaginal temperature has been monitored using insertion type wireless sensor, and results showed good correlation with rectal temperature [32,43]. A few vaginal thermometers, Vel'Phone (Medria, Châteaubourg, France) and Mobile Gyuonkei (Remote Inc., Oita, Japan), were commercialized. These sensors are used to precisely predict the moment of calving. Accurate calving detection is vital for successful calving and safety for the calf and cow. In conventional calving, farmers predict calving timing to avoid the risk of a dystocia event or dead birth and ensure greater success, however it is very difficult to know when exactly the cow will calve. The need to continuously check on the cow to see if it has started calving becomes a large burden to farmers. Hence, there is a need for sensors that will provide alerts on when a cow is ready to birth. These sensors are packaged with a stick-type package of about 100–160 mm long and 20 mm in diameter, and the package wears a stopper. The vaginal thermometer is held in the vagina by hooking the stopper. In a previous study, vaginal temperature showed high correlation with rectal temperature (coefficient determination, R^2 value = 0.90) [43]. The vaginal temperature a day before calving shows lower temperature than those of earlier days. The sensors inform this data to farmers, and the farmers can do the necessary preparation. In addition, the sensor is thrown out from the vagina by premature rupture of membrane, and the temperature of the sensor changes rapidly to ambient temperature, and can send an alert to the farmer signaling the start of calving. This system can relieve farmers of stress and worry when it comes to animal parturition and help in smart farming management. However, there are challenges and limitations, such as not being able to be inserted for a long time because it gives stress to the cow and easily dropping from the vagina, which need to be addressed.

14.6.2.6 Implantable sensors

Monitoring of skin surface or subcutaneous temperature using wearable or implantable wireless sensors has been studied the most [29–31,35,44]. Lefcourt and Adams conducted an early trial of monitoring body temperature using implantable wireless thermometer. They implanted large wireless thermometer sealed with cylinder that was 10 cm long and 3.5 cm in diameter behind the ribcage on the right side, and monitored the core body temperature in the summer and winter seasons [30,31]. For the last decade, researches with implantable wireless thermometer have become active because of the ability to fabricate small wireless thermometers. Reid et al. demonstrated monitoring subcutaneous temperature using commercialized injectable Bio-Thermo microchip (Digital Angel Corp., Saint Paul, MN, USA) [29]. Its size is only 12 mm long and 2 mm in diameter because it is a batteryless system based on radio frequency identifier (RF-ID). The system requires farmers to read data from the microchip by placing the reader near the implanted site. Therefore, this system was insufficient in terms of reducing labor of farmers. In addition, they reported that subcutaneous temperature obtained with Bio-Thermo microchip did not demonstrate positive correlation with rectal temperature when cattle had a fever that was induced by lipopolysaccharide (LPS) injection. Thus, it is difficult to obtain core body temperature from subcutaneous temperature.

14.6.2.7 Wireless thermometers attached to skin surface

A few studies, which obtained valuable data even from skin surface, have reported. Miura et al. demonstrated measurement of skin surface temperature at ventral tail base, and the potential to detect estrus and calving by using the difference between skin temperature and mean skin temperature of the last 3 days [35]. The sensor size is $25.0 \times 25.0 \times 9.6$ mm, which can be placed in a small pit behind the tail base, and its weight is only 10.7 g. The temperature data was wirelessly transferred with 920 MHz band, in which the transmitting distance is about 100 m, if there isn't any obstacle. The sensor can work for more than 6 months using the 220 mAh coin cell battery CR2032 when measurement and transmission interval is set to 2 minutes.

Although these various sensors were developed and their potential to detect estrus and calving were demonstrated, there are both merits and demerits in terms of accuracy of measured signal, long-term reliability, lifetime, and so on. In addition, the accuracies of the detection and probability of false-positive and false-negative detections were not discussed. For example, Rutten et al. concluded that the commercialized wearable wireless thermometer SensOor (Agis Automatisering BV, Harmelen, the Netherlands), which is attached to the ear and can measure activity, rumination, feeding, and temperature by acceleration and temperature sensors, has the potential to accurately predict the start of calving, and also mentioned that false-positive alerts were observed in the last 12 hours before calving started [36].

14.7 Conclusion

Various sensors for livestock were introduced in this chapter. There is the trend for almost all sensors to reduce volume, weight, and power consumption for long-term monitoring using various techniques and approaches. Development of MEMS technology and large-scale integration (LSI) technology enabled miniaturization of volume and reduction of weight and power consumption. As a result, various wearable sensors for IoT in agriculture have already been developed and commercialized. These sensors can measure biological signals continuously or intermittently for the long term, and detect some symptoms of disease, estrus, calving, and health events. However, there still exist many false-positive and false-negative detections because such biological signals have large noise signal affected by various external factors, for example, ambient temperature, behavior of other surrounding livestock, and many more. Therefore, current requirements for IoT in agriculture are not only for the reduction of their volume, weight, and power consumption but also on how to detect these symptoms from biological data measured with such wearable sensors by applying artificial intelligence technology.

Sensor technology has been used in the field of agriculture for the last three decades. In particular, it has been rapidly developing in the past decade with the influence of society's interests and rising expectations for IoT technology. However, future efforts to improve the accuracy of measurement are needed by adapting AI technology as well as by focusing on physiological and biological considerations. To do this, we would need to make a closer coordination and a methodological linkage among engineering, animal, and computer sciences.

References

[1] P. Harrop. Wearable technology for animals 2017-2027: technologies, markets, forecasts. IDTechEx, 2016.

[2] T.T. Hien, N.T. Liem, N.T. Dung, L.T. San, P.P. Mai, N.V.V. Chau, et al., Avian influenza A (H5N1) in 10 patients in Vietnam, New Engl. J. Med. 350 (2004) 1179–1188.

[3] A.M. Kilpatrick, A.A. Chmura, D.W. Gibbons, R.C. Fleischer, P.P. Marra, P. Daszak, Predicting the global spread of H5N1 avian influenza, Proc. Natl. Acad. Sci. USA 103 (2006) 19368–19373.

[4] K.S. Li, Y. Guan, J. Wang, G.J.D. Smith, K.M. Xu, L. Duan, et al., Genesis of a highly pathogenic and potentially pandemic H5N1 influenza virus in eastern Asia, Nature 430 (2004) 209–213.

[5] D.L. Suarez, M.L. Perdue, N. Cox, T. Rowe, C. Bender, J. Huang, et al., Comparisons of highly virulent H5N1 influenza A viruses isolated from humans and chickens from Hong Kong, J. Virol. 72 (1998) 6678–6688.

[6] K. Subbarao, A. Klimov, J. Katz, H. Regnery, W. Lim, H. Hall, et al., Characterization of an avian influenza A (H5N1) virus isolated from a child with a fatal respiratory illness, Science 279 (1998) 393–396.

[7] K. Ungchusak, P. Auewarakul, S.F. Dowell, R. Kitphati, W. Auwanit, P. Puthavathana, et al., Probable person-to-person transmission of avian influenza A (H5N1), New Engl. J. Med. 352 (2005) 333–340.

[8] M.D. De Jong, B. Van Cam, P.T. Qui, V.M. Hien, T.T. Thanh, N.B. Hue, et al., Brief report: Fatal avian influenza A (H5N1) in a child presenting with diarrhea followed by coma, New Engl. J. Med. 352 (2005) 686–691.

[9] Y. Guan, J.S.M. Peiris, A.S. Lipatov, T.M. Ellis, K.C. Dyrting, S. Krauss, et al., Emergence of multiple genotypes of H5N1 avian influenza viruses in Hong Kong SAR, Proc. Natl. Acad. Sci. USA 99 (2002) 8950–8955.

[10] K. Nishihara, W. Iwasaki, M. Nakamura, E. Higurashi, T. Soh, T. Itoh, et al., Development of a wireless sensor for the measurement of chicken blood flow using the laser doppler blood flow meter technique, IEEE Trans. Biomed. Eng. 60 (2013) 1645–1653.

[11] H. Nogami, T. Kobayashi, H. Okada, N. Makimoto, R. Maeda, T. Itoh, Piezoelectric MEMS switch to activate event-driven wireless sensor nodes, Smart Mater.Struct. 22 (2013) 10.

[12] H. Okada, T. Itoh, K. Suzuki, K. Tsukamoto. Wireless sensor system for detection of avian influenza outbreak farms at an early stage. 2009 IEEE Sensors, 25-28 Oct. 2009, pp. 1374–1377.

[13] K. Suzuki, H. Okada, T. Itoh, T. Tada, M. Mase, K. Nakamura, et al., Association of increased pathogenicity of Asian H5N1 highly pathogenic avian influenza viruses in chickens with highly efficient viral replication accompanied by early destruction of innate immune responses, J. Virol. 83 (2009) 7475–7486.

[14] R. Bonner, R. Nossal, Model for laser doppler measurements of blood-flow in tissue, Appl. Opt. 20 (1981) 2097–2107.

[15] E. Higurashi, R. Sawada, T. Ito, An integrated laser blood flowmeter, J.Lightwave Technol. 21 (2003) 591–595.

[16] C.J. Rutten, A.G.J. Velthuis, W. Steeneveld, H. Hogeveen, Invited review: sensors to support health management on dairy farms, J. Dairy Sci. 96 (2013) 1928–1952.

[17] C.J.A.M. De Koning. Automatic Milking - Common Practice on Dairy Farms. The First Notrh American Conference on Precision Dairy Management 2010, 2010 Toronto, Canada, pp. 52–67.

[18] K.J. Hassan, S. Samarasinghe, M.G. Lopez-Benavides, Use of neural networks to detect minor and major pathogens that cause bovine mastitis, J. Dairy Sci. 92 (2009) 1493–1499.

[19] W. Steeneveld, L.C. Van Der Gaag, W. Ouweltjes, H. Mollenhorst, H. Hogeveen, Discriminating between true-positive and false-positive clinical mastitis alerts from automatic milking systems, J. Dairy Sci. 93 (2010) 2559–2568.

[20] D. Cavero, T. Lle, K.H. Buxad, C. Buxadé, J. Krieter, Mastitis detection in dairy cows by application of fuzzy logic, Livestock Sci. 105 (2006) 207–213.

[21] C.H. JøRgensen, A.R. Kristensen, S. Østergaard, T.W. Bennedsgaard, Use of inline measures of L-lactate dehydrogenase for classification of posttreatment mammary *Staphylococcus aureus* infection status in dairy cows, J. Dairy Sci. 99 (2016) 8375–8383.

[22] D. Boichard, Estimation of the economic value of conception rate in dairy-cattle, Livestock Prod. Sci. 24 (1990) 187–204.

[23] J.R. Pursley, R.W. Silcox, M.C. Wiltbank, Effect of time of artificial insemination on pregnancy rates, calving rates, pregnancy loss, and gender ratio after synchronization of ovulation in lactating dairy cows, J. Dairy Sci. 81 (1998) 2139–2144.

[24] G.W. Salisbury, N.L. Vandemark, J.R. Lodge, Insemination of the Cow, Physiology of Reproduction and Artificial Insemination of Cattle, 2nd ed, W. H. Freeman Co, 1978.

[25] K. Maatje, S.H. Loeffler, B. Engel, Predicting optimal time of insemination in cows that show visual signs of estrus by estimating onset of estrus with pedometers, J. Dairy Sci. 80 (1997) 1098–1105.

[26] W.L. Walker, R.L. Nebel, M.L. Mcgilliard, Time of ovulation relative to mounting activity in dairy cattle, J. Dairy Sci. 79 (1996) 1555–1561.

[27] F.J. White, R.P. Wettemann, M.L. Looper, T.M. Prado, G.L. Morgan, Seasonal effects on estrous behavior and time of ovulation in nonlactating beef cows, J. Anim. Sci. 80 (2002) 3053–3059.

[28] P.J. Hansen, Embryonic mortality in cattle from the embryo's perspective, J. Anim. Sci. 80 (2002). E33-E$$.

[29] E.D. Reid, K. Fried, J.M. Velasco, G.E. Dahl, Correlation of rectal temperature and peripheral temperature from implantable radio-frequency microchips in Holstein steers challenged with lipopolysaccharide under thermoneutral and high ambient temperatures, J. Anim. Sci. 90 (2012) 4788–4794.

[30] A.M. Lefcourt, W.R. Adams, Radiotelemetry measurement of body temperatures of feedlot steers during summer, J. Anim. Sci. 74 (1996) 2633–2640.

[31] A.M. Lefcourt, W.R. Adams, Radiotelemetric measurement of body temperature in feedlot steers during winter, J. Anim. Sci. 76 (1998) 1830–1837.

[32] L.M. Andersson, H. Okada, R. Miura, Y. Zhang, K. Yoshioka, H. Aso, et al., Wearable wireless estrus detection sensor for cows, Comput. Electr. Agric. 127 (2016) 101–108.

[33] H. Nogami, S. Arai, H. Okada, L. Zhan, T. Itoh, Minimized bolus-type wireless sensor node with a built-in three-axis acceleration meter for monitoring a cow's rumen conditions, Sensors 17 (2017) 687–696.

[34] S. Sato, A. Kimura, T. Anan, N. Yamagishi, K. Okada, H. Mizuguchi, et al., A radio transmission pH measurement system for continuous evaluation of fluid pH in the rumen of cows, Vet. Res. Commun. 36 (2012) 85–89.

[35] R. Miura, K. Yoshioka, T. Miyamoto, H. Nogami, H. Okada, T. Itoh, Estrous detection by monitoring ventral tail base surface temperature using a wearable wireless sensor in cattle, Anim. Reprod. Sci. 180 (2017) 50–57.

[36] C.J. Rutten, C. Kamphuis, H. Hogeveen, K. Huijps, M. Nielen, W. Steeneveld, Sensor data on cow activity, rumination, and ear temperature improve prediction of the start of calving in dairy cows, Comp. Electr. Agric. 132 (2017) 108–118.

[37] J.B. Roelofs, F.J.C.M. Van Eerdenburg, N.M. Soede, B. Kemp, Pedometer readings for estrous detection and as predictor for time of ovulation in dairy cattle, Theriogenology 64 (2005) 1690–1703.

[38] M.J. Cooper-prado, N.M. Long, E.C. Wright, C.L. Goad, R.P. Wettemann, Relationship of ruminal temperature with parturition and estrus of beef cows, J. Anim. Sci. 89 (2011) 1020–1027.

[39] J.M.D. Enemark, The monitoring, prevention and treatment of sub-acute ruminal acidosis (SARA): a review, Vet. J. 176 (2008) 32–43.

[40] J.L. Kleen, G.A. Hooijer, J. Rehage, J.P. Noordhuizen, Rumenocentesis (rumen puncture): a viable instrument in herd health diagnosis, Dtsch Tierarztl Wochenschr 111 (2004) 458–462.

[41] F.J. Mulligan, M.L. Doherty, Production diseases of the transition cow, Vet. J. 176 (2008) 3–9.

[42] L. Zhang, J. Lu, H. Nogami, H. Okada, T. Itoh, S. Arai, Solid-state pH sensor prototype for real-time monitoring of the rumen pH value of Japanese cows, Microsyst. Technol. Micro Nanosyst. Inform. Storage Process. Syst. 24 (2018) 457–463.
[43] P.E. Hillman, K.G. Gebremedhin, S.T. Willard, C.N. Lee, A.D. Kennedy, Continuous measurements of vaginal temperature of female cattle using a data logger encased in a plastic anchor, Appl. Eng. Agric. 25 (2009) 291–296.
[44] H. Nogami, H. Okada, T. Miyamoto, R. Maeda, T. Itoh, Wearable and compact wireless sensor nodes for measuring the temperature of the base of a calf's tail, Sens. Mater. 25 (2013) 577–582.

Compact disc-type biosensor devices and their applications

15

Izumi Kubo[1] *and Shunsuke Furutani*[2,3]
[1]Graduate School of Bioinformatics, Soka University, Tokyo, Japan, [2]Biomedical Research Institute, National Institute of Advanced Industrial Science and Technology (AIST), Osaka, Japan, [3]Advanced Photonics and Biosensing Open Innovation Laboratory (PhotoBIO-OIL), National Institute of AIST, Osaka, Japan

15.1 Introduction

Recently, microfluidic devices for biosensing have been used in various fields such as analytical chemistry and biochemistry, and various technologies based on microfluidic devices have been developed by many researchers [1]. Such microfluidic devices are referred to as "lab-on-a-chip" because biological reactions, which are traditionally carried out in laboratories, can be performed on one chip. The advantages of these microfluidic devices include a shortened reaction time, integration of multiple reactions, and a reduced amount of sample and reagent on one chip [2,3]. In particular, the handling of a small amount of reagent on microfluidic devices have enabled the manipulation of single cells, which are difficult to handle with human hands [4–7]. However, because pumps and valves are required to control reagents in many microfluidic devices, the miniaturization of the entire apparatus is difficult even if the reaction chip itself is small.

To overcome the drawbacks caused by a requirement for pumps and valves, compact disc (CD)-shaped microfluidic devices have been developed using a centrifugal force as the driving force [8–10]. One of the advantages of CD-shaped microfluidic devices is to achieve multiple parallel reactions easily on one chip because reagents are transported by a centrifugal force. Furthermore, since the power to control the transfer of reagents comes only from a motor that rotates the CD-shaped microfluidic device, downsizing of the whole system is possible. In analytical chemistry, the detection of genes and proteins using such CD-shaped microfluidic devices was also possible. For example, Schuler et al. reported the digital droplet polymerase chain reaction (PCR), which enabled the absolute quantification of genes to be performed on their CD-shaped microfluidic device [11]. They created an integrated system that is compatible with mass fabrication and that combines emulsification, PCR, and fluorescence readout in a single chamber within a disposable CD-shaped microfluidic device to perform digital droplet PCR. In

Chemical, Gas, and Biosensors for the Internet of Things and Related Applications.
DOI: https://doi.org/10.1016/B978-0-12-815409-0.00016-4

addition, Park et al. developed a fully automated enzyme linked immunosorbent assay (ELISA) system for point-of-care testing (POCT) using a CD-shaped microfluidic device. Their system could simultaneously detect three kinds of proteins within 20 minutes [12]. Thus, CD-shaped microfluidic devices allow simple and rapid analyses to be conducted.

Our group has been developing CD-shaped microfluidic devices for biosensing. In this chapter, we introduce some of these CD-shaped microfluidic devices, focusing on the devices for cell isolation [13,14] and PCR [15–18] aimed at the detection of genes in single cells, cell staining [19], and ELISA intended for rapid biomarker detection [20–22].

Since signals from such devices are obtained optically, they may be available through the Internet as the next hopeful stage in the development of this technology.

15.2 CD-shaped microfluidic devices for cell isolation and single cell PCR

15.2.1 Single cell isolation

To understand characteristics such as the gene or expressed gene of each cell, the analysis of single cells is indispensable because they are the smallest units of life. In general, some techniques to isolate cells include flow cytometry [23,24], micromanipulation [25], and limiting dilution. However, large equipment and multiple skills are needed for flow cytometry or micromanipulation. Although limiting dilution is a simple method, accurate handling of many single cells is difficult because the amount of liquid that can be handled manually exceeds 1 μL, which is a much larger volume than that of cells. Therefore, as shown in Fig. 15.1, we have developed a CD-shaped microfluidic device for single cell isolation that enables the simple and rapid separation of a cell suspension into a small number of microchambers

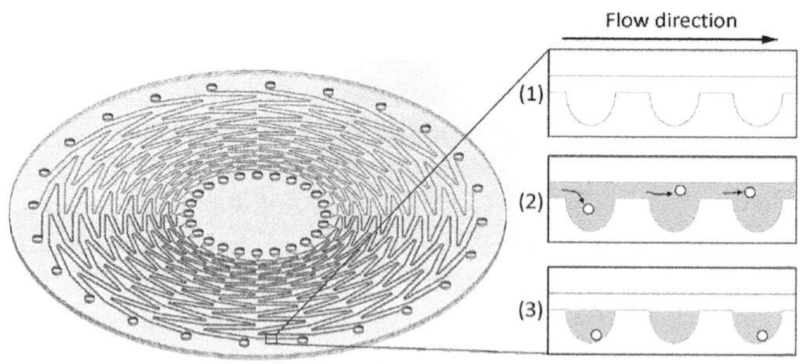

Figure 15.1 The design of a CD-shaped microfluidic device for cell isolation and schematic diagram of single cell isolation. (1) Many U-shaped microchambers are aligned along zigzag shaped microchannels. (2) Cells flow through the microchannel. (3) Single cells are trapped in microchambers.

using a microfluidic technique [13]. In that experiment, to observe the separation of single cells into microchambers, microchannels were prepared from a polydimethylsiloxane (PDMS) elastomer by soft photolithography. A PDMS layer and glass plate were attached and used as the CD-shaped microfluidic device for cell isolation. Microchannels on the CD-shaped microfluidic device, each with a width of 100 μm, were designed in a zigzag shape. Twenty-four microchannels are arranged on one device, each with 530 U-shaped microchambers. Microchambers are arrayed on the periphery to trap liquid and particles, such as cells, by centrifugal force. The dimensions of a microchamber are 180 μm (width) × 120 μm (height) × 40 μm (depth) and the diameter of the CD-shaped microfluidic device for cell isolation is 95 mm.

To isolate cells, 1 μL of cell suspension is injected to the inlet of a microchannel near the center of the device. As the device spins, the cell suspension is introduced into the microchannel by centrifugal force. Throughout this procedure, simple and rapid single cell isolation is possible by maintaining the volume of cell suspension in each microchamber at 500 pL. Fig. 15.2A shows a Jurkat cell, a human leukemia T cell line, isolated in a microchamber on this device. The efficiency of single cell isolation follows a Poisson distribution. Using this device, we succeeded in the simple and rapid cell isolation of 400 cells/μL of a suspension of Jurkat cells with 98.5% probability by spinning the device at 3000 rpm for 30 seconds. Furthermore, in this device, not only Jurkat cells but also bacterial cells such as *Escherichia coli* and *Salmonella enterica* were isolated into microchambers under similar conditions, regardless of cell size. Moreover, after cells were isolated in microchambers, we succeeded in proliferating Jurkat cells from single cells, as nonadherent cells (Fig. 15.2B) and NIH 3T3 cells as adherent cells [14]. Therefore, the activity of single cells was confirmed by examining their proliferation rate.

15.2.2 Single cell PCR of S. enterica

To detect genes in single cells, PCR was performed on our CD-shaped microfluidic device after cell isolation. At that time, by mixing a fluorescent probe and PCR

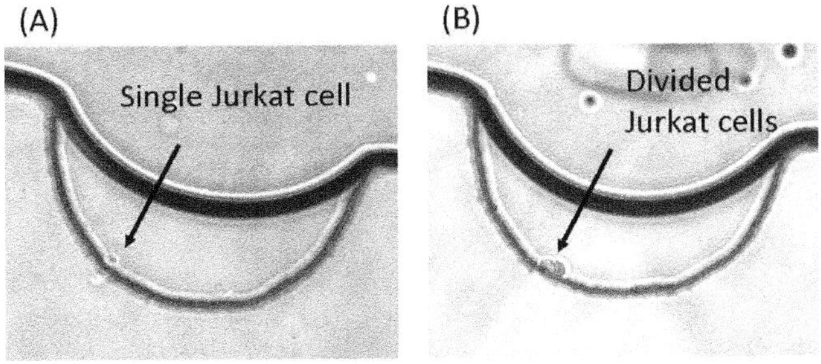

Figure 15.2 Cell isolation by the CD-shaped microfluidic device and cultivation from a single cell in the microchamber. (A) Jurkat cell isolated in a microchamber. (B) Jurkat cells divided from a single cell after cultivation for 2 days at 37°C with 5% CO_2.

reagent with cells, the presence or absence of the target gene in single cells was confirmed by the difference in fluorescence intensity before and after PCR. In this study, to realize single cell PCR with a minute amount of reagent separated into microchambers without the evaporation of reagents, the microchannels were fabricated on a silicon wafer by deep reactive ion etching (deep-RIE). These silicon wafers with microchannels were anodically bonded to glass and used as the CD-shaped microfluidic device for cell isolation and PCR [15]. The dimensions of the microchamber of the CD-shaped microfluidic device for PCR are 300 μm (width) × 200 μm (height) × 46 μm (depth). Twenty-four microchannels are arranged on one device, each of which has 313 U-shaped microchambers. One μL of *S. enterica* suspension at a concentration of 100–400 cells/μL is injected into the inlet. Then, *S. enterica* is isolated into the microchambers by spinning the device. After cell isolation, PCR for the *invA* gene as a specific *S. enterica* gene was carried out on the device. Thermal PCR cycling was initiated at 95°C for 2 minutes to lyse the cells, followed by 50 cycles of 95°C for 5 seconds, 55°C for 10 seconds, and 72°C for 10 seconds. As shown in Fig. 15.3, by measuring the fluorescence intensity of the microchambers before and after PCR, a microchamber with increased fluorescence intensity was confirmed. The number of microchambers with increased fluorescence intensity was almost consistent with the number of microchambers containing *S. enterica* calculated from the Poisson distribution. Furthermore, 50 cells/μL of *S. enterica* mixed with 1000 cells/μL of *E. coli* cells could also be detected, and 30 cfu/g of *S. enterica* from ground meat was successfully detected within 12 hours by combining with 8 hours culture before isolation [18].

15.2.3 Discrimination of microbes

A mixture of various single cells can be isolated using our CD-shaped microfluidic device for single cell isolation and the isolated cells can be distinguished by PCR of specific genes. The discrimination of two different strains of microbes (archaea

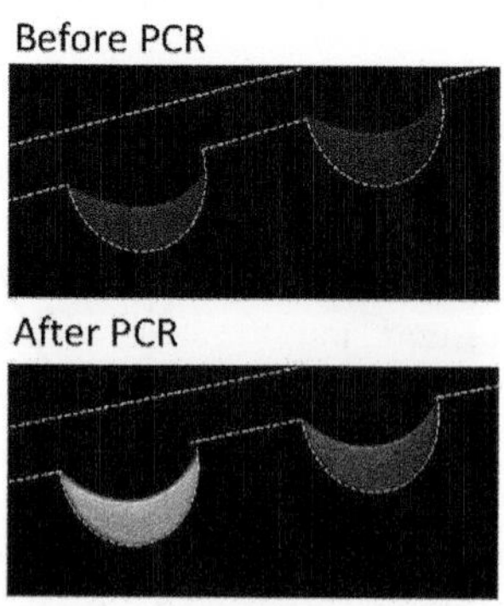

Figure 15.3 Fluorescence image of microchambers on the CD-shaped microfluidic device before and after PCR.

and bacteria) was possible using our CD-shaped device after PCR of specific genes. To achieve this, differences in the sequences of the 16S rRNA gene can usually be targeted. Two different forward primers and TaqMan probes targeting the 16S rRNA gene were designed and applied to discriminate *Metallosphaera sedula* TH-2 (archaea) and *E. coli* (bacteria) as model organisms. To detect archaea and bacteria, two TaqMan probes (FAM probe for *M. sedula* and TAMRA probe for *E. coli*), respectively were employed [26]. The cell suspension (3×10^5 cells/μL of *M. sedula* or *E. coli*) and PCR reagent is mixed, and the mixture is injected into the microfluidic device. Cell lysis by heat treatment (95°C, 5 minutes) and 30 PCR cycles at 95°C for 30 seconds, 58°C for 30 seconds, and 72°C for 30 seconds are performed on the device. The fluorescence intensities of FAM and TAMRA were analyzed sequentially. Detection of archaea and bacteria on the device was confirmed after single cell isolation. The limit of detection was 950 cells/μL *M. sedula* and 3000 cells/μL of *E. coli* on our device.

15.2.4 Single cell RT-PCR for Jurkat cells

There may be large variation in gene expression and behavior in genetically identical cells [27,28]. Therefore, detecting the expressed gene in a large number of single cells would allow for understanding the behavior of individual cells. We isolated Jurkat cells as single cells, and applied reverse transcription (RT)-PCR for mRNA of the glyceraldehyde-3-phosphate dehydrogenase (GAPDH) gene to single Jurkat cells on the CD-shaped microfluidic device. To detect the expressed gene in single cells on the device, RT-PCR must be performed without adding reverse transcriptase after cell lysis by heat treatment. However, since reverse transcriptase generally has no heat resistance, we achieved single cell RT-PCR in the device by using thermostable Tth DNA polymerase, which has reverse transcriptase activity and can be used for RT and PCR [16]. We named this method, in which RT-PCR is possible directly from cells, as hot cell-direct RT-PCR. The CD-shaped microfluidic devices for cell isolation and hot cell-direct RT-PCR were the same as those used for single cell PCR. Thermal cycling for hot cell-direct RT-PCR was initiated at 95°C for 10 minutes as heat treatment for cell lysis and 42°C for 15 minutes as the RT step, followed by 40 cycles of 94°C for 20 seconds, 54°C for 20 seconds, and 72°C for 20 seconds, and a further 72°C for 5 minutes as the final extension step. Using an original detection system, in which the presence or absence of cells in the microchambers could be observed, hot cell-direct RT-PCR on the CD-shaped microfluidic device was performed and the fluorescence intensity of the microchambers before and after hot cell-direct RT-PCR was measured [17]. Fig. 15.4 shows a photo of a Jurkat cell isolated in a microchamber using the CD-shaped microfluidic device, and a fluorescence picture of the microchamber before and after RT-PCR. Fluorescence intensity increased in the microchamber containing the cell by RT-PCR, and the gene expression of single cell was detected. In addition, variations in expression levels of individual cells were confirmed with mRNA of the GAPDH gene, which is a house-keeping gene.

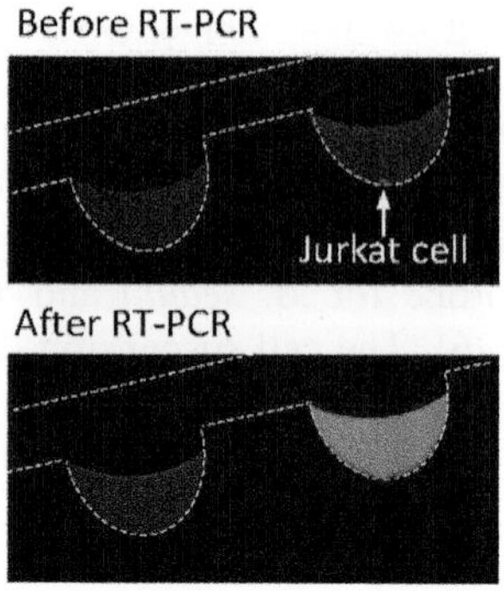

Figure 15.4 RT-PCR for GAPDH of a single Jurkat cell in the CD-shaped microfluidic device. Fluorescence images of the microchambers before and after RT-PCR.

15.3 CD-shaped microfluidic device for cell staining

To investigate characteristics of each cell, staining is an informative method. PCR of genes in a cell reveals its genetic information, while staining of a cell reveals its genetic product. Designing a microfluidic device for cell staining of single nonadherent cells, such as lymphocytes, has been more challenging than adherent cells [29]. Micropatterning to fix cells has been used to immobilize a limited number of cells on a patterned patch. However, micropatterning is limited to adherent cells [30,31]. In those studies, to investigate a single cell's behavior, cells were perfused on the culture medium after they were trapped in microchambers. Flushing a solution to the trapped cell tended to displace not only the solution around the cell, but also the cell itself. Thus, it was not easy to observe a nonadherent cell in a microchamber.

In our study, we proposed a novel CD-shaped device with a microchannel and chambers to trap particles and nonadherent cells, which are maintained in loco during the exchange of solution around the cells and also during observation. As a novel trapping device, a microfluidic device with cup-like chambers was designed and fabricated using PDMS by soft photolithography to trap cells from cell suspension and to exchange liquid around them to allow them to be stained (Fig. 15.5). A cup-like chamber is arranged on each channel. Basically, the main radial channel is arranged on the disc and a cup (200–500 μm wide; 280–400 μm deep) is placed at the bottom of the channel. The bottom of cups is arranged in the direction of centrifugal force, and each cup has a main channel, through which a cell suspension is introduced into the cup, while side channels drain overflowing liquid from the cup. In this way, trapped cells remain in the cup due to centrifugal force.

This device does not need any specific treatment or addition of reagents to retain cells in the trapping chamber and is easy to operate. The proposed device was designed to trap and keep cells in the chamber during the exchange of solution around each cell during staining. Using this device, cells were trapped with liquid. Keeping cells in the chamber, the liquid was exchanged when the device was spun.

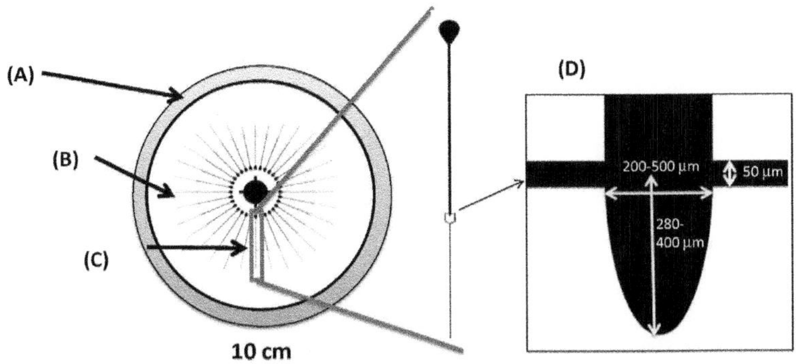

Figure 15.5 Schematic diagram of a microfluidic disc for cell trapping. (A) Glass plate. (B) Polydimethylsiloxane. (C) Microchannel. (D) Microchamber.

This microchannel has two advantages. First, it can trap cells with solution into a chamber and exchange solution around the cell, so that the trapped cells can be stained. Second, the amount of reagent used to stain cells is much less than that required for a conventional microtube-based method. Using this device, we trapped T lymphocytes and stained living and dead cells as mentioned above.

Nonadherent Jurkat cells were examined in this device. After cup size was optimized, it was fixed at a width of 500 μm and a depth of 400 μm. After 1 μL of cell suspension is applied to the inlet, the device is spun at 3000 rpm for 30 seconds to trap cells in the cup-like chamber. After cells have been trapped in the chamber, a solution can be applied to wash and stain them. During the procedure, almost all cells remain in the chamber. The entrapped Jurkat cells are stained with Cellstain (Dojindo Lab., Kumamoto, Japan) to observe cell viability. Living cells were observed in microchamber after staining (Fig. 15.6). Approximately 90% of the cells were alive [19].

Although we successfully entrapped and stained T lymphocyte cells, the chamber size used to trap cells is inappropriate to observe small cells, such as microbes, under a microscope. It is desirable to observe trapped cells in one chamber at a glance under a microscope. In other words, under the same magnification, it is necessary to simultaneously recognize and observe the chamber and the trapped cell. The size of the chamber was reduced (200 μm wide; 280 μm deep) to simultaneously observe one microchamber and microbe cells at a glance under same magnification.

Microbe cells were entrapped into this downsized cup-like chamber and living or dead cells were stained with carboxyfluorescein diacetate (CFDA) or propidium iodide (PI) by exchanging the solution. We used only 1 μL of reagent. We examined *E. coli* incubated overnight at 37°C as living cells and heat-treated living cells as dead cells. We confirmed that living and dead cells could be stained in the chamber. After staining the mixture of living and dead cells with a mixed reagent, CFDA and PI, respectively, we confirmed the presence of both cell types (Fig. 15.7). This technique is promising as it can be applied to other staining methods such as immunostaining. Staining of foodborne bacteria, *S. enterica*, was

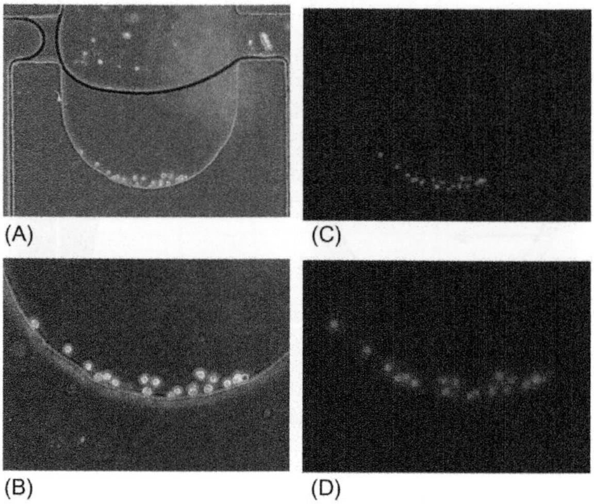

Figure 15.6 Trapped Jurkat cells in a microchamber. (A) Bright images of the entrapped cells. (B) Enlarged image of (A). (C) Fluorescent image of (A). (D) Enlarged image of (C).

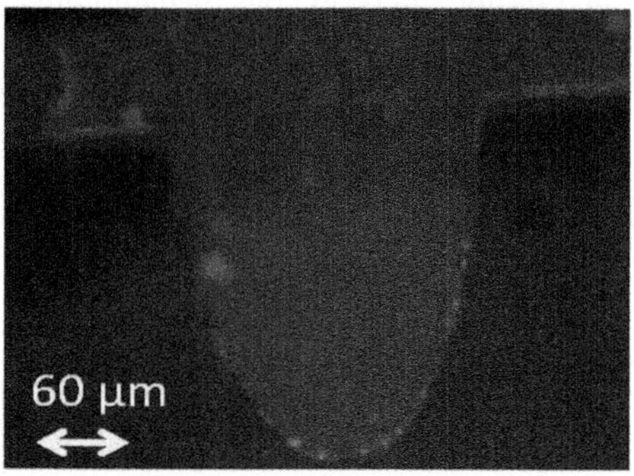

Figure 15.7 Trapped microbial cells in a microchamber.

possible by immunostaining on the device. The use of our microfluidic device shortens the detection time and enables specific detection of *S. enterica* [32].

Living cells (green/bright in monochrome) and dead cells (red/dark in monochrome) were stained by CFDA and PI.

15.4 CD-shaped microfluidic device for ELISA

15.4.1 Detection of bioactive chemicals based on ELISA

ELISA is a highly sensitive and selective analytical technique that is widely used to determine trace amounts of chemicals in diagnostic applications, and in environmental and food analyses. CD-shaped microfluidic devices made of glass and PDMS were developed to perform rapid and high-throughput detection of bioactive chemicals based on ELISA. There are 32 microchannels and a microchamber on each microchannel for the ELISA reaction in the device. No valves or micropumps are necessary because liquid can flow through each microchannel by centrifugal force (Fig. 15.8). The simultaneous ELISA of 32 samples is possible using one device, allowing high-throughput ELISA to be performed easily [21].

Bisphenol A, an endocrine disruptor, is a major constituent of the production pathway of polycarbonate, which is widely used in food containers and feeding bottles because of its endurance to heat. There are fears related to the contamination of bisphenol A in food or drinks and the broad range of adverse effects it could have on public health [33]. To determine the intake of bisphenol A with food or drinks in an animal body, a sensing system of bisphenol A with a small amount of biological sample is necessary. A rapid and easy method to monitor and analyze bisphenol A from a small amount of sample is required.

We investigated the analysis of bisphenol A based on ELISA using a microfluidic device for ELISA. Bisphenol A in rat serum was determined in less time with a much smaller volume of sample than conventional ELISA on a microtiter plate.

A competitive immunoassay of bisphenol A in a small amount of sample was investigated using antibisphenol A antibody and HRP-conjugated bisphenol A. The

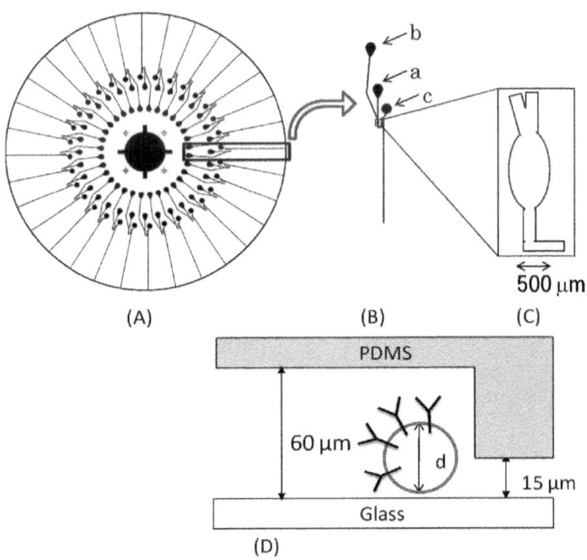

Figure 15.8 Schematic diagram of CD-shaped microfluidic device for ELISA. (A) Top view of CD-shaped microfluidic device for ELISA in which 32 microchannels are arranged radically on a device and samples are introduced into a reaction chamber by simultaneous rotation of the device. (B) A microchannel on the device: (a) an inlet for sample injection, (b) inlet for substrate of enzyme reaction, (c) vent. (C) Reaction chamber. (D) Cross section of a microchamber: (d) antibody-immobilized microbeads.

bisphenol A antibody was immobilized on microbeads and 2000 bead particles were introduced into a microchamber on the device. Sample and reagent were easily introduced into the microchamber by spinning the device for 30 seconds. A competitive immunoassay was performed using HRP-conjugated bisphenol A with a small amount of sample within 20 minutes. After the immunoreaction, HRP activity was determined by luminol chemiluminescence. Sensitive detection of bisphenol A was possible with a small volume of sample using our device.

The CD-shaped device for ELISA was also used to determine two cancer markers, carcinoembryonic antigen and alpha-fetoprotein, using HRP-conjugated antibody via sandwich ELISA [21,34].

15.4.2 Multiple ELISA for diagnosis of diabetes

By quantifying specific proteins in blood, that is, via biomarkers, it is possible to examine various health conditions. In particular, the quantification of a plurality of biomarkers is important to achieve early and accurate diagnosis [35–37]. We developed a rapid ELISA system for multiple biomarkers aimed at the early diagnosis of diabetes [22]. As shown in Fig. 15.9, the CD-shaped microfluidic device for rapid and multiple ELISA consists of three parts: a polymethyl methacrylate (PMMA) disc with main microchannels (main disc), a PMMA disc with a reagent reservoir (reagent cartridge), and PMMA parts for the reaction (reaction parts). By separating the reaction parts from the main disc, it is easy to bind the solid phase antibody to the reaction site. In addition, by separating the reagent reservoirs

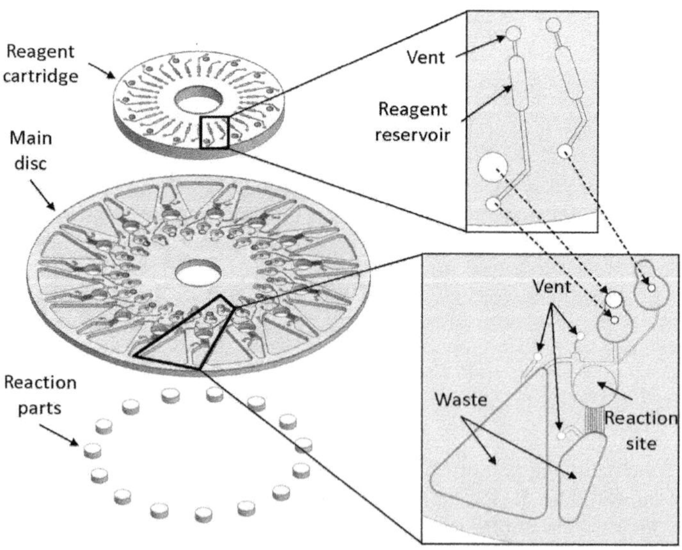

Figure 15.9 Schematic of the CD-shaped microfluidic device for rapid and multiple ELISA. The device consists of three parts: reagent cartridge, main disc, and reaction parts.

Table 15.1 Data comparing the concentration of insulin, adiponectin, and leptin in standard human serum

	Insulin (ng/mL)	Adiponectin (μg/mL)	Leptin (ng/mL)
Guarantee	1.75 ± 0.11	10.20 ± 0.61	2.37 ± 0.19
Rapid ELISA	1.84 ± 0.28	9.60 ± 0.45	2.02 ± 0.43

Guarantee values from suppliers indicate the concentration of each protein contained in the standard human serum.
The values of Rapid ELISA are the measured values of the same serum by our rapid ELISA system.
Data indicate average ± standard deviation (SD).

containing the detection antibody and the luminescent substrate as the reagent cartridge, many microchannels can be integrated on one main disc. The CD-shaped microfluidic device for ELISA can detect 15 samples. Furthermore, by providing a vent at the top of the reagent reservoir in the reagent cartridge, control of movement of the reagent was enabled at an arbitrary timing by making a hole in the vent. By setting the distance between the reaction parts and the upper surface of the microchannels to 100 μm, the required time for each reaction of ELISA was reduced, because the diffusion distance in the reaction site was shortened. Therefore, both the time to bind the protein with the solid phase antibody and the time to bind the detection antibody with the protein were 5 minutes in the CD-shaped microfluidic device, respectively.

As a result, multiple quantifications of insulin, adiponectin, and leptin required for the early diagnosis of diabetes was achieved in 16 minutes. As shown in Table 15.1, the quantitative values of standard human serum were comparable to the concentrations described in the guaranteed insulin, adiponectin, and leptin samples. In addition, we developed a detection system that is able to control the spinning of the CD-shaped microfluidic device and perform washing operations of the reaction site during each reaction step to realize a fully automatic rapid ELISA. Since this rapid ELISA system would be applied to the detection of various biomarkers by changing the type of solid phase antibody and detection antibody, it is expected to be applied to early and accurate diagnosis for not only diabetes but also other diseases.

15.5 Conclusion

Using CD-shaped microfluidic devices for cell isolation and single cell PCR, microbial cells such as *S. enterica* and nonadherent eukaryotic cells such as Jurkat cells were isolated in microchambers easily and detected after hot cell-direct PCR without the need to remove the gene after cell lysis. By using CD-shaped microfluidic device for cell staining, these cells were detected after they were entrapped in microchambers and stained in loco in the chamber. Bisphenol A, an endocrine disrupting chemical, cancer markers, insulin, adiponectin, and leptin as markers of diabetes, could be rapidly determined on our CD-shaped devices based on ELISA. The potential of these CD-shaped devices as a biosensor was shown. As the next logical step to the application of these devices, obtained data will be available through the Internet because signals after PCR, staining and ELISA are obtained optically.

Acknowledgment

A part of this work was financially supported by a grant from the Saitama Leading-edge Industry Design Project. This work was supported in part by funding from MEXT, the Matching Fund for Private Universities, S1001013 and in part by JSPS KAKENHI Grant Numbers JP24510171, JP17K07825.

References

[1] S.J. Trietsch, T. Hankemeier, H.J. van der Linden, Chemom. Intell. Lab. Syst. 108 (2011) 64.
[2] G.M. Whitesides, Nature 442 (2006) 368.
[3] P.S. Dittrich, K. Tachikawa, A. Manz, Anal. Chem. 78 (2006) 3887.
[4] W.-H. Tan, S. Takeuchi, Lab Chip 8 (2008) 259.
[5] L.-S. Jang, M.-H. Wang, Microdevices 9 (2007) 737.
[6] D.D. Carlo, N. Aghdam, L.P. Lee, Anal. Chem. 78 (2006) 4925.
[7] Y. Tanaka, et al., Biosens. Bioelectron. 23 (2007) 449.
[8] M.J. Madou, L.J. Lee, S. Daunert, S. Lai, C.H. Shih, Microdevices 3 (2001) 245.
[9] M. Madou, et al., Annu. Rev. Biomed. Eng. 8 (2006) 601.
[10] D. Duffy, H. Gillis, J. Lin, N. Sheppard, G. Kellogg, Anal. Chem. 71 (1999) 4669.
[11] F. Schuler, et al., Lab Chip 16 (2016) 208.
[12] J. Park, V. Sunkara, T.H. Kim, H. Hwang, Y.K. Cho, Anal. Chem. 84 (2012) 2133.
[13] S. Furutani, H. Nagai, I. Kubo, Sensor Lett. 6 (2008) 961.
[14] I. Kubo, S. Furutani, H. Nagai, ECS Trans. 16 (17) (2009) 1.
[15] S. Furutani, H. Nagai, Y. Takamura, I. Kubo, Anal. Bioanal. Chem. 398 (2010) 2997.
[16] S. Furutani, H. Nagai, Y. Takamura, Y. Aoyama, I. Kubo, Analyst 137 (2011) 2951.
[17] S. Furutani, N. Shozen, H. Nagai, Y. Aoyama, I. Kubo, Sens. Mater. 26 (2014) 623.
[18] S. Furutani, M. Kajiya, N. Aramaki, I. Kubo, Micromachines 7 (2016) 10.
[19] I. Kubo, S. Furutani, K. Matoba, J. Biosci. Bioeng. 112 (2011) 98.
[20] I. Kubo, T. Kanamatsu, S. Furutani, Sens. Mater. 26 (2014) 615.
[21] N. Matsunaga, S. Furutani, I. Kubo, ECS Trans. 16 (2008) 123.
[22] S. Furutani, et al., Anal. Sci. 34 (2018) 379.
[23] J.C. Martin, D.E. Swartzendruber, Science 207 (1980) 199.
[24] J.P. Nolan, L.A. Sklar, Nat. Biotechnol. 16 (1998) 633.
[25] Z.E. Perlman, et al., Science 306 (2004) 1194.
[26] I. Kubo, Y.H. Itoh, S. Furutani, Polymerase Chain Reaction for Biomedical Applications, InTech, Vienna, 2016, pp. 39–56.
[27] C.V. Rao, D.M. Wolf, A.P. Arkin, Nature 420 (2002) 231.
[28] Y. Kuang, I. Biran, D. Walt, Anal. Chem. 76 (2004) 6282.
[29] S. Lindström, H. Andersson-Svahn, Biochim. Biophys. Acta 308 (1810) 2010.
[30] M. Veiseh, B.T. Wickes, D.G. Castnerb, M. Zhang, Biomaterials 25 (2004) 3315.

[31] A. Pulsipher, M.N. Yousaf, Langmuir 26 (2010) 4130.
[32] I. Kubo, H. Hashimoto, Lab on a Chip & Microfluidics Asia 2017 6 (2017).
[33] Y. Ikezuki, O. Tsutsumi, Y. Takai, Y. Kamei, Y. Taketani, Human Reprod. 17 (2002) 2839.
[34] T. Saito, S. Furutani, I. Kubo, Chem. Sens. 29 (2012). s. A, 141.
[35] A. Umeno, et al., PLoS One 10 (2015) e0130971.
[36] J. Wang, et al., J. Geriatr. Cardiol. 14 (2017) 135.
[37] N.L. Henry, D.F. Hayes, Mol. Oncol. 6 (2012) 140.

A CMOS compatible miniature gas sensing system

16

Ting-I Chou, Shih-Wen Chiu and Kea-Tiong Tang
National Tsing Hua University, Hsinchu, Taiwan

16.1 Introduction

Numerous systems have been developed for acquiring odor information. Traditional gas-sensing systems consist of pieces of laboratory equipment that are expensive and large [1,2]; thus, such systems are not suitable for personal use by individuals. An electronic nose system is a suitable alternative to laboratory equipment because such a nose system consumes lower power and has a smaller size. An electronic nose system is a biomimetic system that contains a gas sensor array, signal processing circuits, and algorithms for identifying different odors. An electronic nose system primarily applies a gas sensor array consisting of different types of gas sensors instead of using a single gas sensor. Because different types of gas sensors have different responses to odors, the gas sensor array forms a unique pattern for each odor. The gas sensor responses are acquired using signal-processing circuits, and algorithms are used to identify the odors by recognizing the unique patterns [3].

For devices intended for personal use, scaling down the size of the electronic nose system is necessary. The signal processing circuit and microprocessor can be integrated into a single chip by using very-large-scale integration technology [4]. The size of the sensor array is a critical problem during the process of scaling down the size of the electronic nose system [5]. Commercial gas sensors are larger than the chip of a sensing system. Thus, this study proposes a complementary metal–oxide–semiconductor (CMOS)-compatible gas sensor for realizing a chip-size sensor array. The gas sensor can be integrated with the circuit without the need for postprocessing. A fully integrated nose-on-a-chip system that can serve as a suitable solution for realizing a miniature gas-sensing system can be realized. The ultralow power consumption and small size of the nose-on-a-chip system reveal the ability of system to provide odor information and prove its suitability for personal use.

Chemical, Gas, and Biosensors for the Internet of Things and Related Applications.
DOI: https://doi.org/10.1016/B978-0-12-815409-0.00017-6

16.2 Complementary metal–oxide–semiconductor-compatible gas sensor

16.2.1 Materials and fabrication

Gas sensors based on different principles [6] such as conductive gas sensors, piezoelectric gas sensors, metal–oxide–semiconductor field-effect transistor–based gas sensors, and optical gas sensors have been used to acquire odor information. Among these types of gas sensors, conductive gas sensors are the most suitable for electronic nose system applications because the interface circuits of the sensors are considerably simpler than those of other sensor types. Metal–oxide (MOX) sensors [7] constitute a category of conductive gas sensors and are often used commercially due to their high sensitivity. However, MOX sensors usually operate at high temperature levels. Heaters are required to maintain MOX sensors at a particular operation temperature. The high power consumption and fabrication complexity of CMOS circuitry and front-end microelectromechanical systems (MEMS) [8] lead to difficulties in applying MOX sensors in electronic nose systems. Conducting polymer (CP) composite sensors [9] constitute another category of gas sensors that can operate at a normal temperature level. Moreover, CP sensors can be fabricated by simply depositing sensing materials between two electrodes. Because of these advantages, CP sensors are more suitable than MOX sensors for use in electronic nose systems intended for personal applications.

Compared with MOX sensors that require a complicated MEMS or specific process, CP gas sensors can be manufactured using standard photolithography. In this study, a microsensor array (Fig. 16.1) was first fabricated using standard

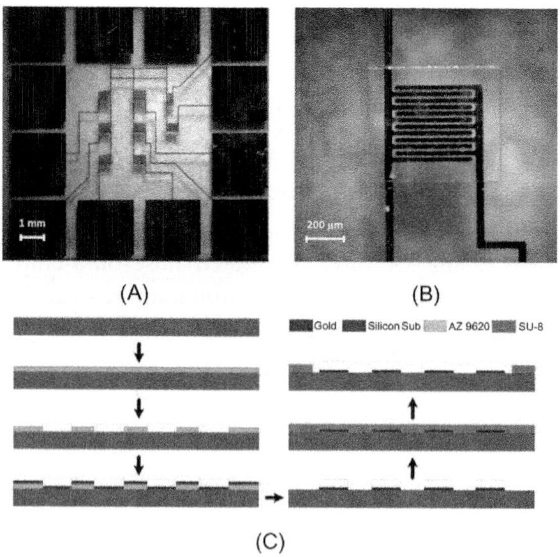

Figure 16.1 Photographs of (A) the microsensor array, (B) a single sensor element, and (C) fabrication process.

Table 16.1 The polymers used in the microsensor array

	Polymer	Solvent
S1	Poly(4-vinyl phenpl-*co*-methyl methacrylate), PVCMM	MEK
S2	Poly(4-vinyl phenpl-*co*-methyl methacrylate), PVCMM	MEK
S3	Poly(vinyl benzyl chloride), PVBC	MEK
S4	Poly(ethylene adipate), PEA	MEK
S5	Poly(methyl vinyl ether-alt-maleic acid), PMVEMA	H_2O
S6	Polyvinylpyrrolidone, PVP	H_2O
S7	Poly(vinylidene chloride-*co*-acrylonitrile), P(VDC-AN)	MEK
S8	Styrene/allyl alcohol copolymer, SAA	MEK

photolithography [10]. Subsequently, CP composite materials were deposited between 100-μm-wide gold leads occupying a small area of $250 \times 250\ \mu m^2$. A 10-μm-thick photoresist AZ9620 was first spin coated on a 700-μm-thick glass substrate, which was then patterned through UV exposure ($\lambda = 350-450$ nm) and to AZ400K developer. Then, an E-gun evaporator was used to sequentially deposit 10 nm of Cr and 200 nm of Au. A lift-off process was executed using acetone to form an interdigital electrode array. Moreover, a 50-μm-thick SU-8 3050 was spun onto the interdigital electrode array substrate after the lift-off process. Seven interdigital electrodes were determined to occupy an area of $400 \times 400\ \mu m^2$, and one interdigital electrode was determined to occupy an area of $500 \times 250\ \mu m^2$. The width of the electrode was 25 μm, and the spacing between the electrodes was 13 μm.

Although CP sensors have some advantages, the poor reproducibility and repeatability of CP sensor fabrication procedures cause major problems. To overcome these problems, a biological two-layer structure was employed [11]. The polymer and the conducting materials in the CP sensors were separated in distinct layers to avoid the aggregation of the conductive material and polymer solutions. The applied two-layer sensing films are fabricated in two steps. First, a conductive path was formed to realize the conducting layer. Multiwalled carbon nanotubes (MWCNTs) [12] were incorporated into the conducting layer because of their advantages such as high sensitivity, low response time, high reproducibility, and long-term stability when used in CP-based gas sensors. The MWCNTs were dissolved in water at 1 mg/mL (1 wt.%). The solution was deposited on the interdigital electrodes by using a nanoliter injector, and the water was pumped out at 70°C, thus forming the conductive film. Subsequently, the polymer layer was deposited on the conducting layer by using the nanoliter injector. Seven polymer composite materials were selected for gas selectivity. Table 16.1 presents the polymers and their solvents.

16.2.2 Gas experimental results

The MWCNT solution was first subject to a 10-minutes homogenous sonication process. Subsequently, the solution was deposited on the microsensor array. The resistance of the sensor array was maintained between 1.0 and 3.0 kΩ after the first layer was deposited. Moreover, the resistance of the sensor array was maintained between 3.0 and 10.0 kΩ after multiple polymer layers were deposited. Fig. 16.2A shows the basic response magnitude of the microsensor array to methanol. The concentration of methanol was maintained between 1000 and 50,000 ppm in dry air. Different polymers possessed different $\Delta R/R_0$ magnitudes to methanol; additionally, the relation curve displayed the linearity between the concentration of

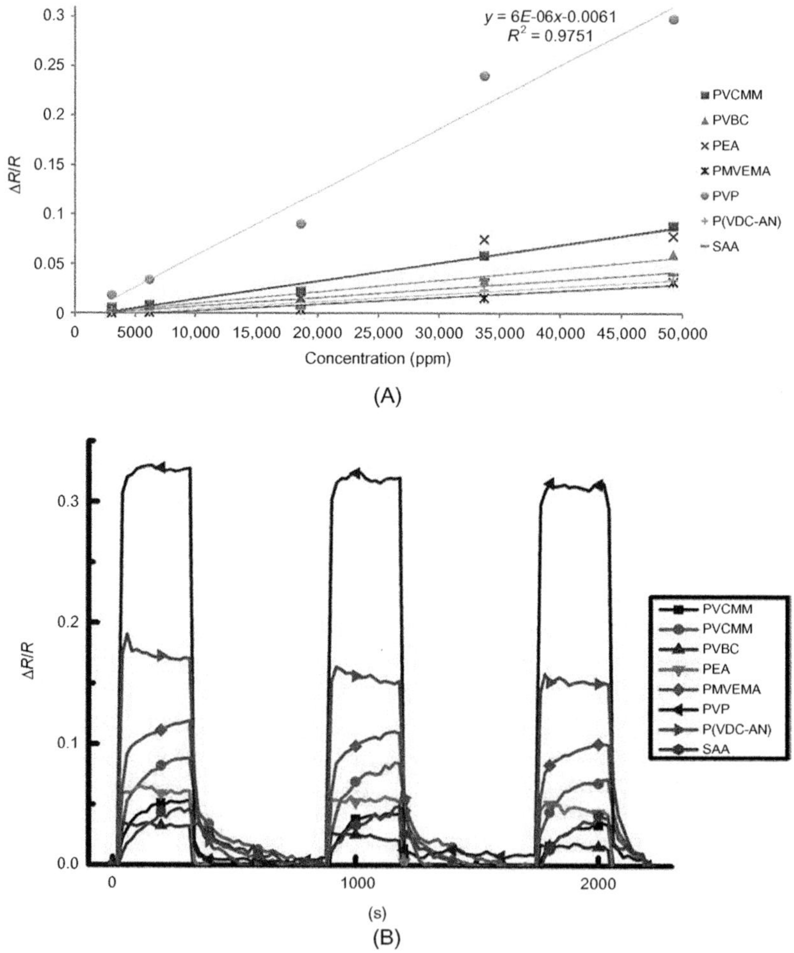

Figure 16.2 (A) Resistance change ratio versus methanol concentration for the polymer-coated microsensors. PEA poly(ethylene adipate), PMVEMA poly(methyl vinyl ether-alt-maleic acid), PVBC poly(vinylbenzyl chloride), PVCMM poly(4-vinylphenol-co-methyl methacrylate), P(VDC-AN) poly(vinylidene chloride-co-acrylonitrile), PVP polyvinylpyrrolidone, SAA styrene/allyl alcohol copolymer (B) Three consecutive gas sensing responses of the microsensor array to methanol gas.

methanol and the change in array resistance. Fig. 16.2B illustrates three consecutive responses of the microsensor array to methanol vapor at a concentration of 4.9%. The absorption and desorption periods in this experiment were set to 5 and 15 minutes, respectively. The magnitude of the change in resistance, the reaction time, and the recovery time were consistent, signifying that the microsensor array exhibits favorable repeatability. The experimental results revealed that the microsensor array was suitable for use in the electronic nose system for personal applications.

16.3 Nose-on-a-chip

16.3.1 System block diagram

The nose-on-a-chip integrates a sensor array, signal processing circuits, and microprocessor on a single chip (Fig. 16.3). Lower metal layers of the proposed nose-on-a-chip are used for the microelectronic circuits, and the top-most metal layer is used as the electrode in the CP sensors [13]. The sensory areas are defined by a pad mask to eliminate the use of silicon nitride in the standard CMOS process. Fig. 16.4 presents a block diagram of the electronic components of the nose-on-a-chip that provide a readout of the gas sensor array information. The proposed nose-on-a-chip comprises an adaptive sensor interface, a 10-bit successive approximation analog-to-digital converter (SAR ADC), a reduced instruction set computing (RISC)-core processor (OR 1200), a continuous restricted Boltzmann machine (CRBM) kernel as a specific digital signal processing unit for data clustering, and an 8K × 32 bit static random access memory (SRAM).

16.3.2 Adaptive interface circuitry

The resistance of a CP sensor usually varies with temperature, humidity, and background odor. The different polymers used in a CP sensor result in different initial sensor resistance levels. Moreover, the baseline resistance shifts after several rounds of absorption and desorption. Therefore, an interface circuit that can adapt to the

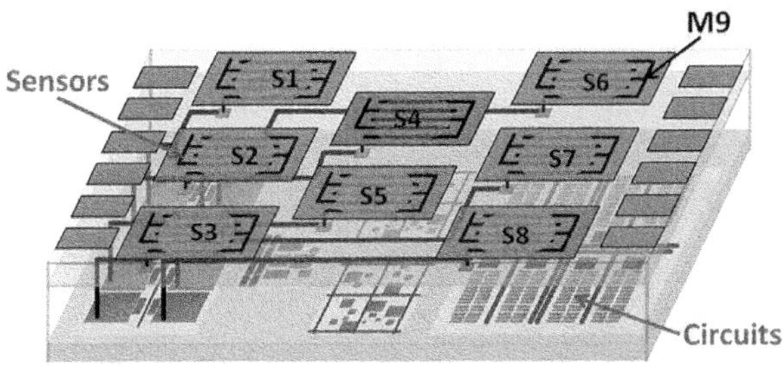

Figure 16.3 The chip architecture of the nose-on-a-chip.

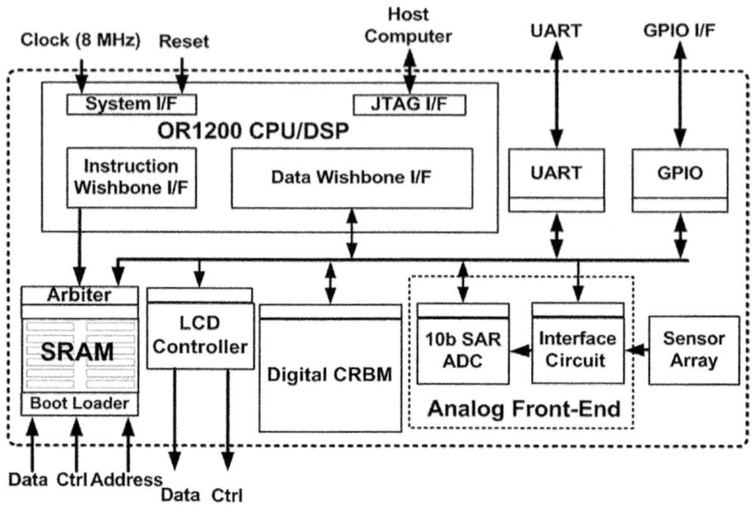

Figure 16.4 System block diagram of the nose-on-a-chip.

shifted sensor resistance is required to nullify nonideal environmental interference [14,15].

The adaptive interface circuit consists of a comparator, a D flip-flop, an 8-bit counter, and an 8-bit current-type digital-to-analog converter (DAC). This circuit is operated in two modes. The first mode is to adapt the circuit to a preset reference voltage. The counter is reset first; thus, the output current of the DAC is zero when the comparator output enables the counter to begin counting. The upcounting output of the counter increases the output current of the DAC. Additionally, the sensor voltage generated by the current from the DAC flows through the gas sensor and thus increases until it reaches the reference voltage, after which the output of the comparator toggles and stops the counter. Thus, the DAC is turned into a constant current source. The circuit is then switched to the second mode and is ready to sense odors. In this mode, a change in the gas sensor resistance is proportional to the voltage change, according to Ohm's law. Specifically, the undesired gas sensor resistance shift is nullified by the first mode of this adaptive interface circuit, and the gas sensor array information is converted to voltage information and is sent to the ADC in the second mode.

16.3.3 SAR ADC

SAR ADCs have high power efficiency in low-voltage operation; thus, they are more suitable for low power applications. For electronic nose system applications, an ADC should be able to operate at hundreds of kilosamples per second to a few megasamples per second with medium resolution. A 0.5-V 10-bit asynchronous SAR ADC is implemented in the proposed nose-on-a-chip. The supply voltage and a charge average switching (CAS) DAC [16] are scaled down to reduce the switching energy of the DAC without using any additional voltage reference or common-mode shift. Moreover, to achieve the 10-bit accuracy requirement, double-boosted sample-and-hold (S—H) circuits and local-boosted switches are utilized in the system.

For the CAS procedure, the MSB capacitor (512 C) should be split into two sub-DAC arrays (128 C to 1 C). One sub-DAC array should be reset to ground, and the other array should be reset to the reference voltage in the sampling phase. The CAS technique is employed to generate the required top-plate voltage shift through charge averaging between the bottom plates instead of the conventional charging or discharging operation. The charge-averaging function is realized by enabling the equalization switches from the control rather than switching to the reference voltages. However, the on-resistance of CAS severely increases in the operation due to the near-threshold operation and insufficient overdrive voltage headroom. Local-boosted switches are used to conduct the averaging function without compromising the speed and accuracy. When the transistor size is optimized and an appropriate layout is prepared, the power overhead of the required boosted control circuit is approximately 3% for the implemented 10-bit ADC.

16.3.4 CRBM kernel

In biomedical applications, raw sensory signals are often high dimensional, noisy, and drifting. The CRBM is a probabilistic neural network that can classify biomedical data [17,18]. The nose-on-a-chip is integrated with the CRBM system to perform dimensional reduction and data clustering for data classification. The CRBM system is implemented in a digital circuit because of the following advantages: immunity to transistor mismatches and noise interference and maintenance of a satisfactory precision level.

The CRBM system consists of a visible layer and a hidden layer of stochastic neurons with interlayer connections. For the CRBM system, let v_i and h_j represent the states of the visible neuron i and hidden neuron j, respectively. The states of the hidden neurons are derived using the following equation:

$$h_j = \varphi\left\{a_j \cdot \left[\sum_i w_{ij} \cdot v_i + N_j(\sigma, 0)\right]\right\},$$

where w_{ij} represents the connection weight between neurons i and j. The term N_j represents zero-mean Gaussian noise with a variance of σ^2, and φ represents the sigmoid function. The CRBM system learns to regenerate training samples in its visible layer. Therefore, the number of visible neurons is equal to the dimension of the data to be modeled, and the number of hidden neurons is selected on the basis of data complexity. During training, each training datum is set as the initial state of the visible neurons, v^0. The initial state of hidden neurons, h^0, is sampled by using v^0. When the visible and hidden neurons are sampled again to obtain the one-step sampled v^1 and h^1, the updated values for all parameters can be calculated on the basis of the following minimizing contrastive divergence (MCD) rule:

$$\Delta w_{ij} = \eta_w \cdot \left(\left\langle v_i^0 \cdot h_j^0 \right\rangle_4 - \left\langle v_i^1 \cdot h_j^1 \right\rangle_4 \right)$$

$$\Delta a_i = \eta_a \cdot \left(\left\langle h_j^0 \cdot h_j^0 \right\rangle_4 - \left\langle h_j^1 \cdot h_j^1 \right\rangle_4 \right)$$

where η_w and η_a are the learning rates and $< \cdot >_4$ denotes the expectation with respect to four data items. The parameter a_j of the visible neurons is updated using the same MCD rule. For the parameter a_j, h_j^0 and h_j^1 are replaced by v_j^0 and v_j^1, respectively.

Because the nose-on-a-chip contains eight on-chip sensors, a CRBM network [19] of eight visible and three hidden neurons is required. The nonlinear sigmoid function in the CRBM system is approximated using piecewise-linear (PWL) curves to reduce the circuit complexity. The noise generator is implemented through a linear-feedback shift register that generates a pseudorandom binary sequence. The amount of 1 seconds in the sequence is subtracted from 16 and divided by $2N$, resulting in a Bernoulli-distributed random value with a zero mean; here, the integer N controls the variance of the random value. The multiplier is designed for 10-bit inputs and 20-bit outputs, and the subtractor or accumulator is also designed for a 20-bit resolution for computations. The register file contains a memory capacity of 138 words $\times$ 20 bits for storing the parameters and the Gibbs sampled states during the training of the digital CRBM system [20].

16.3.5 Memory

For minimizing power consumption, the supply voltage of the nose-on-a-chip is reduced. The proposed nose-on-a-chip adopts an L-shaped 7T (L7T) SRAM cell, which comprises a 6T STAM cell and a one-transistor read-decoupled read port (RP) for preventing the problem of read disturbance [21] that occurs at a low voltage. However, the read bitline (RBL) voltage swing of this L7T cell is limited due to the bitline clamping current from the unselected RPs of the accessed column. To solve this problem, the gate bias of the one-transistor RP is increased by boosting the voltage supply of the cell in the read cycle through parasitic capacitors between the metal lines on the top of the SRAM cells. This technique increases the area overhead by only 1% and reduces the power consumption by 10%. The required cell voltage supply levels differ according to whether the row address is selected. OFS-CVDD [22] technology is utilized to overcome the aforementioned problem. The technology switches the cell voltage supply to a higher voltage and lower voltage for selected and unselected rows, respectively. In the precharge phase, the RBL precharge signal is activated to precharge the RBL to V_{DD}. After the selected read word line is confirmed, the precharged RBL is boosted. This not only increases the sensing margin for read 1 but also increases the cell current of RP for read 0.

16.3.6 RISC core

The proposed nose-on-a-chip adopts an OpenRISC 1200 core microprocessor [23,24], which is an open-source Verilog intellectual property. The OpenRISC 1200 core is adjusted such that a low-power low-voltage operation is realized. A multiply accumulate (MAC) unit is implemented for power efficiency and design flexibility. Furthermore, GPIOs and UART are included in the data interface. A JTAG interface provides debugging and memory-programming capabilities.

Typically, the firmware is first loaded into an external device such as a flash drive. Subsequently, the firmware is loaded into the L7T SRAM through a boot loader. The microprocessor executes a complete program by using the on-chip sensor array signals and processes the received data through the digital CRBM system. Because there are eight on-chip sensors in the nose-on-a-chip, an eight-to-one multiplexer (CH-MUX) controlled using a 3-bit signal from the microprocessor is used to select the desired channel. To change the signal of the sensor array, the microprocessor first attains the baseline value after the interface circuit has adapted to a preset reference voltage. When the target gas flows through the sensor, the microprocessor waits for all the sensors to reach the steady state. Subsequently, the reaction changes are collected by the processor. Normalization is applied to the collected data to reduce the influence of concentration. The data are then set to the digital CRBM system for data dimension reduction and data clustering. The returned data from the digital CRBM system are classified through algorithms in the firmware. Finally, the results are displayed through the LCD–UART–GPIO system.

16.3.7 Chip measurement results

The proposed nose-on-a-chip was fabricated using Taiwan Semiconductor Manufacturing Company 90-nm 1P9M CMOS technology (Fig. 16.5). The chip area was determined to be $3254 \times 3223\ \mu m^2$. A summary of the nose-on-a-chip and subblocks is provided in Table 16.2. The adaptive interface was determined to be able to adapt to the resistance changes in the range of 5–100 kΩ, and the reference voltage could be set between 0.15 and 0.25 V. A higher reference voltage could obtain a higher resolution for the resistance change, whereas a lower reference voltage could obtain a larger signal change. The adaptive interface exhibited favorable linearity when the resistive change ratio was in the range of 0%–76% and consumed 215 μW at 0.5 V. The CAS SAR ADC achieved a 10-bit resolution for a sampling rate in the range of 0.5–4 MS/s. The ADC consumed 1.15 μW at 0.5 V. The digital CRBM system was verified by comparing the measurement results with simulation results obtained using MATLAB; the measurement results exhibited a

Figure 16.5 Chip micrograph of the proposed nose-on-a-chip.

Table 16.2 Summary of the nose-on-a-chip and its subblocks

Chip summary	
Technology	1P9M 90 nm CMOS
Die size	3254 μm × 3223 μm
Supply voltage	0.5 V
Power dissipation	1.27 mW
Operation frequency	8 MHz
Subblocks	
On-chips	8 Polymer-carbon composites
Adaptive interface	8 Channels, range: 5k–100kΩ
SAR-ADC	10b
Processor	32b RISC
SRAM	L7T, 8Kx32b
Learning kernel	CRBM

similar pattern distribution to the simulation after 5000 learning cycles, and the learning rate was 0.25 for {wj} and 0.5 for {ai} and {aj}. This nose-on-a-chip was determined to consume 1.27 mW at 0.5 V and an operating frequency of 8 MHz. Table 16.3 presents a comparison of the proposed nose-on-a-chip with previously developed nose-on-a-chip systems. The proposed nose-on-a-chip achieved the highest level of integration and the highest computational capability with the lowest power dissipation among the state-of-the-art nose-on-a-chip systems that are suitable for personal applications.

Table 16.3 Comparison arts

	JSSC'02 [25]	TBio Eng.'11 [26]	ISSCC'12 [27]	TBCAS'14 [22]
CMOS technology	0.8 μm	0.18 μm	0.25 μm	90 nm
Supply voltage (V)	>5	1.8	3.3	0.5
Sensor type	Capacitive/ calorimetric/ mass-sensitive	Resistive	Frequency	Resistive
Channels	3	8	1	8
Power dissipation (mW)	>7.25	2.81	1.35	1.27
On-chip sensor	Post-MEMS	–	–	Standard CMOS
Sensor interface	Not adaptive	Self adaptive	Not adaptive	Self adaptive
Embedded processor	–	8b 8051	–	32b RISC
On-chip memory	–	6T, 16Kx8b	–	L7T, 8Kx32b
Learning kernel	–	–	–	CRBM
Scalability	–	–	–	YES
Automatic gain control	NO	NO	YES	NO

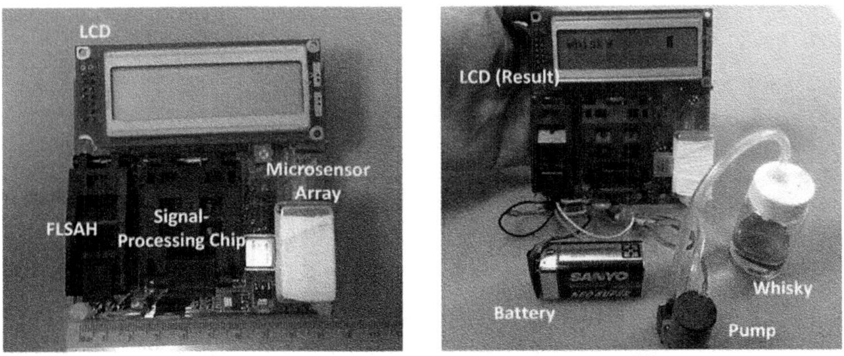

(A) (B)

Figure 16.6 The miniature e-nose device prototype.

16.4 Miniature electronic nose system prototype

Fig. 16.6 shows a miniature gas sensing system prototype including the CMOS-compatible CP microsensor array and a low-power signal processing chip based on an 8051 microprocessor. The prototype contained an LCD (LMC-SSC2A16-01), a flash memory component (AT49SV322D), power and bias commercial integrated circuits (L7805CV and LD1117A), and a 10-MHz crystal oscillator (MCO-1500A). All the discrete components were assembled with the signal processing chip in the

PCB measuring $9.5 \times 9.5 \times 2.2\ cm^3$ and weighing 137.8 g. This prototype was driven by a 9-V battery, and the power module was converted to 5 and 1.8 V for supplying power to all other components and the signal processing chip. For the gas flow path, the testing chamber was connected to a small pump (THOMAS, 2002 model, USA) that could pump the target gas in the sample bottle to the chamber. Computations performed through the K-nearest neighbor algorithm and data in the odor database provided in the firmware revealed that the miniature gas sensing system prototype exhibited 100% accuracy in identifying tincture, sorghum wine sake, whiskey, and vodka.

16.5 Application example

An application of the nose-on-a-chip system is diagnosis [28,29]. Ventilator-associated pneumonia (VAP) lacks a rapid diagnostic strategy. A nose-on-a-chip was installed at the proximal end of an expiratory circuit of a ventilator to monitor and detect the metabolite of pneumonia in the early stage. Fig. 16.7 illustrates the experimental setup involving patients with VAP. The setup was approved by the Institutional Review Board of Taipei Medical University, Taiwan. The sampled gas was collected at the proximal end of the expiratory device by using a bypass approach through the tube. For safety purposes, the sampling time was restricted to 10 seconds. For a pilot test, raw sensor data were obtained using a commercial electronic nose, namely Cyranose, which has been applied in various studies for disease identification [30–33]. The raw data acquired from the sensor array can be used as a standard in the proposed system. Breath samples from each patient were collected

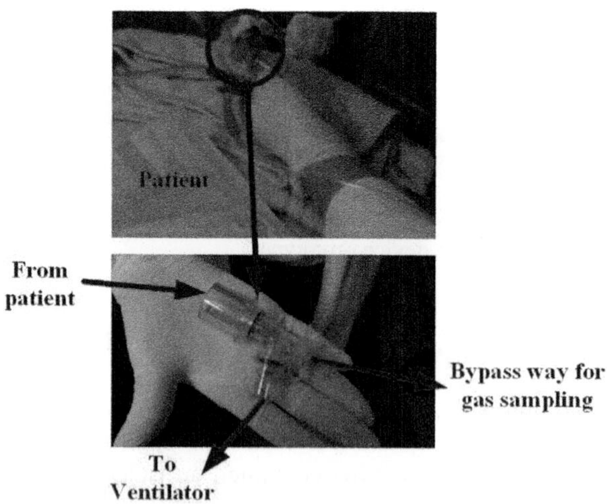

Figure 16.7 Clinical experiment setup of the VAP patients at the proximal end of the expiratory device.

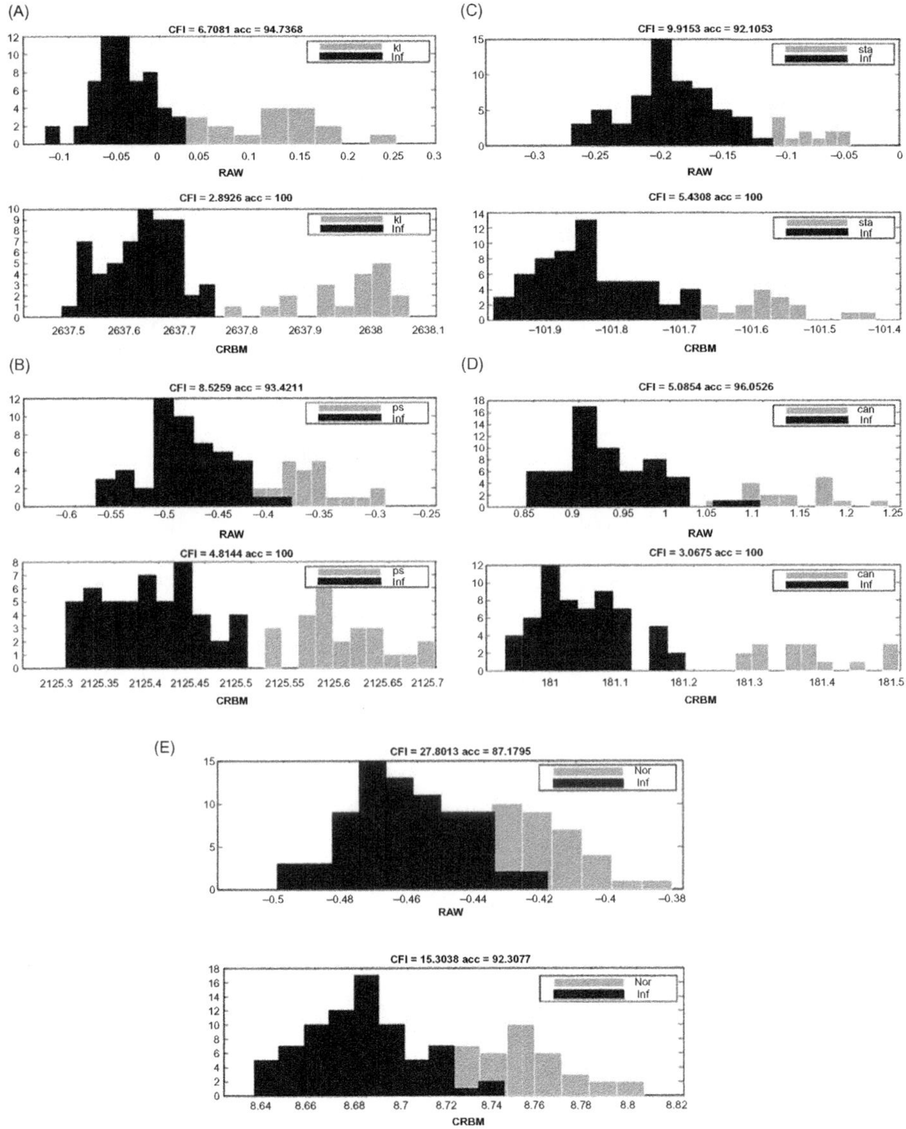

Figure 16.8 Results of VAP identification with/without CRBM: (A) *Klebsiella* vs. other infected patients (*Pseudomonas aeruginosa*, *Staphylococcus aureus*, *Candida*); (B) *Pseudomonas aeruginosa* vs. other infected patients (*Klebsiella*, *Staphylococcus aureus*, *Candida*); (C) *Staphylococcus aureus* vs. other infected patients (*Klebsiella*, *Pseudomonas aeruginos*, *Candida*); (D) *Candida* versus other infected patients (*Klebsiella*, *Pseudomonas aeruginosa*, *Staphylococcus aureus*); (E) normal vs. infected patients (*Klebsiella*, *Pseudomonas aeruginosa*, *Staphylococcus aureus*, *Candida*).

using a 1000-mL gas sampling bag before they were inserted into the device through a gas pumper at 10 mL/s. The readout data containing $\Delta R/R_0$ values served as the odor data for further analysis.

The system's VAP identification ability was verified through the clinical data. A total of 117 samples were collected: 76 samples were from patients with pneumonia (19 pertaining to *Klebsiella*, 25 pertaining to *Pseudomonas aeruginosa*, 16 pertaining to *Staphylococcus aureus*, and 16 pertaining to *Candida*) and 41 samples were obtained from uninfected individuals. The samples were categorized into an experimental group and a control group. First, the classification method identified whether the test sample was infected. If the sample was infected, the method identified the microorganism in the sample. For the classification method, the CRBM system was first trained to model raw data. Subsequently, a linear discriminant analysis (LDA) was applied to project any hidden-neuron space in the trained CRBM. Fig. 16.8 presents a comparison of the LDA projections of the raw data with the corresponding hidden-neuron responses of the trained CRBM. The clustering Fisher index (CFI) [34] is typically used to estimate clustering performance. A lower CFI signifies a lower degree of overlap between two clusters, which implies high classification performance. The results revealed that the original raw data of the experimental group and the control group overlapped. The CRBM system reduced the CFI from 27.8 to 15.3 and improved the classification accuracy from 87.18% to 92.31%. For distinguishing between a specific microorganism and other microorganism, the CRBM system reduced the CFI from 6.71 to 2.89, from 8.53 to 4.81, from 9.92 to 5.43, and from 5.09 to 3.07 and improved the classification accuracy from 94.74% to 100%, from 93.43% to 100%, from 92.11% to 100%, and from 96.05% to 100% when identifying *Klebsiella*, *P. aeruginosa*, *S. aureus*, and *Candida*, respectively. These results thus demonstrate that the miniature gas sensing system is highly efficient for diagnosing VAP.

16.6 Conclusion

A fully integrated nose-on-a-chip is an effective solution for reducing the size and power consumption of an odor-sensing system. A two-layer deposit method not only exhibits compatibility with the standard CMOS process but also enhances the reproducibility and repeatability of a CP gas sensor. A microsensor array was designed for identifying volatile organic compounds and odors that have complex compositions. The nose-on-a-chip occupies an area of 10.49 mm^2 at a 0.5-V supply voltage and consumes only 1.27 mW. VAP diagnosis was conducted with a very high identification rate of 92.31% for identifying healthy and infected patients. Moreover, *Klebsiella*, *P. aeruginosa*, *S. aureus*, and *Candida* were identified with 100% accuracy in patients with VAP. The small size and ultralow power consumption of the nose-on-a-chip render it a promising solution as a miniature gas-sensing system intended for personal use.

Acknowledgments

The authors would like to acknowledge the research team members in National Tsing Hua University, National Chiao Tung University, Taipei Medical University, and National Chung-Shan Institute of Science and Technology, in Taiwan.

Dr. Hsin Chen, Dr. Chih-Cheng Hsieh, Dr. Meng-Fan Chang, and Dr. Yi-Wen Liu from National Tsing Hua University, and Dr. Chia-Hsiang Yang, Dr. Herming Chiueh from National Chiao Tung University contribute for the chip design.

Dr. Chung-Hung Shih from Taipei Medical University contributes for clinical trials.

Dr. Li-Chun Wang from Chung-Shan Institute of Science and Technology contributes for sensor design.

References

[1] O. Laakso, et al., Screening of exhaled breath by low-resolution multicomponent FT-IR spectrometry in patients attending emergency departments, J. Anal. Toxicol. 28 (2) (2004) 111–117.

[2] T.C. Pearce, et al., Handbook of Machine Olfaction: Electronic Nose Technology, John Wiley & Sons, 2006.

[3] B. Hock, Bioresponse-Linked Instrumental Analysis, Springer Science & Business Media, 2012.

[4] D.F. Lemmerhirt, K.D. Wise, Chip-scale integration of data-gathering microsystems, Proc. IEEE 94 (6) (2006) 1138–1159.

[5] J. Hatfield, et al., Towards an integrated electronic nose using conducting polymer sensors, Sens. Actuat. B: Chem. 18 (1–3) (1994) 221–228.

[6] K. Arshak, et al., A review of gas sensors employed in electronic nose applications, Sens. Rev. 24 (2) (2004) 181–198.

[7] M.Y. Afridi, et al., A monolithic CMOS microhotplate-based gas sensor system, IEEE. Sens. J. 2 (6) (2002) 644–655.

[8] B. Guo, et al., A monolithic integrated 4 × 4 tin oxide gas sensor array with on-chip multiplexing and differential readout circuits, Solid-State Electr. 51 (1) (2007) 69–76.

[9] Y.S. Kim, et al., Portable electronic nose system based on the carbon black–polymer composite sensor array, Sens. Actuat. B: Chem. 108 (1–2) (2005) 285–291.

[10] S.-W. Chiu, et al., A miniature electronic nose system based on an MWNT–polymer microsensor array and a low-power signal-processing chip, Anal. Bioanal. Chem. 406 (16) (2014) 3985–3994.

[11] L. Wang, et al., A bio-inspired two-layer multiple-walled carbon nanotube–polymer composite sensor array and a bio-inspired fast-adaptive readout circuit for a portable electronic nose, Biosens. Bioelectron. 26 (11) (2011) 4301–4307.

[12] Y.-W. Lin, T.-M. Wu, Synthesis and characterization of externally doped sulfonated polyaniline/multi-walled carbon nanotube composites, Compos. Sci. Technol. 69 (15–16) (2009) 2559–2565.

[13] K.-T. Tang, et al. 24.5 A 0.5 V 1.27 mW nose-on-a-chip for rapid diagnosis of ventilator-associated pneumonia. in Solid-State Circuits Conference Digest of Technical Papers (ISSCC), 2014 IEEE International, 2014, IEEE.

[14] K.-T. Tang, Goodman, R.M. Towards a wearable electronic nose chip. in Custom Integrated Circuits Conference, 2006. CICC'06. IEEE, 2006.

[15] X. Mu, et al., CMOS monolithic nanoparticle-coated chemiresistor array for micro-scale gas chromatography, IEEE. Sens. J. 12 (7) (2012) 2444–2452.

[16] C.-Y. Liou, C.-C. Hsieh. A 2.4-to-5.2 fJ/conversion-step 10b 0.5-to-4MS/s SAR ADC with charge-average switching DAC in 90nm CMOS. in Solid-State Circuits Conference Digest of Technical Papers (ISSCC), 2013 IEEE International, 2013. IEEE.

[17] H. Chen, A.F. Murray, Continuous restricted Boltzmann machine with an implementable training algorithm, IEE Proc. Vision, Image Signal Process. 150 (3) (2003) 153–158.

[18] H. Chen, P.C. Fleury, A.F. Murray, Continuous-valued probabilistic behavior in a VLSI generative model, IEEE Trans. Neural Networks 17 (3) (2006) 755–770.

[19] J.-H. Wang, K.-T. Tang, H. Chen. An embedded probabilistic neural network with on-chip learning capability. in Biomedical Circuits and Systems Conference (BioCAS), 2013 IEEE, 2013. IEEE.

[20] G.E. Hinton, Training products of experts by minimizing contrastive divergence, Neural Comput. 14 (8) (2002) 1771–1800.

[21] M.-F. Chang, et al., A Sub-0.3 V Area-Efficient L-Shaped 7T SRAM With Read Bitline Swing Expansion Schemes Based on Boosted Read-Bitline, Asymmetric-V $ _ {\rm TH} $ Read-Port, and Offset Cell VDD Biasing Techniques, IEEE J. Solid-State Circ. 48 (10) (2013) 2558–2569.

[22] S.-W. Chiu, et al., A fully integrated nose-on-a-chip for rapid diagnosis of ventilator-associated pneumonia, IEEE Trans. Biomed. Circ. Syst. 8 (6) (2014) 765–778.

[23] T.-J. Chen, et al., The implementation of a low-power biomedical signal processor for real-time epileptic seizure detection on absence animal models, IEEE J. Emerg. Selected Topics Circ. Syst. 1 (4) (2011) 613–621.

[24] S.-F. Liang, et al., A hierarchical approach for online temporal lobe seizure detection in long-term intracranial EEG recordings, J. Neural. Eng. 10 (4) (2013) 045004.

[25] C. Hagleitner, et al., CMOS single-chip gas detection system comprising capacitive, calorimetric and mass-sensitive microsensors, IEEE J. Solid-State Circ. 37 (12) (2002) 1867–1878.

[26] K.-T. Tang, et al., A low-power electronic nose signal-processing chip for a portable artificial olfaction system, IEEE Trans. Biomed. Circ. Syst. 5 (4) (2011) 380–390.

[27] V. Petrescu, et al. Power-efficient readout circuit for miniaturized electronic nose. in Solid-State Circuits Conference Digest of Technical Papers (ISSCC), 2012 IEEE International, 2012. IEEE.

[28] F. Nayeri, et al., Exhaled breath condensate and serum levels of hepatocyte growth factor in pneumonia, Respir. Med. 96 (2) (2002) 115–119.

[29] M. Corradi, et al., Nitrate in exhaled breath condensate of patients with different airway diseases, Nitric Oxide 8 (1) (2003) 26–30.

[30] M.A. Markom, et al., Intelligent electronic nose system for basal stem rot disease detection, Comp. Electr. Agric. 66 (2) (2009) 140–146.

[31] S. Dragonieri, et al., An electronic nose in the discrimination of patients with asthma and controls, J. Allergy Clin. Immunol. 120 (4) (2007) 856–862.

[32] S. Dragonieri, et al., An electronic nose in the discrimination of patients with non-small cell lung cancer and COPD, Lung Cancer 64 (2) (2009) 166–170.

[33] R. Dutta, et al., Bacteria classification using Cyranose 320 electronic nose, Biomed. Eng. Online. 1 (1) (2002) 4.

[34] R.A. Fisher, The use of multiple measurements in taxonomic problems, Ann. Hum. Genet. 7 (2) (1936) 179–188.

Visualization of odor space and quality

17

Fumihiro Sassa[1], *Chuanjun Liu*[2] *and Kenshi Hayashi*[1]
[1]Department of Electronics, Graduate School of Information Science and Electrical Engineering, Kyushu University, Fukuoka, Japan, [2]Research Laboratory, U.S.E. Co., Ltd, Tokyo, Japan

17.1 Introduction

We humans take action in the environment based on the information of five senses: sight, sound, touch, smell, and taste. These senses work in different ways to gain us information about the world we live in. Although every sense is vital for us, it has been demonstrated that the information amount obtained varies from different senses: for example, approximately 83% through vision, 11% through sound, 3.5% through smell, 1.5% through touch, and 1% through taste [1]. The sense of sight is the capability to focus and detect images of visible light. As a long distance physical sense, vision is of the greatest importance in human perception in view of the dominant amount of information that we can obtain by our eyes. Conversely, the sense of smell involves the detection and discrimination of odorous chemicals in air. The information amount obtained by our nose is far less than that by the eyes. In addition, the sense of smell is considered as the most primitive and sophisticated one compared with other senses [2–4]. This is because the olfactory bulb, which is responsible to transmit smell stimulations, is part of the brain's limbic system, which is an area closely associated with our memory and emotion processed by hippocampus and amygdala. Due to the association with memory and emotion, the smell is a very subjective feeling and varies among individuals. Therefore it is difficult to present and transmit odor information objectively. There is still a lack of an effective method to digitalize olfaction.

Since the diffusion coefficient of odorant molecules is far less than the velocity of wind, odors generally exist in environment in the form of a "plume." The odor plume carried by the wind shows complicated structure with instantaneous temporal and spatial distribution of concentrations [5]. The odor information we can perceive includes the odor quality, odor quantity, and the structure of odor plume in the environment. Compared with the development of sensors for the detection of odor quality and quantity, less attention has been paid to the detection of spatiotemporal distribution of odorants in the environment. Actually, the spatiotemporal information of spatial odor distribution is very important for the activity of humans and other species, for example, animals and insects locate food, identify mates and predators, or find habitats by perceiving the spatiotemporal structure of the odor

Chemical, Gas, and Biosensors for the Internet of Things and Related Applications.
DOI: https://doi.org/10.1016/B978-0-12-815409-0.00018-8

plume. They sniff and trace the odor source according to the concentration distribution as well as the direction of odor current [6–10]. Humans have been considered to have a poor reputation in olfactory perception; however, more and more research proves that our noses hold the ultimate ability in odor discrimination [11] and scent-tracking [12].

Sensors with abilities to detect the spatiotemporal structure of odor plumes may have many important applications. For example, such kind of sensors can be loaded on mobile robots to imitate the odor-locating behavior of animals or insects, and thus used in the searching for toxic gas leak, hazardous chemicals, and pollutant sources [13–22]. The olfactory video camera, a sensor array system that helps locate a source of gas or odor, has been proposed by Nakamoto and Ishida to visualize the images of odor flow reaching the array from the source location. Large scale packed sensor arrays based on metal oxide semiconductor (MOS) or quartz crystal microbalance (QCM) have been used in the system [14,22]. Based on the mapping of the response signals, the odor flow is successfully visualized to provide the spatiotemporal odor information. However, the response and recovery time of MOS and QCM sensors are reported from several seconds to several minutes [23,24], which cannot catch up with the instantaneous and random variation of an odor plume in the environment. The visualization effect of the conventional sensor devices is thus limited due to the poor spatiotemporal resolution ability.

Recently, a great deal of attention has been paid to odor detection based on optical sensors. Zhang et al. constructed a cross-reactive chemiluminescence sensor array based on catalytical nanomaterial for the discrimination and identification of flavors and volatile organic compounds [25,26]. Walt et al. propose a fluorescent microbead-based optical fiber array for artificial olfaction, in which the high-density arrays of microbeads make the sensor array much smaller, more reproducible, and with significantly improved data processing ability compared with the conventional electronic nose [27–33]. Suslick and colleagues developed an optoelectronic nose based on colorimetric sensor array using chemoresponsive dyes [34–44]. The advantages of the sensor are rapid, sensitive, portable, and inexpensive for identification of a wide range of gases and vapors. Basically speaking, all of the optical sensors can convert the olfactory-like information into a visual output, that is, making the unseen odors visible, which has been referred to by Suslick and colleagues as "smell-seeing" [35,43]. However, all of the "smell-seeing" technologies put emphasis upon the visual discrimination of odors instead of the visualization of spatiotemporal distribution. A true meaning visualization sensor has been proposed by Mitsubayashi and colleagues, which is fabricated by enzyme catalyzed chemiluminescence reaction. The system is available for the gas components not only for spatial but also for temporal analysis. The sensor system has been applied in the visualization of alcohol metabolism from breath and skin as well as the distribution of alcohol released from wine in a wineglass [45–49].

This chapter introduces the sensor technologies developed by us to visualize both the odor space and the odor quality. The odors in air can only be perceived by our nose. However, the visualization of odor space can allow the unseen odor in the environment to be seen by our eyes. At the same time, the visualization of odor quality can convert the analogous odor perception into a digitalized form and thus help to the transformation and communication of odor information. Sensors with the ability to visualize both the odor quality and odor spatiotemporal distribution will explore new applications that cannot be achieved by conventional sensors.

17.2 Fluorescence imaging for odor visualization

17.2.1 Principle and system of fluorescence imaging

Fluorescence is a well-known phenomenon and has wide application in many fields as a detection probe. The emission spectrum of fluorescent probes is determined by its molecular character and influenced by environmental parameters. The odor visualization of fluorescence imaging is based on the interactions between fluorescent probes and odorant molecules. The possible interactions include solvent effect, polarity effect, pH effect, concentration effect, and fluorescence resonance energy transfer (FRET). All of these effects can be utilized in the odor visualization. Investigations by fluorescence spectrophotometer demonstrate that fluorescence quenching is observed for most odorants when they are mixed into the solution of fluorescent probes such as quinine sulfate and tryptophan (Fig. 17.1). In the case of

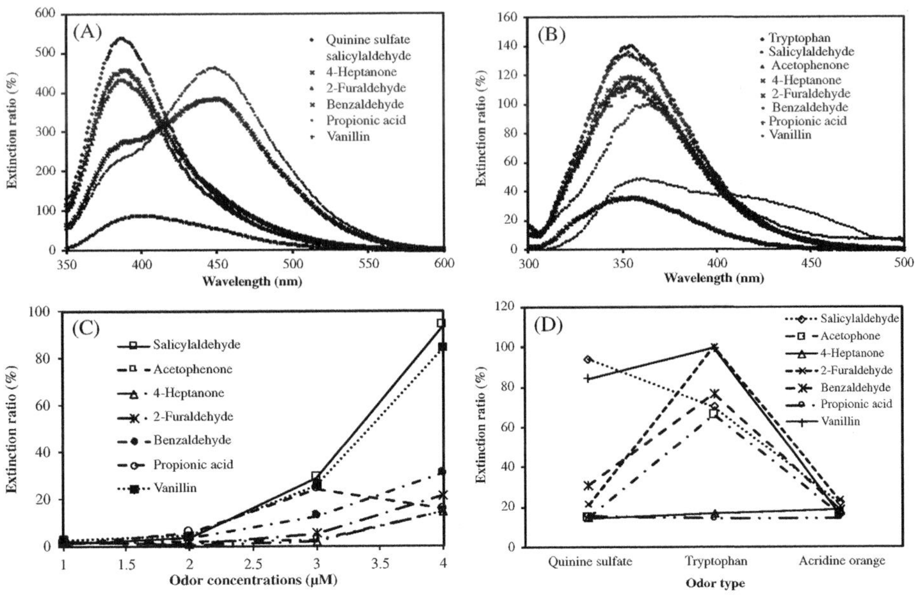

Figure 17.1 Fluorescence interactions between fluorophore and odorants. Odorant induced fluorescence intensity change of quinine sulfate (A) and tryptophan (B); odorant concentration dependent extinction of quinine sulfate (C), and response pattern shown by different fluorescent probes on various odorants (D).

quinine sulfate, an emission shift is shown in the reaction with propionic acid, which should be attributed to the pH-dependent fluorescence character of quinine sulfate. It is also found that the extinction degree is depended on the concentrations of both the probes and the odorants (Fig. 17.1A–C). In addition, the response patterns of the odorants with different probes are different from each other, indicating the possibility to discriminate the odorants by using fluorescence reactions.

17.2.2 Fabrication of the visualization system

The odor visualization system basically consists of an excitation light source, a fluorescent sensing film, and a high sensitive charge-coupled device (CCD) camera (Fig. 17.2A). A compact xenon light source combined with bandpass filters is used to provide excitation light with desired wavelengths. The CCD camera records the fluorescence change of the sensing film after reacted with odorants. The fluorescent sensing film is prepared by dissolving fluorescent dyes in agarose gel. This system can be used in two kinds of odor visualization: one is the visualization of odor flows injected onto the film surface; the other is the visualization of odor marks remained on substrate by odor exposure. The odor flows are provided by purging activated charcoal filtered, odorless air into a glass sample chamber containing odor source liquid chemicals. In the odor exposure experiment, the odor marks are prepared by using odor-soaked black papers with different shapes. The fluorescent sensing film is set above the odor-marked substrate with a distance of approximately 5 mm (we call this process "odor exposure"). After the exposure, changes of the fluorescence images are analyzed by an image processing software.

17.2.3 Visualization based on single fluorescent probe

Based on the fluorescence interactions between typical odorants and probes, we have tried the odor visualization by using quinine sulfate as fluorescent probes. Fig. 17.2B shows the images of two odor marks (acetophenone: AP and hexanoic acid: HA) with different shapes of triangle and round, separately. After the odor exposure with a time (90 seconds), the two shapes are clearly visualized. In addition, the two odorants show different fluorescence reactions: fluorescence quenching by AP and enhancing by HA [50,51]. Thus both the shape and type of odor marks can be visualized by the fluorescence imaging. The odor exposure is a novel approach to record invisible odorants remaining in the environment, which cannot be realized by using conventional sensor technologies.

One precondition for the factual record of the odor shape is that the diffusion process of odors is not influenced by wind and the air convection is too small to be negligible. However, the air convection always exists in real situations and the velocity of wind is far larger than the diffusion coefficient of odorant chemicals. Therefore the shape of odor source should be determined by the wind conditions. In Fig. 17.2B, the image of a round-shape odor source (HA) under a wind stream is demonstrated. During the process of odor exposure, a microfan (airflow 0.015 L/min) is set at one side of the channel to generate the airflow. Compared with the image

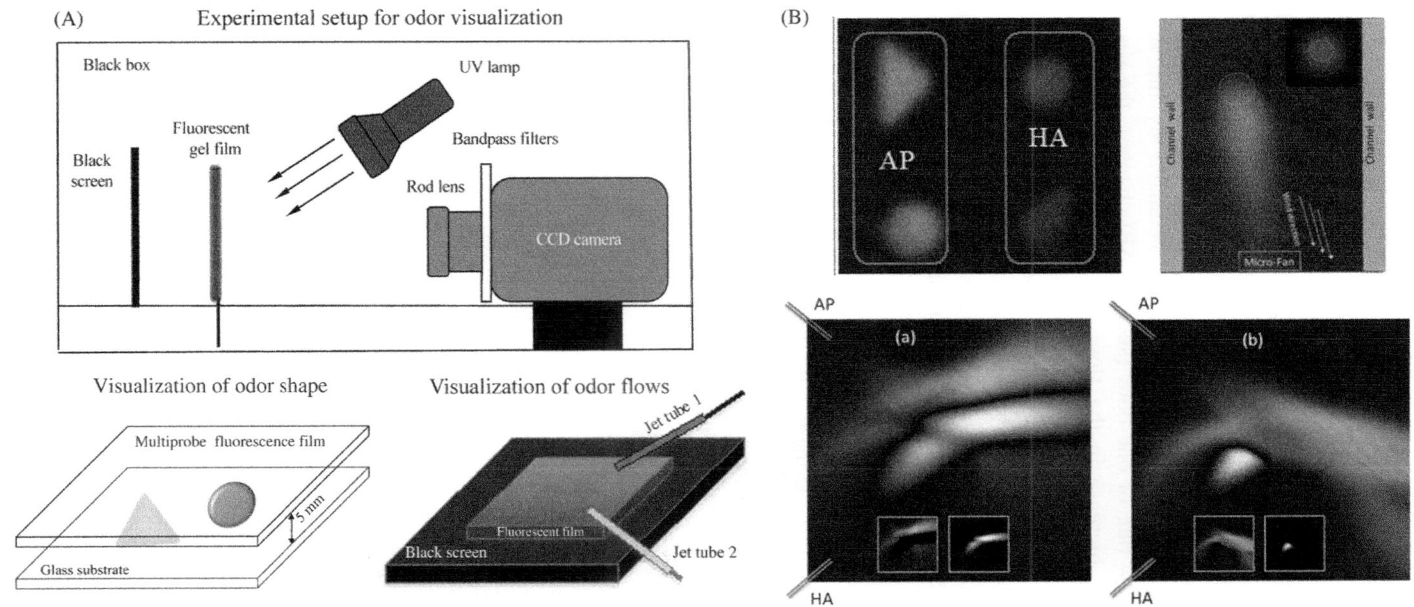

Figure 17.2 Experiment setup (A) used for the visualization of odor shape and odor flows (B).
Source: Reproduced with permission from C. Liu, R. Yokoyama, S. Uchisa, K. Nakano, K. Hayashi, Odor spatial distribution visualized by a fluorescent imaging sensor, IEEE Sensor (2013), 1506–1509 [50].

without airflow, a comet-shape image is visualized with the tail along the direction of the wind stream. Moreover, the initial point of the image deviates from the real position of the odor mark (indicated by a dash line), which agrees with the real diffusion situation of the odor plume in the wind channel.

Fig. 17.2B also shows the images of odor flows of AP and HA, injected onto the film surface at vertical angle. The red area represents the fluorescence enhanced by HA and the green area represents the fluorescence quenched by AP, separately. Both the images and the merged RGB images are presented. It can be seen that two domains with a clear boundary are successfully visualized by the fluorescence imaging. When the flow rate of one odor (such as AP) changes, the shape and the boundary also change. Increase in the flow rate of AP (from 0.6 to 1.2 L/min) greatly suppresses the spatial distribution of HA, and thus only a small red spot is visualized. Animation videos based on the sequence images of the two odor flows demonstrate that the shape and boundary of the two odor domains change dynamically in the scale of second, which in certain degree reflects the high spatiotemporal resolution of the developed fluorescence imaging sensor.

17.2.4 Visualization based on multispectral fluorescence imaging

Odors in the environment are generally mixtures of various chemical compounds. Therefore the discrimination ability of the single fluorescence probe is limited. To overcome this problem, we propose a multiple probe–based multispectral fluorescence imaging technique for the visualization of complicated odorant mixtures [52]. Multispectral imaging is a technique that can obtain images of an object simultaneously in a number of discrete spectra bands. It shows a high accuracy in target discrimination and identification due to its characteristics of multiple channel, narrow bandwidth, and large amount of information [53].

The multispectral fluorescence imaging is realized by mixing multiple fluorescent probes in the agarose gel to increase the discrimination ability. The mixing of fluorescent species will result in complicated fluorophore–fluorophore interactions, such as concentration or polarity-induced fluorescence quenching, FRET, the excimer-derived emission spectra alternation, and the spectral overlap-caused bleed through (crossover or crosstalk) [54]. Generally, the interactions of adjacent fluorophores may make the spectral and image analysis more complex. From another point of view, odorants can be discriminated in view of the multichannel characteristics of the sensing film caused by the complicated interactions such as odors-fluorophore and fluorophore–fluorophore (as shown in Fig. 17.3A). Fig. 17.3B shows the multispectral images of 12 odorant molecules under different wavelength bands. The result of principle component analysis on the images is presented in Fig. 17.3C. It can be seen that the 12 odorant molecules can be discriminated with separated distribution in the principal component (PC) space according to their molecular structures. For example, the molecules in cluster A (1-hexanol and 1-nonanol) belong to long chain alcohols. The molecules in cluster B (salicylaldehyde, acetophenone, hexanone, *o*-cresol, and anisole) hold similar aromatic structure (benzene ring). The molecules in cluster C (ethanol, phenethyl alcohol) belong to short chain or aromatic alcohol. The molecules of

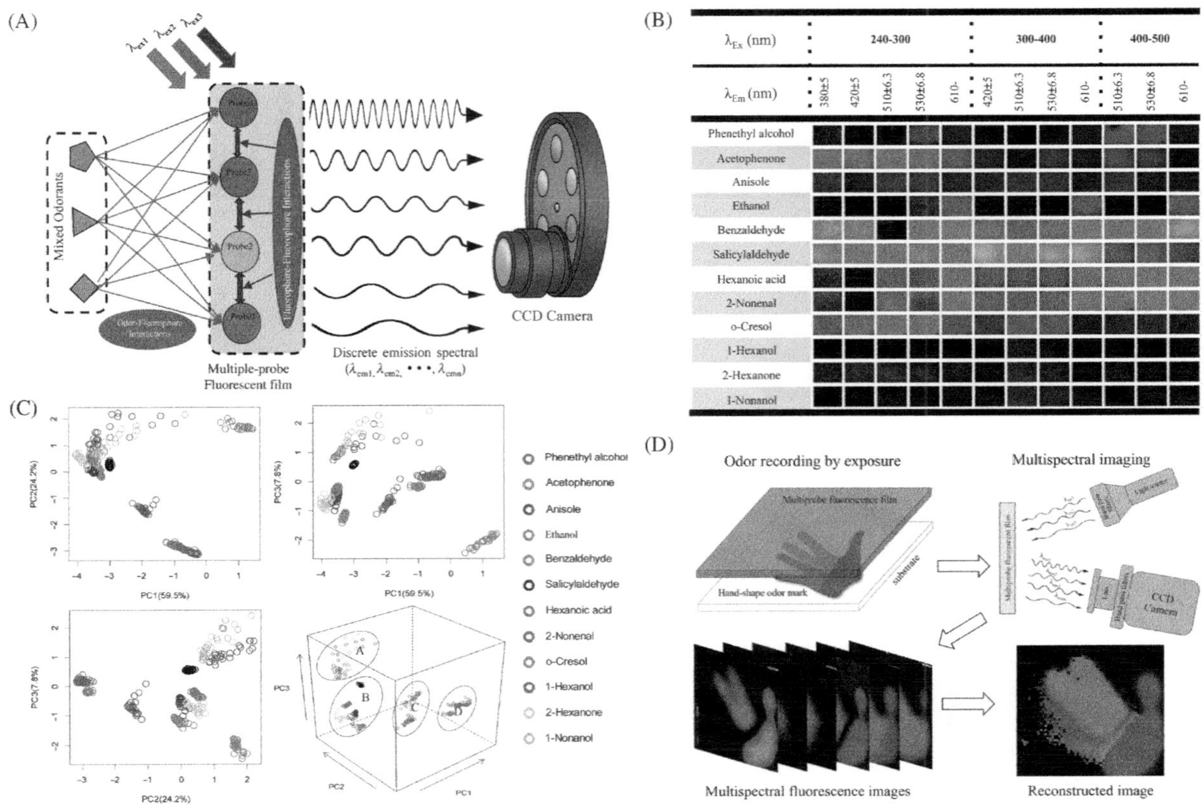

Figure 17.3 Schematic illustration of multispectral fluorescence imaging (A); multispectral images of 12 different odorants (B); discrimination of the 12 odorants by principal component analysis of the multispectral images (C); visualization of region segmentation of odor source with multiple components (D).

Source: Reproduced with permission from H. Yoshioka, C. Liu, K. Hayashi, Multispectral fluorescence imaging for odorant discrimination and visualization. Sens. Actuat. B-Chem., 220 (2015), 1297–1304 [52].

cluster D (hexanoic acid, 2-nonenal, and benzaldehyde) hold the similar carbonyl groups. This result demonstrates that the multispectral fluorescence imaging can be used as a powerful tool in the molecular structure-related odorant discrimination. Fig. 17.3D demonstrates the multispectral images of a hand-shape odorant mark containing three components: salicylaldehyde, hexanoic acid, and benzaldehyde. The discrete multispectral images are reconstructed by image analysis such as principal component analysis (PCA). It can be seen that both the shape and the segmented regions are visualized successfully.

Compared with the existing odor sensor techniques, the multispectral fluorescence imaging can be used not only in the odorant discrimination but also in the visualization of odorant spatial distribution. Therefore the invisible odors in our living environment can become visible by the imaging technique. This new technique has been applied for the visualization of human body odor [55], as well as the odor release from the fragrance inclusion complexes [56].

17.3 Localized surface plasmon resonance sensor for odorant visualization

In the real environment, the spread of an odor plume is patchy, intermittent, and time-variant [57]. The real-time capture of the odor plume requires sensors having rapid response and recovery ability. The images visualized by the fluorescent gel film are a time-averaged spatial distribution of the volatilized odorants in the environment [51]. Although the fluorescence imaging technique shows advantages in recording the shape and distribution of odorants traces via the exposure process, it is not able to real-time capture the spatial distribution of odorants due to the lack of sufficient time resolution. In view of this, we pay attention to odor sensors based on localized surface plasmon resonance (LSPR) phenomenon of metal nanoparticles (MNPs). LSPR is a well-known optical phenomenon of MNPs. The interaction between the incident light and surface electron of MNPs in conduction band results in coherent localized plasmon oscillation with a resonant frequency. The frequency is highly sensitive to the refractive index of environment medium, which can be used to fabricate gas or vapor sensors (Fig. 17.4A). Since the sensor response depends only on the change of refractive index of medium surrounding on the NPs, the LSPR sensors show a high-speed response and recovery, which make them very suitable for the real-time visualization of odor space.

AuNPs used in LSPR sensors are fabricated by vacuum deposition on glass substrate. It is well known that the LSPR spectral depends strongly on the structure of the NPs such as size, geometry, dielectric environment, as well as particle–particle separation distance of NPs (Fig. 17.4B). The performance of the LSPR sensors is thus influenced by the preparation conditions (deposition current and time, annealing times and temperatures) [58]. The rapid response character of the LSPR sensor is proved by the result shown in Fig. 17.4C. The times of response and recovery of the LSPR sensor on ethanol vapor are shown within a second, which are superior to

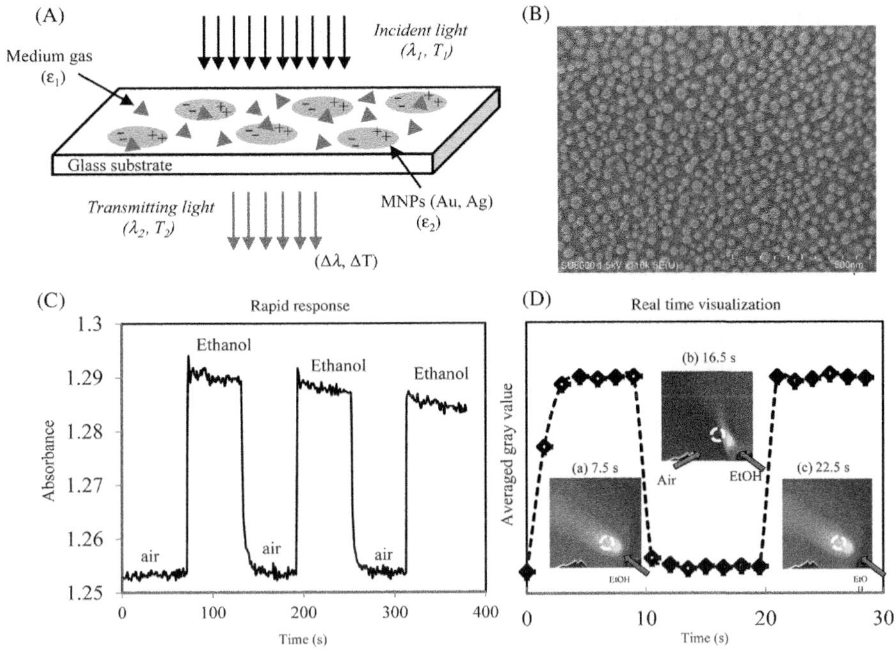

Figure 17.4 Illustrated mechanism of LSPR sensor (A); Structure of AuNPs used in the fabrication of LSPR sensors (B); Rapid response character of an LSPR sensor on ethanol vapor (C); A demonstrated visualization of odor flow by an LSPR sensor (D).

other sensor transducers such as MOS or QCM sensors [59]. The rapid response and recovery characteristics are utilized for the real-time visualization of vapor flows. As shown in a demonstration in Fig. 17.4D, an ethanol vapor is blown from the nozzle in the lower right of the LSPR substrate. Due to the higher refractive index of the ethanol than air, the vapor flow can be visualized by a CCD camera [60]. When air is blown from the lower left of the substrate, the deviation of the ethanol flow is observed immediately. The visualized images agree with the change of gray value recorded (area marked with white circle), which proves the real-time character of the LSPR sensor in the visualization of vapors and odors.

The characteristic of rapid response and recovery makes the LSPR sensor suitable for high-speed odor sensing purposes. The next section describes application of such high-speed plasmonic sensors for robotics.

17.4 Collecting spatial odor information from on-ground odor sources with a robot system

In contrast to acoustic waves or light, information from a chemical substance is recorded as an attached substance on the place where an event has happened.

Here, identification of things or events with volatile chemical substances (odor information) has great potential from the view point of noncontact, nondestructive, and remote detection. To utilize the odor information, currently we need to depend on people or animals who underwent special training in most cases. In many commercial applications or disaster response scenes, a dog's olfactory system is still the best odor detection system. Digitizing the creature's olfactory system is a great challenge for science and industry and can dramatically reduce the large cost and amount of time needed for maintenance and training. If we have a sensor robot with high ability for odor identification, we could access and utilize the odor information throughout the world.

In the past, much research on odor sensing robots has been carried out [15,61–63]. Ishida's group demonstrated a gas plume tracing robot for finding the source of odor drifting in the atmosphere [64].

Although this research represents a big achievement in the engineering of odor, this system could not obtain the information of a past event due to the nature of the plume as sensing target that was mobilized moment by moment via the wind. Developing usage of "odor trace information" such as footprint tracking or touch history or odor trace communication, which are utilized by animals [65,66] but not utilized as engineering technology, is a frontier of this field.

We developed an odor sensor robot that can collect odor information from on-ground odor sources such as odor traces or spatially corded odor information. The detection target of this robot is not a gas plume but volatile compound attached on the floor or solid surface [67]. The concept and a photograph of the robot are shown in Fig. 17.5A. The robot was sampling gases from on-ground odor sources by a sensor probe tip that was placed close to the floor surface. Here, high-speed gas sensor is required for obtaining geometrical information for the odor trace with high spatial resolution. We developed a high-speed gas sensor module based on LSPR [58,59] for the robot instead of a semiconductor gas sensor, which is usually used by gas sensor robots (Fig. 17.5B). The module consisted of LSPR substrate based on AuNPs, LED light source (peak wave length = 630 nm), photo diode, suction pump, and fine stainless tubing for gas sampling with 0.5 mm opening that is 1 mm separated from detecting surface. Total mass and dimensions of the module are 50 g and 30 mm $\times$ 60 mm $\times$ 42 mm respectively. Response time and recovery time of this sensor module show less than 50 ms (25 Hz) in experiment with ethanol gas (Fig. 17.5C). It is around 100 times shorter than typical commercial semiconductor gas sensors commonly used. The robot equipped with the gas sensor module can readout an on ground odor trace with 2.5 cm special resolution.

As a demonstration of advanced odor trace reading performance of the developed robot, we performed a reading of digital odor trace information coded on the floor by alignment of an odor marker to represent the symbol 0 or 1. The obtained data was decoded to ASCII code as O, D, O, R (Fig. 17.5D). In this experiment ethanol was used as odorant as clean and biocompatible and rewritable volatile substance. This way provides a new asynchronous communication pathway to multirobot systems that works as a digitized pheromone.

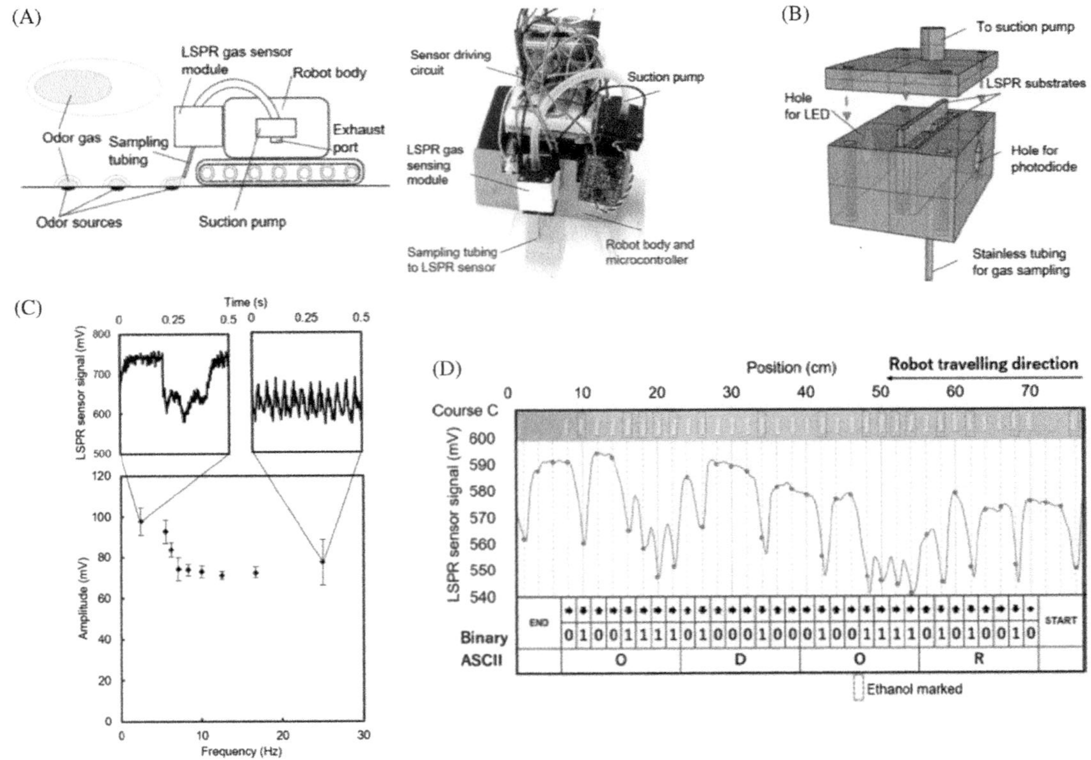

Figure 17.5 A robot system for collecting spatial odor information from on-ground odor sources. (A) Concept and photograph of the sensor robot. (B) LSPR based high-speed gas sensor module. (C) Frequency response of the LSPR gas sensing module to ethanol gas signal ($n = 5$). (D) Demonstration of the reading of digital odor trace information with the robot.

Reproduced with permission from Z. Yang, F. Sassa, K. Hayashi, A robot equipped with a high-speed LSPR gas sensor module for collecting spatial odor information from on-ground invisible odor sources, ACS Sens. (2018), doi:10.1021/acssensors.8b00214 [67]. Copyright 2018 American Chemical Society.

This result shows a developed sensor robot system enabled to the collection of information from volatile substances attached in the past on a site in a systematic manner. This concept could bring a new kind of application combined with emerging robot technology to gas sensor research.

17.5 Visual odor representation of a volatile molecular based on chemical property by network diagram

As described above, much research has achieved a variety of ways to obtain odor information with the advancement of sensing techniques and applications. On the other hand, even if gas sensing systems provide much odor information (e.g., chemical species or amount of the substance) to humans, intuitive understanding of the huge and complicated odor information [11,68] is difficult. One of the reasons is that a way of visual expression for odor has not been established yet.

We are developing a method to make an intuitive visual representation of odor based on the relationship of molecule parameters, which shows a unique chemical feature of a molecule such as boiling point or pH. Especially, we introduce a project that uses a network diagram to show odor feature, data on human feelings about an odor (odor descriptors: sweet, green, etc. [69,70]), and relationship between other odorants at once.

To make the network diagram, we chose 321 typical odor molecules and used its 1006 molecule parameters. A network diagram of odor molecules using 1006 of all molecular parameters based on cosine similarity as a distance between molecules is shown in Fig. 17.6A. Each node shows each molecule, and a green-colored node represents a molecule that was related to "green" odor descriptors. Each node is connected at the edge if cosine similarity of these two nodes is larger than 0.98.

Each odor descriptor has a unique set of molecular parameters that has strong correlation to the odor descriptor [70]. Fig. 17.6B shows the network around a typical odor molecule generated with a set of molecular parameters correlated to odor descriptors of sweet, green, fruity, and herbaceous respectively. Each network shows that the focused molecule has a neighbor molecule that is classified to same odor descriptor and it can help to estimate the odor of the focused molecular.

We performed quantitative evaluation to the generated odor network by using network parameters used in graph theory. A result with evaluation by the number of cut edges between a set of nodes that are classified to a focused odor descriptor and a set consisting of other nodes, which indicate degree of clustering, is shown in Fig. 17.6C. It shows ratio of the cut number of a network generated with selected molecule parameter and cut number of a network generated with all 1006 molecule parameters. This result shows that the network generated with parameters related to the odor descriptors sweet and fruity are more clustered compared with the one generated with all 1006 parameters. On the other hand, it did not work in the case of odor descriptors green and herbaceous. The other network parameters stress and closeness centrality were also used for evaluation. However, making a strict model

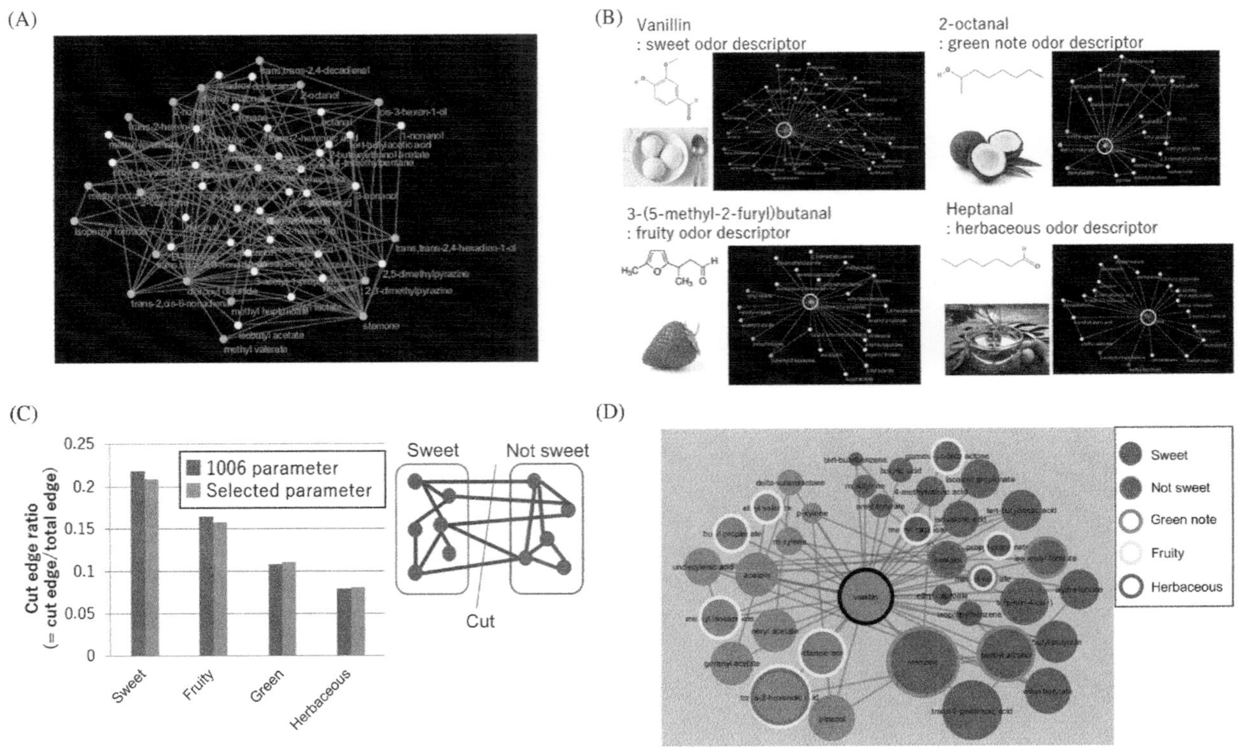

Figure 17.6 Visual odor representation of a volatile molecule. (A) A network diagram of odor molecules using all 1006 molecular parameters based on cosine similarity as a distance between molecules. Green node represent molecule related to "green node" odor descriptors. (B) Network diagrams around a typical odor molecule generated with a set of molecular parameters correlated to odor descriptor. (C) Ratio of the cut number of a network generated with selected molecule parameter and cut number of a network generated with 1006 of all molecule parameters. (D) Visual representation of vanillin. A node highlighted with black circle on center shows vanillin molecule.

for evaluation of a parameter's effect on efficiency of the visual representation is difficult. So, we defined a figure of merit (FOM) as a testbed for evaluation. The FOM values of network generated with parameters related with the odor descriptor sweet and with all 1006 parameters and with a random 200 parameters are 3.8, 3.4, and 3.3 respectively. Although the difference of each FOM values and the corresponding visual representations are not significant yet, this approach could be a way to improve the efficiency of visualization of odor information. Consequently, visual representation of vanillin is realized in Fig. 17.6D; node size shows closeness centrality as importance of the network.

References

[1] O.O.o.T.a.E. U.S. Department of Labor, Presenting Effective Presentations With Visual Aids, May 1996. <http://www.rufwork.com/110/mats/oshaVisualAids.html>, (accessed 28.02.19).

[2] S. Firestein, What the nose knows—the science of scent in everyday life, Science 321 (2008) 774.

[3] R.C. Araneda, A.D. Kini, S. Firestein, The molecular receptive range of an odorant receptor, Nat. Neurosci. 3 (2000) 1248–1255.

[4] H. Yahya, The Miracles of Smell and Taste, 1999. <https://www.researchgate.net/publication/320616479_The_Miracles_of_Smell_and_Taste>, (accessed 28.02.19).

[5] M.A. Reidenbach, M.A.R. Koehl, The spatial and temporal patterns of odors sampled by lobsters and crabs in a turbulent plume, Integr. Comp. Biol. 49 (2009) E141.

[6] J.L. DeBose, G.A. Nevitt, The use of odors at different spatial scales: comparing birds with fish, J. Chem. Ecol. 34 (2008) 867–881.

[7] Q. Gaudry, K.I. Nagel, R.I. Wilson, Smelling on the fly: sensory cues and strategies for olfactory navigation in Drosophila, Curr. Opin. Neurobiol. 22 (2012) 216–222.

[8] P. Moore, J. Crimaldi, Odor landscapes and animal behavior: tracking odor plumes in different physical worlds, J. Marine Syst. 49 (2004) 55–64.

[9] M.J. Weissburg, M.C. Ferner, D.P. Pisut, D.L. Smee, Ecological consequences of chemically mediated prey perception, J. Chem. Ecol. 28 (2002) 1953–1970.

[10] J.P. Martin, J.G. Hildebrand, Innate recognition of pheromone and food odors in moths: a common mechanism in the antennal lobe? Front. Behav. Neurosci. 4 (2010).

[11] C. Bushdid, M.O. Magnasco, L.B. Vosshall, A. Keller, Humans can discriminate more than 1 trillion olfactory stimuli, Science 343 (2014) 1370–1372.

[12] J. Porter, B. Craven, R.M. Khan, S.J. Chang, I. Kang, B. Judkewicz, et al., Mechanisms of scent-tracking in humans, Nat. Neurosci. 10 (2007) 27–29.

[13] A. Gomez-Marin, B.J. Duistermars, M.A. Frye, M. Louis, Mechanisms of odor-tracking: multiple sensors for enhanced perception and behavior, Front. Cell. Neurosci. 4 (2010).

[14] H. Ishida, T. Tokuhiro, T. Nakamoto, T. Moriizumi, Improvement of olfactory video camera: gas/odor flow visualization system, Sens. Actuat. B-Chem. 83 (2002) 256–261.

[15] H. Ishida, Y. Wada, H. Matsukura, Chemical sensing in robotic applications: a review, IEEE. Sens. J. 12 (2012) 3163–3173.

[16] K. Song, Q. Liu, Q. Wang, Olfaction and hearing based mobile robot navigation for odor/sound source search, Sensors 11 (2011) 2129–2154.

[17] T. Asada, Y. Takei, H. Nanto, T. Oyabu, Y. Iwasaki, Y. Yoshie, et al., Mobile robot with early fire detection system using odor sensor array, Chem. Senses 33 (2008) J25.

[18] P. Batog, A. Wolczowski, Odor markers detection system for mobile robot navigation, 26th Eur. Conf. Solid-State Transducers Eurosensor 2012 47 (2012) 1442–1445.

[19] H. Ishida, T. Nakamoto, T. Moriizumi, Remote sensing of gas/odor source location and concentration distribution using mobile system, Sens. Actuat. B-Chem. 49 (1998) 52–57.

[20] H. Ishida, H. Tanaka, H. Taniguchi, T. Moriizumi, Mobile robot navigation using vision and olfaction to search for a gas/odor source, Auton. Robots 20 (2006) 231–238.

[21] J.G. Li, Q.H. Meng, Y. Wang, M. Zeng, Odor source localization using a mobile robot in outdoor airflow environments with a particle filter algorithm, Auton. Robots 30 (2011) 281–292.

[22] H. Ishida, T. Yamanaka, N. Kushida, T. Nakamoto, T. Moriizumi, Study of real-time visualization of gas/odor flow image using gas sensor array, Sens. Actuat. B-Chem. 65 (2000) 14–16.

[23] K.J. Albert, N.S. Lewis, C.L. Schauer, G.A. Sotzing, S.E. Stitzel, T.P. Vaid, et al., Cross-reactive chemical sensor arrays, Chem. Rev. 100 (2000) 2595–2626.

[24] E.M.K. Arshak, G.M. Lyons, J. Harris, S. Clifford, A review of gas sensors employed in electronic nose applications, Sens. Rev. 24 (2004).

[25] N. Na, S. Zhang, S. Wang, X. Zhang, A catalytic nanomaterial-based optical chemosensor array, J. Am. Chem. Soc. 128 (2006) 14420–14421.

[26] D. Liu, M.Y. Liu, G.H. Liu, S.C. Zhang, Y.Y. Wu, X.R. Zhang, Dual-channel sensing of volatile organic compounds with semiconducting nanoparticles, Anal. Chem. 82 (2010) 66–68.

[27] T.A. Dickinson, K.L. Michael, J.S. Kauer, D.R. Walt, Convergent, self-encoded bead sensor arrays in the design of an artificial nose, Anal. Chem. 71 (1999) 2192–2198.

[28] S.E. Stitzel, K.J. Albert, D.R. Walt, Organic vapor detection with fiber-optic bead arrays, Adv. Mater. Opt. Syst. Chem. Biol. Det. 3858 (1999) 92–97.

[29] K.J. Albert, D.R. Walt, D.S. Gill, T.C. Pearce, Optical multibead arrays for simple and complex odor discrimination, Anal. Chem. 73 (2001) 2501–2508.

[30] C.L. Schauer, S.E. Stitzel, D.R. Walt, Cross-reactive optical sensing arrays, Chem. Taste: Mech. Behav. Mimics 825 (2002) 318–329.

[31] D.R. Walt, Imaging optical sensor arrays, Curr. Opin. Chem. Biol. 6 (2002) 689–695.

[32] K.L. Brogan, D.R. Walt, Optical fiber-based sensors: application to chemical biology, Curr. Opin. Chem. Biol. 9 (2005) 494–500.

[33] D.R. Walt, Bead-based optical fiber arrays for artificial olfaction, Curr. Opin. Chem. Biol. 14 (2010) 767–770.

[34] N.A. Rakow, K.S. Suslick, A colorimetric sensor array for odour visualization, Nature 406 (2000) 710–713.

[35] K.S. Suslick, N.A. Rakow, A colorimetric nose: "Smell-seeing", Artif. Chem. Sens. Olfact. Electr. Nose (Isoen 2001) 2001 (2001) 8–13.

[36] K.S. Suslick, N.A. Rakow, A. Sen, Colorimetric sensor arrays for molecular recognition, Tetrahedron 60 (2004) 11133–11138.

[37] N.A. Rakow, A. Sen, M.C. Janzen, J.B. Ponder, K.S. Suslick, Molecular recognition and discrimination of amines with a colorimetric array, Angew. Chem. Int. Ed. 44 (2005) 4528–4532.

[38] C. Zhang, K.S. Suslick, A colorimetric sensor array for organics in water, J. Am. Chem. Soc. 127 (2005) 11548–11549.
[39] M.C. Janzen, J.B. Ponder, D.P. Bailey, C.K. Ingison, K.S. Suslick, Colorimetric sensor Arrays for volatile organic compounds, Anal. Chem. 78 (2006) 3591–3600.
[40] S.H. Lim, L. Feng, J.W. Kemling, C.J. Musto, K.S. Suslick, An optoelectronic nose for the detection of toxic gases, Nat. Chem. 1 (2009) 562–567.
[41] B.A. Suslick, L. Feng, K.S. Suslick, Discrimination of complex mixtures by a colorimetric sensor array: coffee aromas, Anal. Chem. 82 (2010) 2067–2073.
[42] J.W. Kemling, K.S. Suslick, Nanoscale porosity in pigments for chemical sensing, Nanoscale 3 (2011) 1971–1973.
[43] K.S. Suslick, Synesthesia in science and technology: more than making the unseen visible, Curr. Opin. Chem. Biol. 16 (2012) 557–563.
[44] J.R. Askim, M. Mahmoudi, K.S. Suslick, Optical sensor arrays for chemical sensing: the optoelectronic nose, Chem. Soc. Rev. 42 (2013) 8649–8682.
[45] X. Wang, E. Ando, D. Takahashi, T. Arakawa, H. Kudo, H. Saito, et al., 2D spatiotemporal visualization system of expired gaseous ethanol after oral administration for real-time illustrated analysis of alcohol metabolism, Talanta 82 (2010) 892–898.
[46] X. Wang, E. Ando, D. Takahashi, T. Arakawa, H. Kudo, H. Saito, et al., Non-invasive spatial visualization system of exhaled ethanol for real-time analysis of ALDH2 related alcohol metabolism, Analyst 136 (2011) 3680–3685.
[47] T. Arakawa, E. Ando, X. Wang, M. Kumiko, H. Kudo, H. Saito, et al., A highly sensitive and temporal visualization system for gaseous ethanol with chemiluminescence enhancer, Luminescence 27 (2012) 328–333.
[48] T. Arakawa, X. Wang, K. Kita, K. Miyajima, H. Kudo, K. Mitsubayashi, Chemiluminescent visualization system for evaluation of alcohol metabolism based on transdermal emission of gaseous ethanol, in: 2012 IEEE Sensors Proceedings, 2012, pp. 235–238.
[49] T. Arakawa, X. Wang, T. Kajiro, K. Miyajima, S. Takeuchi, H. Kudo, et al., A direct gaseous ethanol imaging system for analysis of alcohol metabolism from exhaled breath, Sens. Actuat. B-Chem. 186 (2013) 27–33.
[50] C. Liu, R. Yokoyama, S. Uchisa, K. Nakano, K. Hayashi, Odor spatial distribution visualized by a fluorescent imaging sensor, IEEE Sensor (2013) 1506–1509.
[51] H. Matsuo, Y. Furusawa, M. Imanishi, S. Uchida, K. Hayash, Optical odor imaging by fluorescence probes, J. Robot. Mechat. 24 (1) (2012) 8.
[52] H. Yoshioka, C. Liu, K. Hayashi, Multispectral fluorescence imaging for odorant discrimination and visualization, Sens. Actuat. B-Chem. 220 (2015) 1297–1304.
[53] C.-I. Chang, Hyperspectral Data Exploitation: Theory and Applications, John Wiley & Sons, 2007.
[54] M.T.Z. Spence, I.D. Johnson, The Molecular Probes Handbook: A Guide to Fluorescent Probes and Labeling Technologies, Life Technologies Corporation, 2010.
[55] C. Liu, Y. Furusawa, K. Hayashi, Development of a fluorescent imaging sensor for the detection of human body sweat odor, Sens. Actuat. B-Chem. 183 (2013) 117–123.
[56] C. Liu, K. Hayashi, Visualization of controlled fragrance release from cyclodextrin inclusion complexes by fluorescence imaging, Flav. Fragr. J. 29 (2014) 356–363.
[57] A. Marjovi, L. Marques, Swarm robotic tracking for intermittent and time-variant odor dispersion, in: Proceedings of the IEEE European Conference on Mobile Robots (ECMR), Barcelona, Spain, 2013, pp. 379–384.

[58] B. Chen, M. Mokume, C. Liu, K. Hayashi, Structure and localized surface plasmon tuning of sputtered Au nano-islands through thermal annealing, Vacuum 110 (2014) 94–101.

[59] B. Chen, M. Ota, M. Mokume, C. Liu, K. Hayashi, High-speed gas sensing using localized surface plasmon resonance of sputtered noble metal nanoparticles, IEEJ Trans. Sens. Micromach. 133 (2013) 90–95.

[60] T. Koga, H. Yoshioka, C. Liu, K. Hayashi, Gas visualization based on localized surface plasmon resonance of Au nanoparticle film, in: Proceedings of IEEE sensors, 2015, pp. 72–55.

[61] A.J. Lilienthal, A. Loutfi, T. Duckett, Airborne chemical sensing with mobile robots, Sensors 6 (2006) 1616–1678.

[62] A. Loutfi, S. Coradeschi, Smell, think and act: a cognitive robot discriminating odours, Auton. Robots 20 (2006) 239–249.

[63] J.A. Farrell, S. Pang, W. Li, Chemical plume tracing via an autonomous underwater vehicle, IEEE J. Ocean. Eng. 30 (2005) 428–442.

[64] H. Ishida, G. Nakayama, T. Nakamoto, T. Moriizumi, Controlling a gas/odor plume-tracking robot based on transient responses of gas sensors, IEEE. Sens. J. 5 (2005) 537–545.

[65] R.E. Johnston, Chemical communication in rodents: from pheromones to individual recognition, J. Mammal. 84 (2003) 1141–1162.

[66] L. Amo, I. Galván, G. Tomás, J.J. Sanz, Predator odour recognition and avoidance in a songbird, Funct. Ecol. 22 (2008) 289–293.

[67] Z. Yang, F. Sassa, K. Hayashi, A robot equipped with a high-speed LSPR gas sensor module for collecting spatial odor information from on-ground invisible odor sources, ACS Sens. (2018). Available from: https://doi.org/10.1021/acssensors.8b00214.

[68] Markus Meister, On the dimensionality of odor space, eLife 4 (2015) e07865.

[69] Sigma-Aldrich. Flavors and Fragrances. Natural and Food Grade Ingredients, 2016. <http://www.sigmaaldrich.com/industries/flavors-and-fragrances.html>, (accessed 06.10.18).

[70] L. Shang, C. Liu, Y. Tomiura, K. Hayashi, Machine-learning based olfactometer: prediction of odor perception from physicochemical features of odorant, Anal. Chem. 89 (2017) 11999–12005.

Bio-sniffer and sniff-cam

18

Takahiro Arakawa, Koji Toma and Kohji Mitsubayashi
Department of Biomedical Devices and Instrumentation, Institute of Biomaterials and Bioengineering, Tokyo Medical and Dental University, Tokyo, Japan

18.1 Introduction: breath and skin gas analysis

In the last decade, measurement of gaseous components from the human body has played an important role in a noninvasive, painless, and attractive diagnostic method of monitoring disease states without risks of blood collection to patients. Besides, several volatile organic compounds (VOCs) in breath have been identified as biomarkers of some diseases. The exhalation diagnosis and human body odor were reported to be helpful in diagnosis of diseases by monitoring for VOCs [1–5]. For example, nitric oxide is well known as one of the important indicators for asthma [6,7]; a higher concentration of breath acetone, which is related to lipid metabolism, was observed in diabetes mellitus patient [8–12]; and breath isopropanol (IPA) concentration related to breast cancer [13]. Searching biomarkers for screening of cancers in an early stage is always an important issue in health care [3,14–16].

Hence, instant and simple methods are strongly desired to discriminate VOCs from the human body, breath, skin, and so on [5,17]. However, gas chromatography methods are time consuming, expensive, and require large-scale equipment. It is difficult for low selective semiconductor type gas sensor to arrest density and a change in the space precisely. Therefore, a continuous and easy monitoring system for various gas components' behavior in spatial localization is strongly required. Some technology has been developed for gas sensors employing enzymatic reactions, such as biochemical gas sensors for ethanol, acetaldehyde, formaldehyde, and various VOCs and a reduced nicotinamide adenine dinucleotide (NADH)-dependent fiber-optic biosensor "bio-sniffer" for determination of gaseous components [18–24]. Also, the enzyme-based biosensors are highly selective and sensitive for target chemicals.

We introduce a biochemical gas sensor "bio-sniffer" by exploiting biochemical reactions and characteristics of various enzymes. Firstly, an optical bio-sniffer (gas sensor) for acetone was applied to measure lipid metabolism in breath [25,26]. A novel bio-sniffer for continuous monitoring of gaseous acetone by the reverse catalytic reaction of secondary alcohol dehydrogenase (S-ADH) was developed.

Chemical, Gas, and Biosensors for the Internet of Things and Related Applications.
DOI: https://doi.org/10.1016/B978-0-12-815409-0.00019-X

When S-ADH catalyzes the reduction of acetone in the presence of NADH, which acts as an electron donor, 2-propanol and oxidized nicotinamide adenine dinucleotide (NAD^+) are produced. This study aims to describe the biosensor design, characteristics, and validation through analysis of breath acetone concentration in volunteer subjects at rest and during exercise according to the lipid metabolism. Secondary, we construct a bio-sniffer for an IPA vapor measurement using a S-ADH immobilized membrane, a high-power UV-LED, a high sensitivity PMT, and an optical fiber with flow-cell [27,28]. We investigate and optimize the factors that may interfere the performance of the bio-sniffer, and apply the bio-sniffer to analyze the IPA concentration in the exhaled breath supplying from healthy subjects [29]. Finally, a two-dimensional visualization system of gaseous ethanol was demonstrated in spatial and temporal imaging of gaseous components in exhaled breath for a noninvasive diagnostic method of monitoring disease states without risk to patients [30–34]. We applied alcohol oxidase (AOD), which catalyzes low molecular weight alcohols by molecular oxygen into aldehydes with the production of hydrogen peroxide, which can lead to the HRP-luminol-H_2O_2 system for chemiluminescence analysis. The chemiluminescence generated by the catalytic reaction of gaseous ethanol was analyzed. In addition, we constructed a visualization system that detects the intrinsic fluorescence of coenzymes generated from alcohol dehydrogenase reaction and images the spatiotemporal distribution of the concentration of ethanol vapor as well [35,36]. We conducted imaging of a body gas, sampling postalcohol consumption breath and transdermal vapor from the human palm.

18.1.1 Construction of bio-sniffer

Many enzymatic bio-sniffer devices were developed for VOCs measurement. The bio-sniffer consists of a UV-LED-based excitation system, a photomultiplier tube (PMT, C9692, Hamamatsu Corporation, Japan), and an optical fiber probe with a flow-cell. Prior to the development of the bio-sniffer, the optical system for NADH determination was constructed. To enhance the sensitivity, the custom-fabricated LED excitation system was employed [21,37–39]. The excitation system utilizes a 335-nm wavelength UV-LED (UVTOP BL335) with a custom-made UV-LED power supply circuit. The LED and the PMT detector were connected to each separated end of a bifurcated optical fiber assembly, respectively. The optical fiber probe was connected to the common end of the optical fiber assembly. To reduce the background signal, two band-pass filters (BPF; $\lambda_{ex} = 340 \pm 10$ nm and $\lambda_{fl} = 490 \pm 10$ nm) were used in excitation and detection, respectively.

The fluorescence response of NADH by enzymatic reactions was investigated using the NADH fluorescence measurement system. The typical fluorescence response to the change in NADH concentration has been shown [26,27]. The fluorescence intensity (Δ intensity) at 490 nm wavelength was decreased as the NADH concentration changed. When phosphate buffer solution (P)B was dropped into the NADH/PB solution, the NADH solution was diluted, resulting in the prompt decrease of NADH fluorescence [26]. A sigmoidal calibration curve for measurement of

NADH was obtained from 100 nmol/L to 10 mmol/L. It is unable to determine a higher NADH concentration than 10 mmol/L because the Δintensity is saturated. Considering our previous works in which the NADH system allowed one to detect increased NADH fluorescence, the results prove that the system is capable to trace any change in NADH concentration with high responsiveness. This ability of the system makes it suitable to be used for evaluating NADH-dependent S-ADH immobilized biosensor. In the next section, we introduce an acetone bio-sniffer using decreasing NADH fluorescent measurement system.

18.1.2 Acetone bio-sniffer

Acetone is a typical VOCs in exhaled breath. It is affected by fasting, exercise, and diabetes mellitus. Biochemical pathways of acetone production have been studied and reported [40–42]. Acetone is metabolized by two pathways: decarboxylation of acetoacetate or dehydrogenation of 2-propanol. Acetoacetate is generated by dextrose metabolism and lipolysis, and is a major source of acetone in the human body. Acetone concentrations in exhaled breath have previously been shown to correlate strongly with acetone concentrations in blood, as well as with other ketones such as beta-hydroxybutyrate. A correlation between the acetone concentration in exhaled breath and the blood glucose concentration has also been reported. The acetone concentration in exhaled breath from healthy people ranges from 200 to 900 ppb in previous research [40,43]. Increases in the breath acetone concentration are usually a sign that cells lack insulin or are unable to effectively use available insulin; this occurs in diabetes. The breath acetone concentrations of diabetic patients are higher than those of healthy people, and can exceed 900 ppb. For this reason, acetone in exhaled breath is a potential biomarker for a noninvasive diagnosis of diabetes.

A rapid, convenient, noninvasive, and highly selective method for gaseous acetone detection was developed. Secondary alcohol dehydrogenase (S-ADH) catalyzes the reduction of acetone in the presence of NADH, which acts as an electron donor. 2-propanol and NAD^+ are produced through the following enzymatic reaction:

$$\text{Acetone} + \text{NADH} \xrightleftharpoons{S\text{-ADH}} \text{2-propanol} + \text{NAD}^+ \tag{18.1}$$

Optical methods are also commonly used for NADH detection. It is known that NADH can be excited by 340 nm ultraviolet radiation and it has a fluorescent emission peak at 490 nm, while NAD^+ has no fluorescent emission or optical properties.

Fig. 18.1 shows the experimental setup for monitoring of standard acetone vapor. In this system, the acetone vapor was supplied from the standard gas generator. The flow-rate of the supplied gas was controlled using a mass flow controller with a needle-bulb regulator. The NADH-containing PB was circulated into the flow-cell by an intelligent pump. S-ADH loses its specific activity if chemical conditions are not adequate, such as in dry conditions, and at an inappropriate temperature or pH.

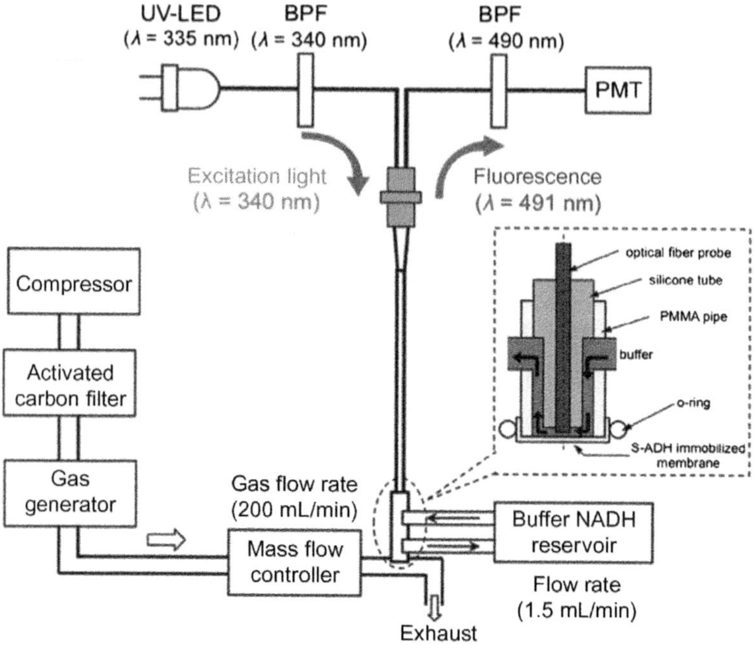

Figure 18.1 Experimental setup for characterization of the fiber-optic acetone bio-sniffer. The bio-sniffer measures acetone vapor as fluorescence of NADH, which is consumed at the enzyme membrane [12]. Biosensors and Bioelectronics permitted.

The PB flow kept the enzyme membrane wet, supplying NADH and also efficient removing of the reaction products (including NAD^+) and excessive substrates. Hence, reaction conditions were optimized by the circulating PB.

The fluorescent response to acetone in the gas phase was investigated using the standard gas measurement system. Fig. 18.2A shows typical responses of fluorescence intensity (Δintensity) to various concentrations of gaseous acetone. When the sensing region was exposed to acetone vapor, the fluorescence intensity decreased and then reached steady state value immediately. The fluorescence intensity returned to the initial state upon cessation of acetone vapor flow. These responses indicate that NADH decreases at the enzyme membrane as it is consumed through the catalytic reaction involving S-ADH and acetone vapor, and immediately removed from the sensing region by the buffer flow. The calibration curve for acetone vapor is shown in Fig. 18.2B. The dynamic range was from 20 to 5300 ppb, which includes breath acetone concentrations of healthy (200–900 ppb) and diabetic ($>$ 900 ppb) individuals. The fluorescent output of the bio-sniffer also demonstrated high reproducibility. For acetone vapor at 500 ppb, a coefficient of variation of 2.6% ($n = 5$) was obtained. These values provide evidence supporting the use of the bio-sniffer to measure acetone in exhaled human breath.

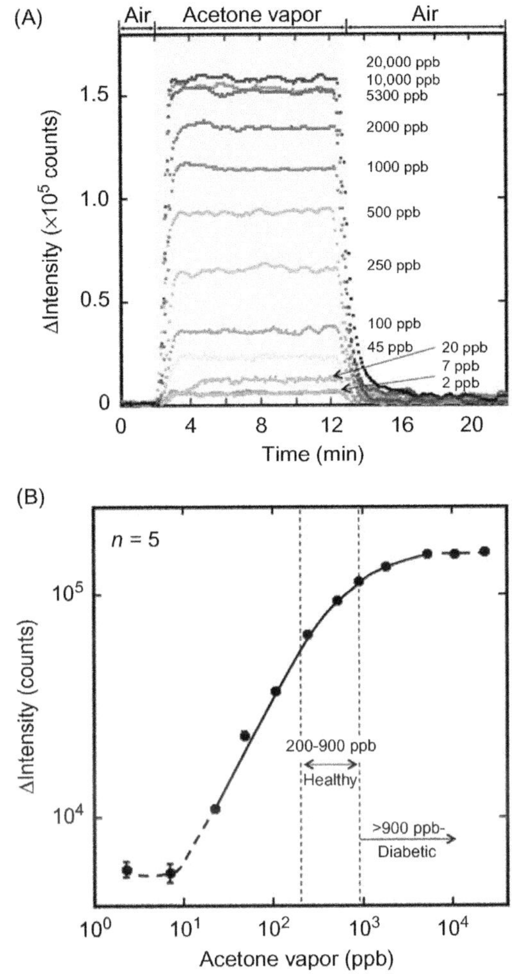

Figure 18.2 (A) Typical responses to various concentrations of standard acetone in the gas phase. (B) Calibration curve of the acetone bio-sniffer for acetone in the gas phase. Biosensors and Bioelectronics permitted.

The acetone bio-sniffer enables a real-time measurement of acetone vapor. We used this function to assess the lipid metabolism based on the breath acetone analysis. Fig. 18.3 represents the time courses of acetone concentration in exhaled breath before, during, and after an aerobic exercise stress test (50 W, 30 minutes). Subjects fasted for at least 6 hours beforehand. The acetone concentrations in exhaled air increased during the exercise, and peaked 15–30 minutes after exercise. At the peaks, the acetone concentrations reached 110%–140% of the initial values. Thereafter, they gradually decreased to steady states. This proof of concept demonstrates that the acetone bio-sniffer can be used to monitor breath acetone related to lipid metabolism.

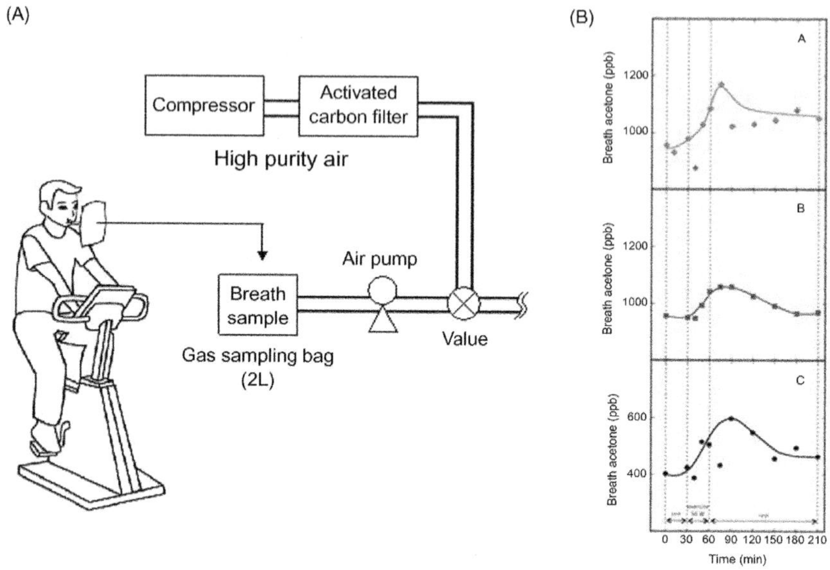

Figure 18.3 (A) Schematic image of the experimental setup for breath gas analysis with aerobic exercise stress testing (50 W, 30 min). (B) Time course of acetone concentration in exhaled breath before, during, and after aerobic exercise. Biosensors and Bioelectronics permitted.

18.1.3 Isopropanol bio-sniffer

Some studies pointed out that concentration of isopropanol (IPA) in the exhaled air could relate with illnesses such as liver disease, chronic obstructive pulmonary (COPD), and lung cancer. In this section, a highly sensitive and selective bio-sniffer for the breath IPA concentration determination is described [44–46]. The IPA bio-sniffer measured the concentration of gaseous IPA according to the fluorescence intensity of increased NADH, which was produced by an enzymatic reaction of S-ADH. As mentioned before, the NADH detection system was employed the UV-LED and the PMT.

This bio-sniffer utilized the S-ADH to transform the concentration of IPA to an optical signal, which presented by the change in the NADH fluorescence intensity (Δintensity). The calibration curve and the typical fluorescence response of the IPA bio-sniffer are presented in Fig. 18.4. When the bio-sniffer contacted to the IPA vapor, the fluorescence intensity increased immediately and reached a steady state within two minutes. After that, the fluorescence intensity recovered to the initial value when the supply of the IPA vapor stopped. The bio-sniffer had an advantages of high sensitivity and wide calibration range. The calibration curve using power approximation curve fitting for 1–9060 ppb determination was confirmed. The correlation coefficient was estimated to 0.997. This bio-sniffer could be used for human exhaled air analysis because the calibration range includes the reported breath IPA concentration in healthy people and some disease patients.

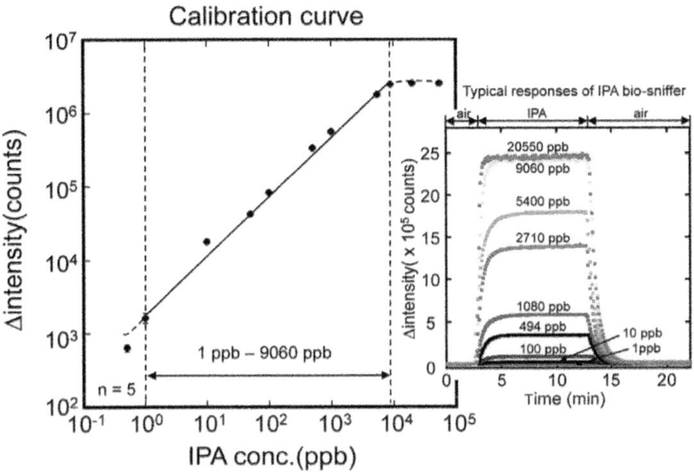

Figure 18.4 Calibration curve of the S-ADH bio-sniffer for IPA in the gas phase. Inset: typical responses of the bio-sniffer to various concentrations of standard acetone vapor. Biosensors and Bioelectronics permitted.

Lung cancer patients' exhaled IPA concentrations were found in a range from 81.2 to 329 ppb with the mean value of 159 ppb [47]. And also, chronic obstructive pulmonary (COPD) patients presented breath IPA concentrations from 167 to 900 ppb with a median value of 523 ppb [44].

The IPA concentration in breath from a healthy human was investigated using the developed IPA bio-sniffer. In total we measured 67 breath samples from 46 healthy volunteers. For the total volunteers, the range of the measured breath was from 2.1 to 54.4 ppb, and the mean concentration was 16.1 ± 11.9 ppb. This obtained result is close in comparison to other studies [48].

In addition, the breath acetone and IPA analysis in healthy subjects shows that the mean values of acetone and IPA were 750.0 ± 434.4 ppb and 15.4 ± 11.3 ppb in Fig. 18.5. Both acetone and IPA did not show a statistical difference among different genders and ages. The breath acetone analysis for diabetic patients shows a mean value of 1207.7 ± 689.5 ppb, which was higher than that of healthy subjects ($P < 1 \times 10^{-6}$). In particular, type-1 diabetic (T1D) patients exhaled a much higher concentration of acetone than type-2 diabetic (T2D) patients ($P < .01$). The breath IPA also had a higher concentration in diabetic patients (23.1 ± 20.1 ppb, $P < .01$), but only T2D patients presented a statistical difference (23.9 ± 21.3 ppb, $P < .01$). The results could be worthwhile in the research of breath biomarkers for diabetes mellitus diagnosis [29].

18.1.4 Sniff-cam system with chemiluminescence

Recently evaluation and examination methods using urine and breath have been put into practice, and research is also being conducted into health monitoring using

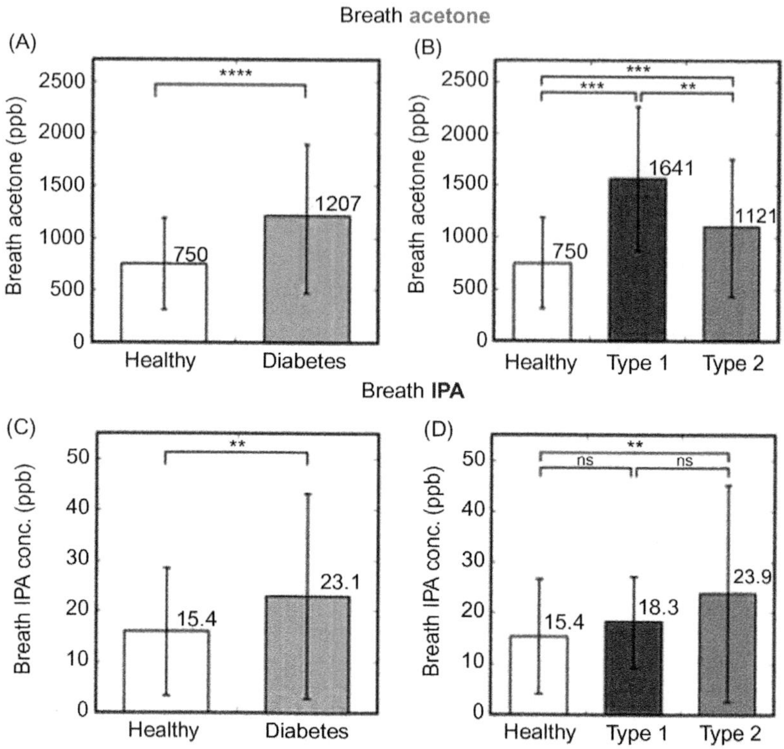

Figure 18.5 (A) Comparison of breath acetone between healthy groups and diabetic patients. (B) Breath acetone b/w healthy groups, diabetes mellitus type-1 (T1D), and type-2 (T2D). (C) Breath IPA b/w healthy subjects and diabetic patients. (D) Breath IPA b/w healthy groups, type-1, and type-2 diabetes mellitus. Only T2D patients showed higher exhaled IPA concentrations. Other comparisons did not observe a significant difference. ($^{****}P < 1 \times 10^{-6}$, $^{***}P < .001$, $^{**}P < .01$, ns: no significant difference). Permission of Analytical Chemistry.

gaseous compounds emitted from the skin surface. However, body-derived volatile chemical compounds see large temporal and spatial variation in concentration, making it difficult to accurately (real time) and selectively evaluate the behavior of these spatiotemporal distributions with conventional analytical methods that use a batch sampling (i.e., gas chromatography), or that show weak sensitivity (i.e., semiconductor gas sensors). Therefore, our group has developed and demonstrated gas measurement methods that use biocatalytic enzymes as sensors to measure gaseous compounds selectively and with high sensitivity using chemiluminescence (CL). CL of luminol reaction is widely used in the fields of analytical chemistry, bioanalytical chemistry, and clinical chemistry, among others [49–51]. The CL methods are characterized by a high intensity of luminescence, rapid reactions, and a wide dynamic range of luminescence. The reaction of luminol–H_2O_2–horseradish peroxidase (HRP) is a valuable and readily available CL reaction that has been applied in the determination of VOCs [52]. We have developed a conventional breath gaseous ethanol imaging system with CL using a gas sampling bag [30,33] (Fig. 18.6A).

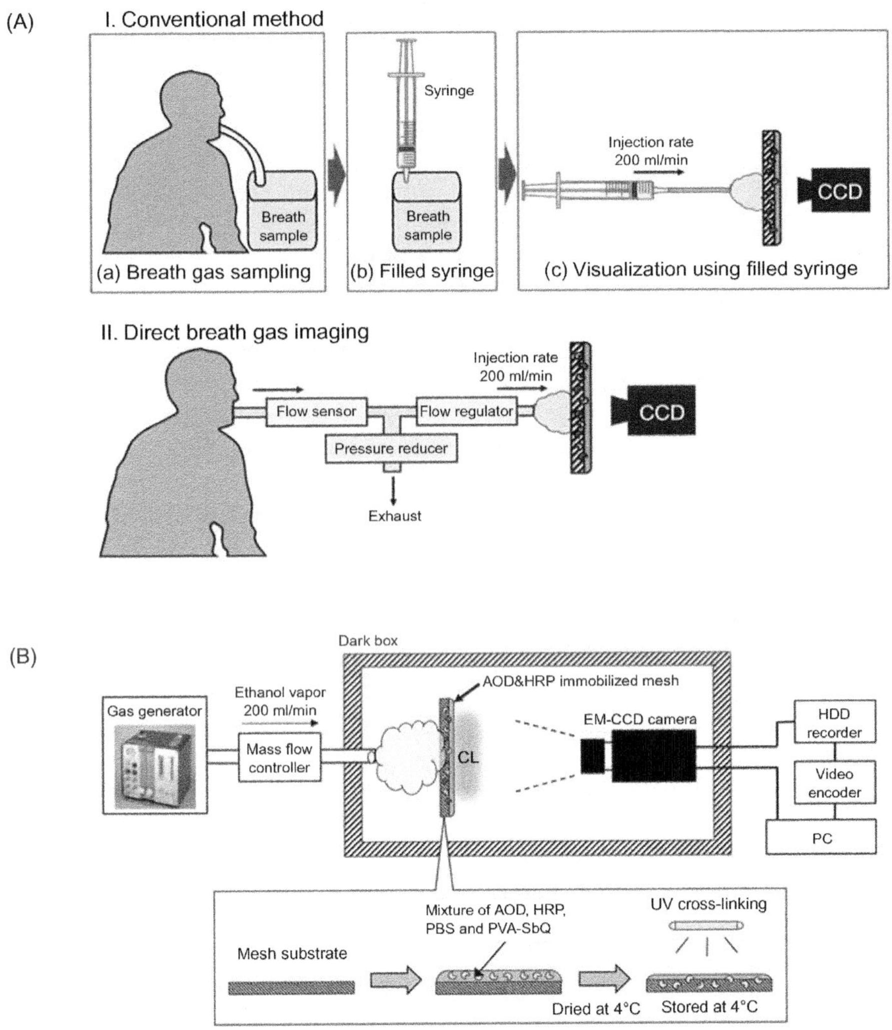

Figure 18.6 (A) Schematic view of sampling methods for imaging of breath chemical. (I: breath-gas sampling bag and syringe, II: Exhaust breath flow regulation system for direct gaseous ethanol imaging without gas-sampling bag.) (B) Schematic of (A) gaseous ethanol imaging system. This system was composed of ethanol gas generator, mass flow controller, enzyme immobilized mesh, and EM-CCD recoding system. Fabrication process of AOD and HRP enzymes and PVA-SVQ immobilized mesh substrate. Permission of Sensors and Actuators B: Chemical.

The developed gas imaging system for ethanol is shown in Fig. 18.6B. In principle, ethanol is oxidized to acetaldehyde with production of hydrogen peroxide by alcohol oxidase (AOD) in the presence of oxygen. The hydrogen peroxide (HRP) reacts with the luminol solution by catalysis of HRP, resulting in the chemiluminescence. These reactions are as follows [53]:

$$\text{Ethanol} + O_2 \xrightarrow{\text{AOD}} \text{acetaldehyde} + H_2O_2 \tag{18.2}$$

$$\text{Luminol} + H_2O_2 + OH^- \xrightarrow{\text{HRP}} \text{3-aminophthate} + 2H_2O + N_2 + h\nu \tag{18.3}$$

AOD and HRP were immobilized at mesh-substrate with photo-cross-linkable poly (vinyl alcohol) (PVA-SbQ) for gaseous ethanol imaging. For the fabrication of the enzyme-immobilized substrate, AOD and HRP were dissolved in phosphate buffer solution (PB, 0.1 mmol/L, pH 7.5), mixed with PVA-SbQ in a volume/weight ratio of 1:2. The enzymes/PVA-SbQ mixture was coated onto the mesh substrate, spread and cured for 3 hours at 4°C in a dark place. And then the substrate was cured using low power ultraviolet irradiation for 5 minutes. A standard gaseous ethanol was supplied at a flow rate of 200 mL/min for 20 seconds from a gas generator unit, which is employed as a standard gas generator for calibration purposes.

The typical responses to various concentrations of standard gaseous ethanol are shown in Fig. 18.7A. Insets in Fig. 18.7A show color images of CL intensity peak at each concentration of gaseous ethanol. These color images indicate that gradation of CL related to the concentration distribution of gaseous ethanol on the mesh substrate. The CL average intensity was increased by injections of standard gaseous ethanol, the CL peaks appeared at 30 seconds after the gas supply and gradually decreased until 60 seconds at 100 ppm gaseous ethanol. The concentration of gaseous ethanol was calibrated from 10 to 400 ppm (Fig. 18.7B). The CL intensity increased rapidly following the injection of gaseous ethanol. The CL average intensities were related to the concentration of gaseous ethanol over the range of 10–400 ppm.

The pharmacokinetic profiles of mean exhaled ethanol for the ALDH2(+) and ALDH2(−) volunteers were demonstrated. The concentration of exhaled ethanol rose rapidly after oral administration, peaking at 30 minutes' postadministration, then gradually decreasing with a linear slope until the end of sample collection at 300 minutes. The ALDH2-deficient (−) volunteer slowly metabolized the acetaldehyde oxidation product, while the breath ethanol concentration in the ALDH2(+) volunteer decreased faster. ALDH2 is the main enzyme involved in acetaldehyde oxidization related to ethanol metabolism. Hence, in volunteers with ALDH2(−), acetaldehyde levels are increased, leading to ethanol being metabolized at a slower rate. This imaging system would be useful and significant for the evaluation of ethanol metabolism using exhaled human breath. In addition, the rate of ethanol metabolism by ALDH2(+) and (−) human subjects was also determined using the breath detection method.

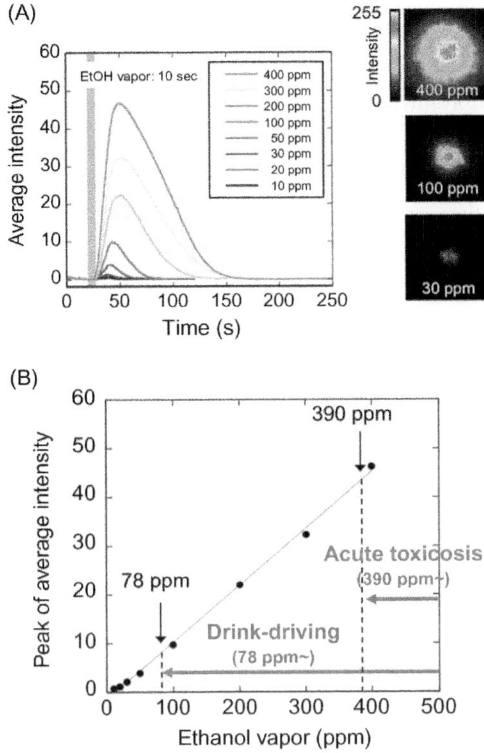

Figure 18.7 (A) Typical responses to 10–400 ppm concentrations of gaseous ethanol and typical images (30, 100, 400 ppm EtOH). (B) Calibration curve of the imaging system for standard gaseous ethanol. The peak of average intensity related to the concentration of gaseous ethanol from 30 to 400 ppm. Permission of Analytical Chemistry.

18.1.5 Biofluorometric "sniff-cam"

Our group have constructed and demonstrated an imaging system that uses biological luminescence in 3.1, and applied it to the imaging measurements and metabolic capacity evaluation of alcohol administration. However, the imaging system have some issues in sensitivity and reactivity because its use of biological luminescence requires two types of enzyme reactions (AOD and HRP). Accordingly, biofluorescence was used instead; gaseous compounds could be induced to biofluorescence with one type of enzyme, with the expectation of increased reactivity and sensitivity.

The principle of the ethanol vapor imaging system based on the biofluorescence of coenzyme nicotinamide adenine dinucleotide (NADH) is shown as follows.

$$\text{Ethanol} + \text{NAD}^+ \xrightarrow{\text{ADH}} \text{acetaldehyde} + \text{NADH} + \text{H}^+ \tag{18.4}$$

Ethanol is oxidized to acetaldehyde by the catalysis of alcohol dehydrogenase (ADH), whereupon it generates reduced NADH with the oxidized NAD (NAD^+) as an electron acceptor. Because of the fluorescent properties of NADH (ex. $\lambda = 340$ nm, fl. $\lambda = 490$ nm), combining an excitation light source with a high-sensitivity camera allows us to detect and image NADH generated by the enzyme reaction when the gaseous ethanol is loaded. Based on this principle, we developed a novel gas imaging system that uses a biofluorescence method for human breath and skin gas (Fig. 18.8A and B). In the optical system used for the NADH bio-

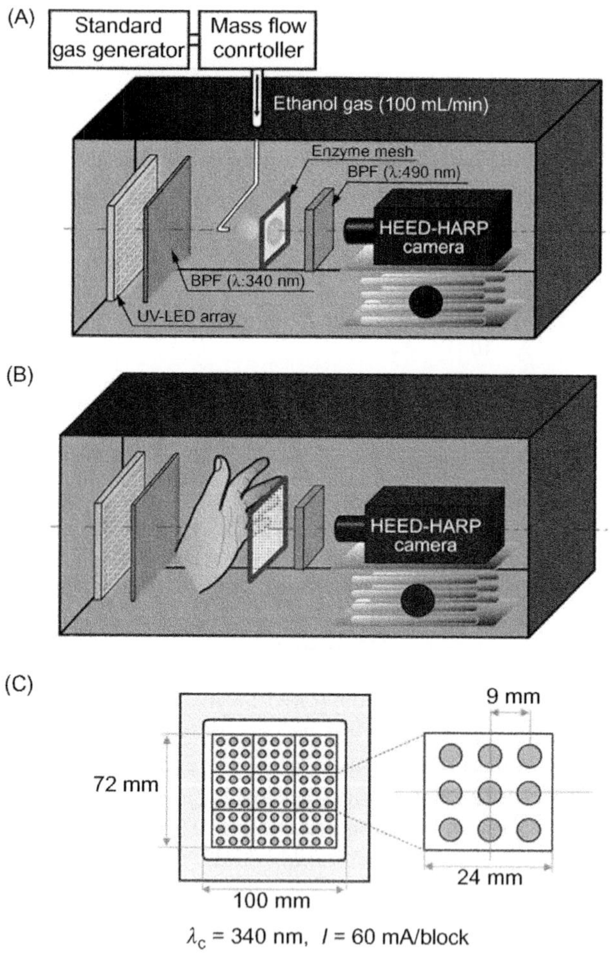

Figure 18.8 (A) Gas imaging system using a biofluorescence method to standard ethanol in the gas phase (B) to skin-derived ethanol. (C) Schematic of UV-LED sheet (9 × 9 LEDs) for excitation of NADH fluorescence on the enzyme mesh. Permission of Analytical Chemistry.

fluorometric approach, the designed UV-LED sheet was positioned opposite a high-sensitivity camera (HEED-HARP camera). An excitation source bandpass filter (BPF) with a central wavelength of 340 nm was set in front of the LED sheet, and another BPF with a central wavelength of 490 nm was set on the imaging side of the camera. We decided to create the enzyme mesh using cotton mesh as the carrier and using enzymes immobilized using glutaraldehyde (GA) cross-linkage because GA has low fluorescence noise and good stability.

Standard ethanol gas samples of various concentrations were loaded into the imaging system, which contained ADH derived from *Saccharomyces cerevisiae*. Biofluorescence was observed centering on the point where the ethanol vapor was loaded (Fig. 18.9A). The biofluorescent intensity of 256 gradations was detected for each of all pixels of a captured image. The NADH generated through the enzyme reaction remained on the enzyme mesh and the fluorescence intensity did not change. Therefore, it was difficult to show increases or decreases in the distribution of the ethanol vapor load. To improve this issue, we used differential analysis to find the change in fluorescence intensity for each unit of time. The average intensity curve calculated by a differential analysis showed increases with loading gas and decreases when loading gas stopped. The differential analysis method can provide better response and analysis time compared with CL imaging. The 90-percentile response time at 50 ppm ethanol was decreased from 35 seconds (CL) to 20 seconds. The inset pictures are the color images at several time points (20, 40, 80 seconds), taken from the differentiated moving images created with this method (Fig. 18.9B). Fig. 18.9C shows these two calibration curves. The quantitative range was 0.5–150 ppm (from *S. cerevisiae*), and similar quantitative performance (1–150 ppm) was obtained for the peak value from differential analysis. In addition, we investigated the selectivity of the system using representative components of breath. Fig. 18.9C shows a comparison of the fluorescence output using two kinds of ADH (from *yeast* and *S. cerevisiae)* between various components in the gas phase. We can find that neither of the enzymes react to breath components other than ethanol, and high selectivity based on the substrate specificity of the enzymes was obtained.

Next, the imaging system using ADH was applied to the biofluorometric imaging of ethanol in breath and skin gas. The results were obtained by inputting the post–alcohol consumption breath of ALDH2(+) and (–) experimental subjects into the created imaging system (data not shown) [35]. Fig. 18.10 shows biofluorometric images of post–alcohol consumption skin gas. As can be seen from this figure, the ethanol contained in the transdermal gas emitted from the palm skin could be imaged, and the spreading of the ethanol emissions across the palm skin was also confirmed by the ADH biofluorometric approach. It would offer improved sensitivity and reactivity compared with the traditional CL method and the biofluorometric system would be used for detailed evaluation of the generation of body gases.

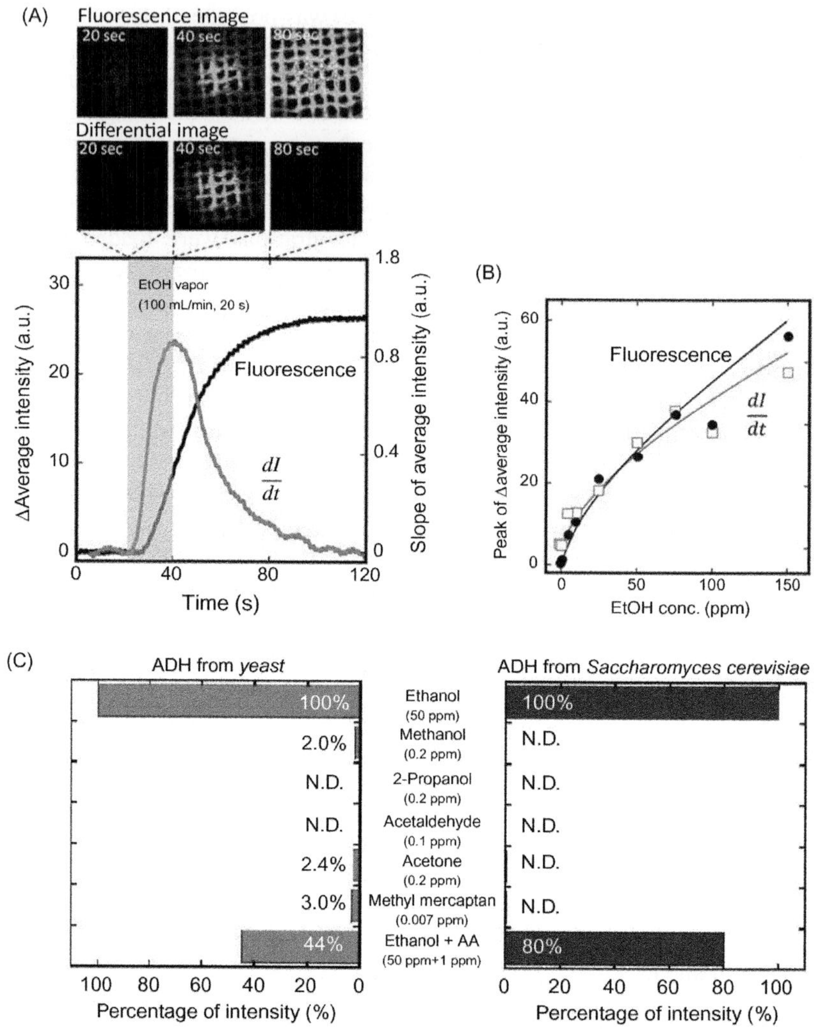

Figure 18.9 (A) Typical response of average fluorescence intensity in the imaging of ethanol vapor (50 ppm) (fluorescence), and calculated curve of differential analysis after creating a running average from them (*dI*/*dt*). Inset images: fluorescence and differential analysis images. (B) Comparison of calibration formulae between fluorescence intensity (●) and differential analysis results (□) against gaseous ethanol in the system. (C) Comparisons of fluorescence intensities in enzyme membranes with ADH derived from *yeast* (left) and *Saccharomyces cerevisiae* (right) b/w several breath components. Permission of Analytical Chemistry.

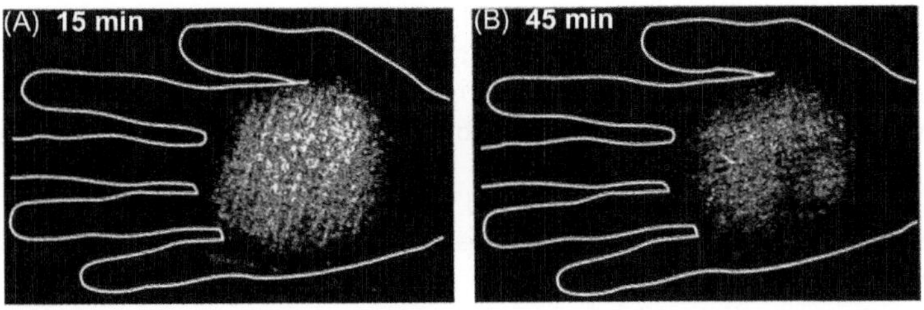

Figure 18.10 Images of skin-derived ethanol vapor on a volunteer subject after alcohol administration (fluorescence intensity magnified 25 times). (A) 15 min, (B) 45 min after drinking. Permission of Analytical Chemistry.

18.2 Summary

We introduce biochemical gas sensor "bio-sniffer" using biochemical reactions and characteristics of various enzymes for determination of acetone and isopropanol in the gas phase. The bio-sniffer devices were successfully applied to measure concentrations of acetone and isopropanol in exhaled air from healthy subjects and diabetes patients as volatile biomarkers of diabetes mellitus and evaluation of a lipid metabolism. In addition, two types (chemiluminescence and biofluorescence) of gas imaging system "sniffer-camera" also developed bio-optical reactions, respectively. These imaging systems measure ethanol concentrations as intensities of optical images by enzyme reaction. In future work, the bio-sniffer and imaging system in the gas phase would be applied for analysis of VOCs information from humans, transdermal analysis, evaluation of metabolic conditions, and noninvasive screening of some diseases.

Acknowledgments

This work is partly supported by Japan Society for the Promotion of Science (JSPS) Grants-in-Aid for Scientific Research System, by Japan Science and Technology Agency (JST) and by MEXT (Ministry of Education, Culture, Sports, Science and Technology) Special Funds for Education and Research "Advanced Research Program in Sensing Biology," and KAKENHI Grant Numbers 26280053, 15H04013, 16KK0143, 17H01759.

References

[1] A. Amann, B.D.L. Costello, W. Miekisch, J. Schubert, B. Buszewski, et al., The human volatilome: volatile organic compounds (VOCs) in exhaled breath, skin emanations, urine, feces and saliva, J. Breath Res. 8 (2014) 034001.

[2] X. Sun, K. Shao, T. Wang, Detection of volatile organic compounds (VOCs) from exhaled breath as noninvasive methods for cancer diagnosis Young Investigators in Analytical and Bioanalytical Science, Anal. Bioanal. Chem. 408 (2016) 2759–2780.

[3] Y. Saalberg, M. Wolff, VOC breath biomarkers in lung cancer, Clin. Chim. Acta 459 (2016) 5–9. Available from: http://dx.doi.org/10.1016/j.cca.2016.05.013.

[4] S. Dragonieri, G. Pennazza, P. Carratu, O. Resta, Electronic nose technology in respiratory diseases, Lung 195 (2017) 157–165. Available from: http://dx.doi.org/10.1007/s00408-017-9987-3.

[5] A. Krilaviciute, J.A. Heiss, M. Leja, J. Kupcinskas, H. Haick, H. Brenner, et al., Detection of cancer through exhaled breath: a systematic review, Oncotarget 6 (2015) 38643–38657.

[6] I. Horváth, et al., Exhaled breath condensate: methodological recommendations and unresolved questions, Eur. Respir. J. 26 (2005) 523–548.

[7] P. Montuschi, N. Mores, A. Trové, C. Mondino, P.J. Barnes, The electronic nose in respiratory medicine, Respiration 85 (2012) 72–84.

[8] A. Reyes-Reyes, R.C. Horsten, H.P. Urbach, N. Bhattacharya, Study of the exhaled acetone in type 1 diabetes using quantum cascade laser spectroscopy, Anal. Chem. 87 (2015) 507–512.

[9] C.N. Tassopoulos, D. Barnett, T. Russell Fraser, Breath-acetone and blood-sugar measurements in diabetes, Lancet 293 (1969) 1282–1286.

[10] M. Righettoni, A. Tricoli, S.E. Pratsinis, Si:WO 3 Sensors for highly selective detection of acetone for easy diagnosis of diabetes by breath analysis, Anal. Chem. 82 (2010) 3581–3587.

[11] C. Deng, J. Zhang, X. Yu, W. Zhang, X. Zhang, Determination of acetone in human breath by gas chromatography-mass spectrometry and solid-phase microextraction with on-fiber derivatization, J. Chromatogr. B Anal. Technol. Biomed. Life Sci. 810 (2004) 269–275.

[12] P. Spaněl, K. Dryahina, D. Smith, Acetone, ammonia and hydrogen cyanide in exhaled breath of several volunteers aged 4-83 years, J. Breath Res. 1 (2007) 011001.

[13] M. Phillips, R.N. Cataneo, C. Saunders, P. Hope, P. Schmitt, J. Wai, Volatile biomarkers in the breath of women with breast cancer, J. Breath Res. 4 (2010) 026003.

[14] F. Buljubasic, G. Buchbauer, The scent of human diseases: A review on specific volatile organic compounds as diagnostic biomarkers, Flavour Fragr. J. 30 (2015) 5–25.

[15] A. Amann, P. Mochalski, V. Ruzsanyi, Y.Y. Broza, H. Haick, Assessment of the exhalation kinetics of volatile cancer biomarkers based on their physicochemical properties, J. Breath Res. 8 (2014) 016003.

[16] M. Gallagher, C.J. Wysocki, J.J. Leyden, A.I. Spielman, X. Sun, G. Preti, Analyses of volatile organic compounds from human skin, Br. J. Dermatol. 159 (2008) 780–791.

[17] B. de Lacy Costello, A. Amann, H. Al-Kateb, C. Flynn, W. Filipiak, T. Khalid, et al., A review of the volatiles from the healthy human body, J. Breath Res. 8 (2014) 014001.

[18] H. Saito, Y. Suzuki, T. Gessei, K. Miyajima, T. Arakawa, K. Mitsubayashi, bioelectronic sniffer (biosniffer) based on enzyme inhibition of butyrylcholinesterase for toluene detection, Sens. Mater. 26 (2014) 121–129.

[19] H. Saito, Y. Hashimoto, T. Minamide, T. Kon, K. Toma, T. Arakawa, et al., Fiber optic biosniffer (biochemical gas sensor) for gaseous dimethyl sulfide, Sens. Mater. 28 (2016) 1295.

[20] H. Kudo, T. Yamashita, K. Miyajima, T. Arakawa, K. Mitsubayashi, NADH-fluorometric biochemical gas sensor (Bio-Sniffer) for evaluation of indoor air quality, IEEE Sens. J. 13 (2013) 2828–2833.

[21] H. Kudo, M. Sawai, Y. Suzuki, X. Wang, T. Gessei, D. Takahashi, et al., Fiber-optic bio-sniffer (biochemical gas sensor) for high-selective monitoring of ethanol vapor using 335 nm UV-LED, Sens. Actuators, B Chem. 147 (2010) 676–680.

[22] K. Mitsubayashi, H. Matsunaga, G. Nishio, S. Toda, Y. Nakanishi, Bioelectronic sniffers for ethanol and acetaldehyde in breath air after drinking. Biosens. Bioelectron., 2005, pp. 1573–1579.
[23] H. Kudo, Y. Suzuki, T. Gessei, D. Takahashi, T. Arakawa, K. Mitsubayashi, Biochemical gas sensor (bio-sniffer) for ultrahigh-sensitive gaseous formaldehyde monitoring, Biosens. Bioelectron. 26 (2010) 854–858.
[24] H. Kudo, T. Goto, T. Saito, H. Saito, K. Otsuka, K. Mitsubayashi, Biochemical sniffer with choline oxidase for measurement of choline vapour. Microchim. Acta, 2008, pp. 421–426.
[25] M. Ye, P.J. Chien, K. Toma, T. Arakawa, K. Mitsubayashi, An acetone bio-sniffer (gas phase biosensor) enabling assessment of lipid metabolism from exhaled breath, Biosens. Bioelectron. 73 (2015) 208–213.
[26] M. Ye, T. Arakawa, P.J. Chien, T. Suzuki, K. Toma, K. Mitsubayashi, Acetone biosensor based on fluorometry of reduced nicotinamide adenine dinucleotide consumption in reversible reaction by secondary alcohol dehydrogenase, IEEE Sens. J. 17 (2017) 5419–5425.
[27] P.J. Chien, M. Ye, T. Suzuki, K. Toma, T. Arakawa, Y. Iwasaki, et al., Optical isopropanol biosensor using NADH-dependent secondary alcohol dehydrogenase (S-ADH), Talanta 159 (2016) 418–424.
[28] P.-J. Chien, T. Suzuki, M. Tsujii, M. Ye, K. Toma, T. Arakawa, et al., Bio-sniffer (gas-phase biosensor) with secondary alcohol dehydrogenase (S-ADH) for determination of isopropanol in exhaled air as a potential volatile biomarker, Biosens. Bioelectron. 91 (2017) 341–346.
[29] P.J. Chien, T. Suzuki, M. Tsujii, M. Ye, I. Minami, K. Toda, et al., Biochemical gas sensors (biosniffers) using forward and reverse reactions of secondary alcohol dehydrogenase for breath isopropanol and acetone as potential volatile biomarkers of diabetes mellitus, Anal. Chem. 89 (2017) 12261–12268.
[30] X. Wang, E. Ando, D. Takahashi, T. Arakawa, H. Kudo, H. Saito, et al., 2D spatiotemporal visualization system of expired gaseous ethanol after oral administration for real-time illustrated analysis of alcohol metabolism, Talanta 82 (2010) 892–898.
[31] T. Arakawa, E. Ando, X. Wang, M. Kumiko, H. Kudo, H. Saito, et al., A highly sensitive and temporal visualization system for gaseous ethanol with chemiluminescence enhancer, Luminescence 27 (2012) 328–333.
[32] T. Arakawa, E. Ando, X. Wang, K. Miyajima, S. Takeuchi, H. Kudo, et al., Chemiluminescent visualization for evaluation of gaseous ethanol distribution during "la France" pear maturation, IEEE Sens. J. 13 (2013) 2842–2848.
[33] X. Wang, E. Ando, D. Takahashi, T. Arakawa, H. Kudo, H. Saito, et al., Non-invasive spatial visualization system of exhaled ethanol for real-time analysis of ALDH2 related alcohol metabolism, Analyst 136 (2011) 3680–3685.
[34] T. Arakawa, X. Wang, T. Kajiro, K. Miyajima, S. Takeuchi, H. Kudo, et al., A direct gaseous ethanol imaging system for analysis of alcohol metabolism from exhaled breath, Sens. Actuators B Chem. 186 (2013) 27–33.
[35] T. Arakawa, T. Sato, K. Iitani, K. Toma, K. Mitsubayashi, Fluorometric biosniffer camera "sniff-Cam" for direct imaging of gaseous ethanol in breath and transdermal vapor, Anal. Chem. 89 (2017) 4495–4501.
[36] K. Iitani, P.J. Chien, T. Suzuki, K. Toma, T. Arakawa, Y. Iwasaki, et al., Improved sensitivity of acetaldehyde biosensor by detecting ADH reverse reaction-mediated NADH fluoro-quenching for wine evaluation, ACS Sens. 2 (2017) 940–946.
[37] T. Koshida, T. Arakawa, T. Gessei, D. Takahashi, H. Kudo, H. Saito, et al., Fluorescence biosensing system with a UV-LED excitation for l-leucine detection, Sens. Actuators B Chem. 146 (2010) 177–182.

[38] T. Gessei, T. Arakawa, H. Kudo, K. Mitsubayashi, A fiber-optic sorbitol biosensor based on NADH fluorescence detection toward rapid diagnosis of diabetic complications, Analyst 140 (2015) 6335–6342.
[39] H. Kudo, T. Yamashita, M. Ye, K. Miyajima, T. Arakawa, K. Mitsubayashi, et al., NADH-fluorometric biochemical gas sensor (bio-sniffer) for assessment of indoor air quality, Proc. IEEE Sens., 2012.
[40] T.D.C. Minh, D.R. Blake, P.R. Galassetti, The clinical potential of exhaled breath analysis for diabetes mellitus, Diabetes Res. Clin. Pract. 97 (2012) 195–205.
[41] W. Miekisch, J.K. Schubert, G.F. Noeldge-Schomburg, Diagnostic potential of breath analysis—focus on volatile organic compounds, Clin. Chim. Acta 347 (2004) 25–39.
[42] T. Mitsui, N. Yuza, N. Kato, T. Kondo, Is acetone a normal breath constituent in Japanese? Clin. Chim. Acta (1999) 205–208.
[43] N. Teshima, J. Li, K. Toda, P.K. Dasgupta, Determination of acetone in breath, Anal. Chim. Acta 535 (2005) 189–199.
[44] J. Rudnicka, M. Walczak, T. Kowalkowski, T. Jezierski, B. Buszewski, Determination of volatile organic compounds as potential markers of lung cancer by gas chromatography-mass spectrometry versus trained dogs, Sens. Actuators B Chem. 202 (2014) 615–621.
[45] A. Wehinger, A. Schmid, S. Mechtcheriakov, M. Ledochowski, C. Grabmer, G.A. Gastl, et al., Lung cancer detection by proton transfer reaction mass-spectrometric analysis of human breath gas, Int. J. Mass Spectrom. 265 (2007) 49–59.
[46] A. Bajtarevic, et al., Noninvasive detection of lung cancer by analysis of exhaled breath, BMC Cancer 9 (2009) 348.
[47] S. Kischkel, W. Miekisch, P. Fuchs, J.K. Schubert, Breath analysis during one-lung ventilation in cancer patients, Eur. Respir. J. 40 (2012) 706–713.
[48] C. Turner, B. Parekh, C. Walton, P. Španěl, D. Smith, M. Evans, An exploratory comparative study of volatile compounds in exhaled breath and emitted by skin using selected ion flow tube mass spectrometry, Rapid Commun. Mass Spectrom. 22 (2008) 526–532.
[49] C.A. Marquette, L.J. Blum, Applications of the luminol chemiluminescent reaction in analytical chemistry, Anal. Bioanal. Chem. 385 (2006) 546–554.
[50] F. Li, C. Zhang, X. Guo, W. Feng, Chemiluminescence detection in HPLC and CE for pharmaceutical and biomedical analysis, Biomed. Chromatogr. 17 (2003) 96–105.
[51] J. Lin, H. Ju, Electrochemical and chemiluminescent immunosensors for tumor markers, Biosens. Bioelectron. (2005) 1461–1470.
[52] H. Cui, C.G. Xie, C.Z. Lai, A novel chemiluminescent method for determination of phloroglucinol, Luminescence 18 (2003) 318–323.
[53] P. Fletcher, K.N. Andrew, A.C. Calokerinos, S. Forbes, P.J. Worsfold, Analytical applications of flow injection with chemiluminescence detection–a review, Luminescence 16 (2001) 1–23.

Part III

Information and Network Technologies for Sensor-Internet of Things Applications

Flexible and printed biosensors based on organic TFT devices

19

Kuniaki Nagamine and Shizuo Tokito
Research Center for Organic Electronics (REOL), Yamagata University, Yonezawa, Japan

19.1 Introduction

19.1.1 Biosensors for the Internet of Things society

The concept of the Internet of Things (IoT), connecting objects to a network via sensors to solve global societal needs, is now spreading to various application fields to realize an emerging "smart lifestyle," as shown in Fig. 19.1. Healthcare is one of the most attractive applications of IoT because it will provide various healthcare services for individuals including daily health monitoring, fitness support, and elderly care. To realize these applications, there is a desire to develop sensor devices that can be worn comfortably on the human body and monitor human health parameters in real time and in a noninvasive manner.

A biosensor is an analytical device composed of biological receptors connected to signal transducers. Owing to its highly selective quantification of analytes in bodily fluids, which include proteins, nucleic acids, and cells, they have been successfully applied in the medical field for the diagnosis of physiological diseases. Recently developed microfabrication technologies based on photolithography methods have enabled integration of the biosensing elements into small, portable devices, opening the opportunity to apply these devices in periodic medical tests such as point-of-care-testing (POCT) and bedside physical examination [1–4]. Also, the appearance of biocompatible materials accelerates development of on-skin (wearable) [5–8] and implantable biosensors [9,10] for real-time monitoring of physiological conditions. The rise of Internet technology allows for the collection of obtained biological data through a network, followed by their utilization in big data analyses for future, advanced medical applications. However, the continuous or temporal sampling of bodily fluids in a noninvasive manner is a challenging issue to be resolved for the potential daily use of these sensors. Additionally, the low-cost mass production of the biosensor systems is another challenge that, if addressed, can enable a spread in the use of biosensors for healthcare [11].

Chemical, Gas, and Biosensors for the Internet of Things and Related Applications.
DOI: https://doi.org/10.1016/B978-0-12-815409-0.00020-6

Figure 19.1 Future "smart lifestyle" potentially realized by various types of sensors and the Internet of Things (IoT).

19.1.2 Printed organic biosensors for human healthcare applications

The omics analyses have revealed that externally secreted biological fluids from the human body such as tears, saliva, drainage from wounds, sweat, and urine include biomarkers that are partitioned from blood. For example, salivary amylase, cortisol, NO_3^-, chromogranin A are indicators for mental stress reflecting sympathetic nervous system activity [12,13], and some salivary metabolites and microRNAs are considered to be biomarkers for cancer [14–16]. Another example is sweat which includes some biomarkers of ischemia (lactate) [17], cystic fibrosis (Cl^- ion) [18], heat stroke (Na^+, K^+) [19], schizophrenia (some proteins) [20], and atopic dermatitis (dermicidin) [21]. These external bodily fluids can be sampled in a noninvasive manner, allowing for not only patients but also healthy subjects to understand their daily physiological conditions without the uncomfortable sampling of their blood.

Recently, various types of wearable biosensors have been developed for the external bodily fluids as follows: soft contact lens type for tear [22,23], mouth guard type for saliva [24,25], bandage type for wound [26,27], and wristband type for sweat [17,28–32]. An integrated sensor system is composed of not only a sensing electrode but also a signal transduction circuit and a wireless network module is necessary to realize a fully wearable device. In particular, the development of a fully flexible integrated sensor system that conforms to the human body is still challenging [17,33,34].

Printing technology has emerged as a low-cost, environmentally friendly mass manufacturing technology for the fabrication of next-generation printed flexible electronics devices, such as flexible displays [35,36], radio frequency identifier tags [37,38], smart labels [39,40], and various types of sensors based on thin-film transistors. Organic semiconducting materials are particularly suitable for printed electronics because they are solution-processable. Organic thin-film transistor (OTFT) devices can be combined with the biosensing elements to create highly sensitive devices that are completely flexible. By recently employing biocompatible semiconductors such as conducting polymers (PEDOT) we have been able to utilize these devices as implantable flexible sensors that can be applied to organs and tissues with complex surface geometries [41]. Employing advanced printing technologies [42] will enable direct and on-demand printing of the biosensing systems on the three-dimensional curved surfaces of conventional wearable medical devices such as mouth guards, soft contact lenses, wound dressings, wristbands, and even human skin in the future [43]. In this chapter, we summarize recent developments in printed organic material–based biosensors with an emphasis on our research results.

19.2 Organic thin-film transistor-based biosensors

19.2.1 Printing techniques for device fabrication

There are a number of printing methods that can be applied to electronic device fabrication, such as inkjet printing [44,45], screen printing [6], gravure offset printing [46], soft blanket gravure printing [47–50], and reverse offset printing [51]. These printing techniques are utilized on-demand depending on device configuration using conventional conductive inks for printed electronics such as silver and copper pastes. For example, inkjet printing is well known as a promising method for digital and straightforward digital-on-demand patterning method without the use of printing plate, but with limited pattern resolution (around 50–300 nm in thickness and 50–300 μm in feature size). Reverse offset printing, which utilizes a printing mask or plate, has the advantage of being able to make precise patterns close to 1 μm in width with thicknesses of about 100 nm, as shown in Fig. 19.2B and C [51,52]. However, utilizing biomaterials such as DNA, RNA, proteins, polymers, and tissues/cells themselves for printing inks is still challenging due to their vulnerability under printing conditions [53,54]. We are now developing printable biomaterial-based inks that can be employed in the fabrication of biosensors.

19.2.2 Organic thin-film transistor-based biosensor principles

We have developed extended-gate type OTFT-based biosensors for the detection of several biomarkers in a sample solution. Fig. 19.3 shows a photograph and structure of the extended-gate type OTFT device used in this study. The basic structure of the sensor consists of an OTFT device connected to an extended gate electrode whose surface is modified with a biorecognition layer. The OTFT acts as an amplifier as well as a transducer for the detected signal (driving portion), and the extended gate electrode is employed as the detection portion. The OTFT device can be printed near the detection portion without sacrificing device flexibility, allowing

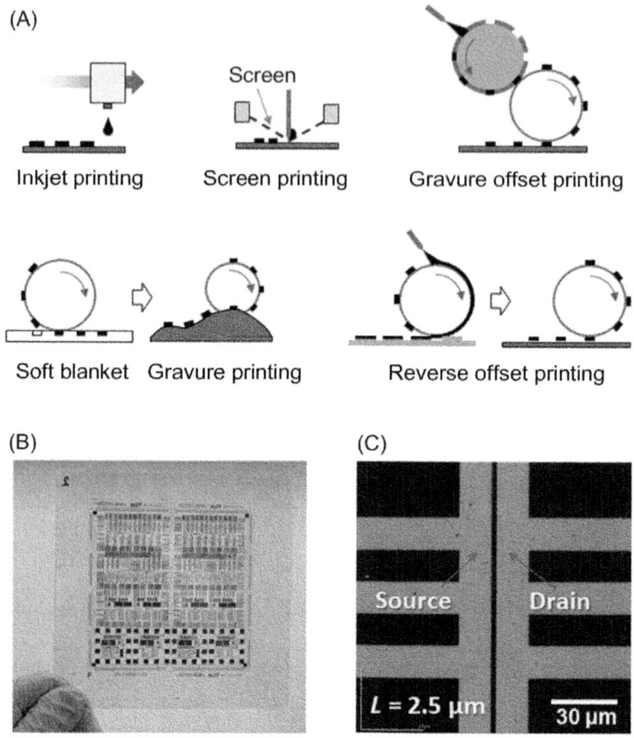

Figure 19.2 (A) Schematic illustration of the printing techniques. (B, C) Microscopic images of an integrated circuit (B) and an organic transistor device (C) created with the reverse offset printing method.

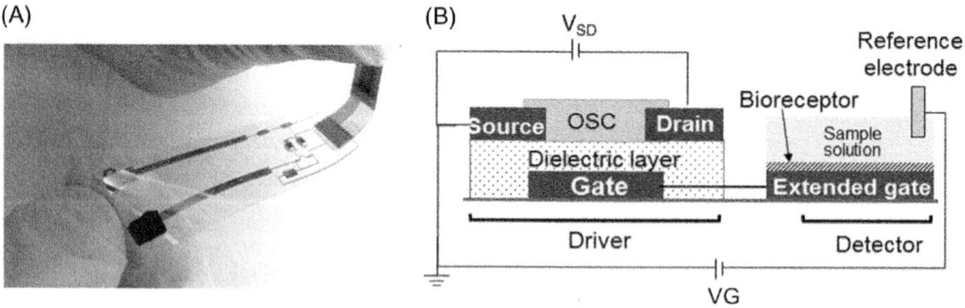

Figure 19.3 (A) Photograph and (B) structure of an extended-gate type OTFT-based biosensor.

the fabrication of highly sensitive biosensors. The extended-gate type transistor is advantageous for stable and reproducible detection of analytes in a sample solution because the sensing portion immersed in the solution can be completely isolated from the driving portion.

In a generic transistor device, the source−drain current flowing in the organic semiconductor (OSC) can be directly controlled by a gate voltage applied between source and gate electrodes, because the gate voltage induces polarization at the interface between the OSC and dielectric layers. In an extended-gate type transistor, the applied gate voltage is distributed in the dielectric layer, the extended-gate electrode, and the reference electrode. The potential of the reference electrode is always constant, whereas the surface potential of extended-gate electrode changes depending on the reaction with analytes. As a result, the potential applied to the dielectric layer changes, followed by inducing a change in source−drain current. Based on this principle, the biorecognition reaction at the extended-gate electrode can be detected as a change of source−drain current or threshold voltage in the OTFT device. We have developed a variety of OTFT-based biosensors utilizing different types of bioreceptors: enzyme [55−57], antibody [58−62], and ionophore [63]. Additionally, artificial receptors have also been synthesized for development of biomaterial-free chemical sensors because they are chemically stable and adaptable to printing techniques, but exhibit less selectivity than bioreceptors [64−74]. In this chapter, three representative biosensors for lactate, IgA, and electrolytes (Na^+, K^+) are introduced.

19.2.3 Enzyme-based biosensors

Fig. 19.4 shows the setup for an extended-gate type OTFT-based biosensor utilizing an enzymatic reaction for biorecognition events.

The OTFT device was fabricated as follows:

- An aluminum (Al) gate electrode was deposited onto a substrate via thermal evaporation (30 nm thickness).
- The surface of Al gate electrode was treated with an oxygen plasma to form an aluminum-oxide layer (5 nm thickness) onto which a self-assembled monolayer (1.7 nm thick) was formed using tetradecylphosphonic acid to obtain a gate dielectric layer.
- Au (gold) source−drain electrode patterns (30 nm thickness) were thermally deposited onto the dielectric layer through a shadow mask.
- After preparing a bank layer using an amorphous fluoropolymer, a semiconductor polymer, poly(2,5-bis(3-hexadecylthiophene-2-yl)thieno[3,2-b]thiophene) (pBTTT-C_{16}), was drop-casted as to bridge the source−drain electrodes, followed by thermal annealing.
- The surface of the OTFT device was passivated using a Cytop layer (100 nm thickness).

The OTFT sensor device was then created as follows:

- The Au (50 nm thick) extended-gate electrode was thermally deposited on a polyethylene naphthalate (PEN) film (125 μm thick) through a metal mask.
- The lead area of the Au extended-gate electrode was insulated with an amorphous fluoropolymer to expose only the active region (15 mm^2) to a sample solution.
- The active area of the extended-gate electrode was modified with a carbon paste containing redox mediator, Prussian blue (PB), onto which the enzyme was immobilized with a chitosan polymer.
- The enzyme/PB-modified extended-gate electrode and Ag/AgCl reference electrode were connected to the gate and source electrodes of the OTFT, respectively.

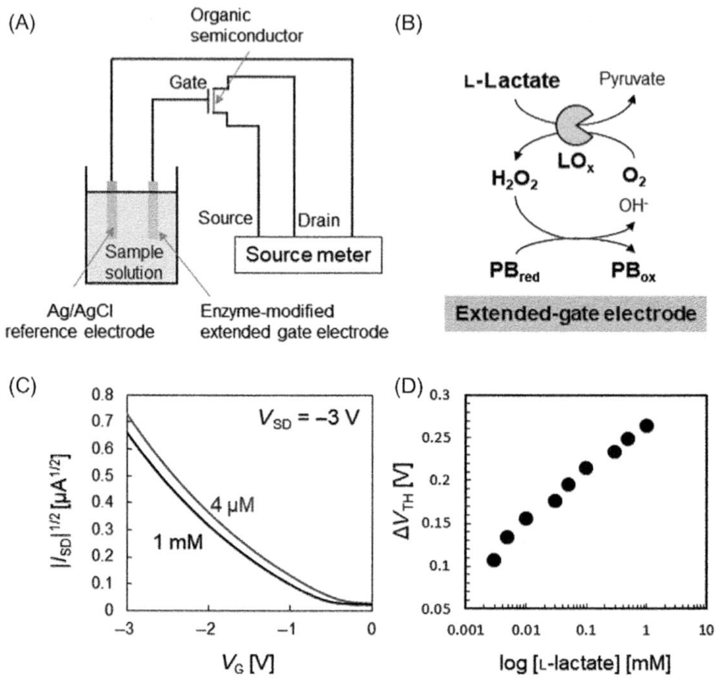

Figure 19.4 (A) Setup for the extended-gate type OTFT-based biosensing system. (B) Enzymatic reaction on an extended-gate electrode. (C) Transfer characteristics of the lactate sensor upon titration with L-lactate. (D) Change in threshold voltage of the L-lactate sensor at various concentration of L-lactate in PBS (−).

A source-meter controlled the voltage between the source−drain and gate−reference electrodes. Fig. 19.4B shows the enzymatic reaction generating on the active area of extended-gate electrode. The target analytes (L-lactate) are oxidized by the enzyme (lactate oxidase, LO_x) generating hydrogen peroxide, followed by oxidation of redox mediator PB from di- to trivalent form with the hydrogen peroxide. As a result, the electric potential of the extended-gate electrode changes according to the Nernstian equation, resulting in a change in source−drain current or threshold voltage of the OTFT as described above. Fig. 19.4C shows the typical transfer characteristics of the OTFT-based L-lactate sensor. The curve was shifted in the leftward direction with increase of L-lactate concentration from 4 μM to 1 mM, suggesting that negative-doping was induced at the interface between the OSC layer and the dielectric layer with reducing PB at the extended-gate electrode. The shift of threshold voltage V_{TH} exhibited a linear increase against the logarithm of L-lactate at a concentration range from 4 μM to 1.0 mM, as shown in Fig. 19.4D. This sensor is appropriate for disposable, temporal biosensing of the analytes in a sample solution, but is unsuitable for real-time monitoring due to an irreversible redox reaction of PB at the extended-gate electrode connecting to the high impedance of dielectric layer. We are now developing the next generation of OTFT-based enzyme sensors, which can induce a reversible redox reaction of the mediator for continuous monitoring of the analytes.

19.2.4 Immunosensors

Immunosensors utilize an immunoreaction (antigen–antibody reaction) to selectively capture target analytes in a sample solution, followed by transducing its response to detectable electric and photonic signals. We established an immunosensor system using the extended-gate type OTFT device that can detect changes in the electrical potential of the antibody-modified extended-gate electrode upon capturing charged analytes [59]. The target analyte of immunoglobulin (Ig) A is a well-known protein that is related to allergies and infectious diseases. Fig. 19.5A shows the structure of extended-gate Au electrode modified with anti-IgA antibody. The Au electrode was firstly treated with 3-mercaptopropionic acid to form its self-assembled monolayer, followed by 1-ethyl-3-(3-dimethylaminopropyl) carbodiimide/*N*-hydroxysuccinimide (EDC/NHS) coupling of an amine group of streptavidin and carboxylic group of the SAM. Biotinylated monoclonal anti-IgA antibody was then immobilized via the avidin–biotin interaction, which allows oriented

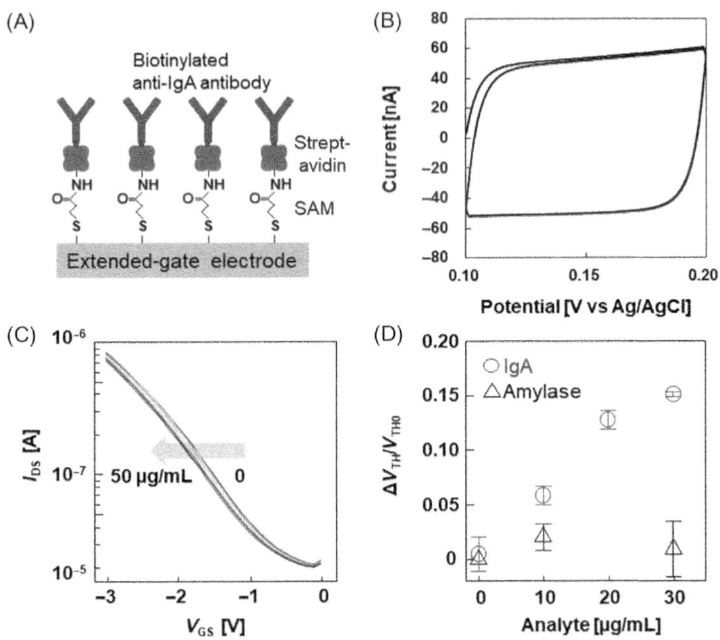

Figure 19.5 (A) Structure of an extended-gate Au electrode modified with anti-IgA antibody. (B) Cyclic voltammogram of 3,3′-Dithiodipropionic Acid SAM-modified extended-gate electrode detected in PBS(−). (D) Transfer characteristics of the IgA sensor upon titration with IgA. (D) Change in threshold voltage of the IgA sensor against various concentration of Human IgA and amylase.

immobilization of the antibody on the electrode. Fig. 19.5B shows the cyclic voltammogram detected using the SAM-modified extended-gate electrode in a phosphate buffer saline (PBS) (−). The electrode exhibited charging/discharging current of an electrical double-layer capacitance independent of the applied potential. This current−voltage relationship is typical behavior for an insulative alkanethiol SAM-modified electrode [75]. The electrical double-layer capacitance of the SAM-modified electrode calculated from Fig. 19.5C was about 10 $\mu F/cm^2$, which is a similar value to that of close-packed SAM reported in the previous manuscript [75]. Fig. 19.5C shows the typical transfer characteristics of the immunosensor upon titration with human IgA. The transfer characteristics clearly shifted in the negative direction due to change of the electric potential of extended-gate electrode upon capturing the charged IgA. Fig. 19.5D shows the titration curve for human IgA dissolved in PBS (−). The y-axis represents the ratio of V_{TH} change divided by the V_{TH} before titration with human IgA (V_{TH0}). This ratio increased with increases in the human IgG concentration in the sample. However, the sensor exhibited almost no response against amylase, which is one of the interferences contained in saliva, suggesting selectivity of the present sensor. The same detection principle was applied to other proteins such as IgG and chromogranin A and its reproducible performance was demonstrated using artificial saliva [58−62]. These preliminary studies suggested the potential of our immunosensor for being applied in the noninvasive management of human health.

19.2.5 Ion-selective sensors

Monitoring the concentration of physiological electrolytes in bodily fluids such as sweat is vital for preventing the excessive exercise by athletes and heat stroke in the elderly [76]. We fabricated a highly sensitive, printed potentiometric ion-selective sensor composed of the ion-selective membrane-modified extended-gate electrode and the amplifier circuits employing OTFT-based pseudo-CMOS inverters [63]. Fig. 19.6A and B shows its photograph and magnified image (A), and a circuit diagram of the entire amplification system (B). The system was composed of two

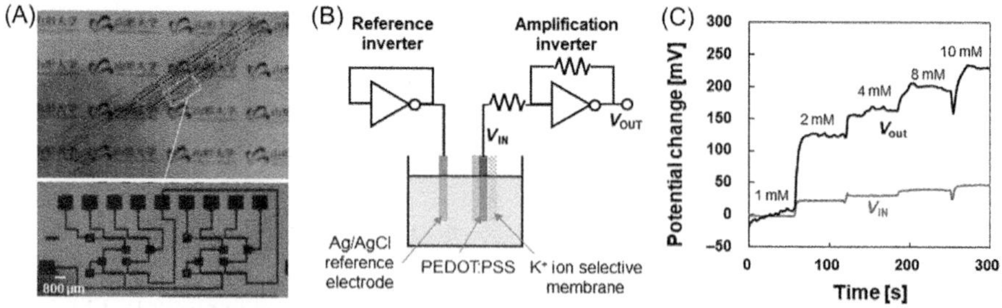

Figure 19.6 (A) A Photograph and magnified image of the printed organic amplification system. (B) Experimental setup and circuit diagram of the amplification system. (C) Time course of zero-adjusted potential change of V_{IN} (*gray line*) and V_{OUT} (*black line*).

inverters designed in a pseudo-CMOS configuration with only p-type OTFTs (2,7-dihexyl-dithieno[2,3-*d*;2′,3′-*d*′]benzo[1,2-*b*;4,5-*b*′]dithiophene (DTBDT-C_6) and polystyrene (PS) blend active layer). The first inverter connected to the extended-gate ion-selective sensor electrode (denoted as "Amplification Inverter" in Fig. 19.6B) amplifies the sensor signal with a tunable gain of 3.1−8.3. The second inverter connected to the Ag/AgCl reference electrode (denoted as "Reference Inverter" in Fig. 19.6B) is used for the self-adjustment of the offset voltage. A K^+ ion sensor was composed of an ion-sensitive membrane laminated on an Ag electrode modified with PEDOT-PSS layer as an ion-to-electron transducer (Fig. 19.6B). The concentration of K^+ was increased from 1 to 10 mM, and the input voltage (V_{IN}) and output voltage (V_{OUT}) of the amplification unit were measured simultaneously. Fig. 19.6C shows the shift of V_{IN} and V_{OUT} relative to those at 1 mM, indicating that the system functions reasonably as an amplifier. The system amplified the of K^+ ion sensor signal from 34 to 160 mV/dec (a factor of 4.6), which exceeds the theoretical sensitivity derived from the Nernst equation (59 mV/dec). This printed potentiometric sensor system is appropriate for amplifying the signals of other types of potentiometric biosensors with enzyme- or antibody-based biorecognition elements described above.

19.2.6 Wearable sensors using microfluidics

A wearable microfluidic system is advantageous for continuous sampling of externally secreted body fluids such as sweat to monitor its composition [76−78]. We are currently developing the prototype of a thin film−assembled wearable microfluidic system integrated with a Na^+ ion selective sensor and an Ag/AgCl reference electrode. Fig. 19.7A shows the design of wearable microfluidic device. The microfluidic device was fabricated by assembling three thin-film layers: a hydrophilic upper planar sheet (100 μm thick), a double-sided adhesive spacer sheet (90 μm thick) with microchannel-formed through hole, and a bottom PEN film (125 μm thick) onto which the Na^+ ion sensor electrode and Ag/AgCl reference electrode were fabricated. The three films were assembled to form a microchannel (length: 1.27 mm, width: 1 mm, height: 25 μm) with a volume of 32 nL. The inlet hole (hole area: 24 mm^2) was created to the bottom film to introduce sweat secreted from the human skin surface. Assuming an average sweat gland density of 200 cm^{-2} [76], the number of sweat glands exposed to the inlet area is calculated to be 48 glands. As the typical human perspiration rate is 1−20 nL/min/gland [76], the detection channel is filled with sweat within 1 minutes. Fig. 19.7B shows potentiometric response of the microfluidic Na^+ ion sensor against the logarithm of Na^+ ion concentration. The sensor was connected to external multimeter to measure the electrical potential difference between the Na^+ ion sensor electrode and the Ag/AgCl reference electrode. An Na^+ ion solution different concentrations of was sequentially introduced into the microchannel. The sensor signal linearly increased with increases on the logarithm of Na^+ ion concentration, suggesting functionally working Na^+ ion sensor and the Ag/AgCl reference electrode in the microfluidic device. This microfluidic Na^+ ion sensor was then placed onto the human skin as

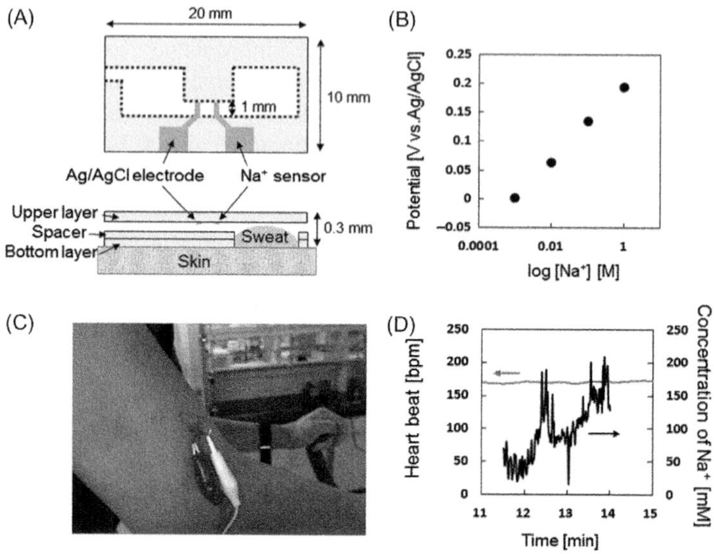

Figure 19.7 (A) Design of the microfluidic Na^+ ion sensor. (B) The Na^+ ion sensor response corresponds to the logarithmic concentration of Na^+ ions. (C) Photograph of the microfluidic Na^+ ion sensor worn on the human skin. (D) Time-course of the sweat Na^+ ion concentration during bicycle ergometer exercise. Heartbeat is also monitored simultaneously using conventional heart rate meter. [This experiment was approved by the institutional review board of Yamagata Prefectural Yonezawa University of Nutrition Science (29-9). Before carrying out these experiments, the purpose of this study was communicated to subjects who provided university-approved informed consent.]

shown in Fig. 19.7C to monitor sweat Na^+ ion concentration during bicycle ergometer exercise. Fig. 19.7D shows the time-course of sweat Na^+ concentration monitored using the microfluidic device. The gray line shows heart beat simultaneously detected with commercially available heart rate monitor. Sweat could be visually observed on the subject's skin after about 10 minutes of exercise. At the same timing, the concentration of sweat Na^+ ions detected with the sensor began to increase from 10 to 150 mM. This concentration range is nearly the same as physiological sweat Na^+ ion [76], suggesting reasonable response for the wearable microfluidic sensor.

19.3 Sensor systems using flexible hybrid electronics

To realize simple, thin, and flexible sensors for IoT applications, total integration of the sensor system, not only for the sensor electrode but also surrounding circuits for amplification, A/D conversion, and wireless transmission of the detected signals, on a thin film substrate is ultimately necessary. In particular, the advanced printing technologies described above will enable the low-cost mass production of the

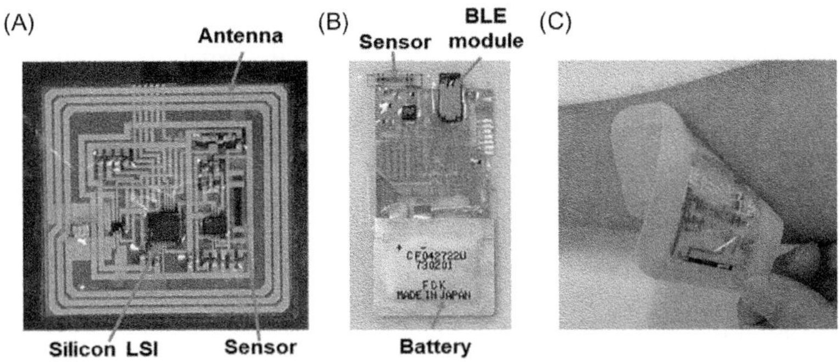

Figure 19.8 Photographs of (A) NFC type and (B) BLE type FHE wireless sensor devices fabricated on plastic film substrates. (C) The FHE-based temperature sensor adhered to human skin using an adhesive bandage.

sensors. Recently, a new approach to realize flexible printed sensors has been adopted, called flexible hybrid electronics (FHE), where LSI die-based silicon circuit technology is utilized together with printing technology on a thin plastic film substrate. In FHE, sensors, interconnects, and antennas are first printed on a flexible plastic film, and then Si-LSI die and resistors are mounted onto the same film substrate. Near-Field Communication (NFC) or Bluetooth Low Energy (BLE) protocols are used for the wireless communication. Although there are few reports on the FHE-based biosensors [79,80], FHE has become the default platform in the field of wearable biosensor development. Fig. 19.8 shows our FHE-based wearable temperature sensor, whereby printed PEDOT:PSS was employed as a temperature-sensitive layer. For the NFC-type device, a coil-shaped antenna was patterned on the film substrate, and a thin Li-battery was used for the BLE-type device. Human body temperature could be continuously monitored by transmitting detected data to a tablet device. Although flexibility, thinness, and cost of FHE devices are remaining issues, this strategy has proved quite useful in enabling the practical use of flexible biosensors. Ultimately, we will replace the Si-LSI die with printed integrated circuits based on OTFT devices, allowing for fully flexible, wearable biosensor devices that can potentially be mass-fabricated at low-cost.

19.4 Conclusion

This chapter describes recent progress in the development of wearable biosensors for noninvasive healthcare applications and details our approach based on the printed electronics. Printed electronics is a promising future technology for low-cost mass fabrication of wearable, flexible, thin film–based biosensor systems. The continuous sampling of slightly excreted fresh bodily fluids such as tears, saliva, urine, and sweat is vitally required for noninvasive monitoring of human health. A microfluidic system is one of the potential solutions for this issue, a configuration that should be simple and flexible to achieve a wearable and conformable device. Besides, some excreted bodily fluids are contaminated. Sweat, for example, can be contaminated with skin surface components such as components of dead skin surface cells and secretions from resident bacteria. Furthermore, the wearable sensor tends to be exposed to dramatically changeable environmental conditions such as skin surface pH and body temperature, which affects sensor performance. For reliable measurements, a multisensing system with not only a biosensor but also pH and temperature sensors will enable reliable monitoring of the analytes in bodily fluids. Printing technology is an attractive near-future option for simple low-cost, mass fabrication of the multisensing systems.

Acknowledgments

We would like to acknowledge each of our laboratory members and our colleagues for productive discussions and their contributions to these experiments. This review involves the results of several projects that are financially supported by Japan Science and Technology Agency (JST).

References

[1] M.U. Ahmed, I. Saaem, P.C. Wu, A.S. Brown, Personalized diagnostics and biosensors: a review of the biology and technology needed for personalized medicine, Crit. Rev. Biotechnol. 34 (2014) 180–196.

[2] C. Dincer, R. Bruch, A. Kling, P.S. Dittrich, G.A. Urban, Multiplexed point-of-care testing-xPOCT, Trends Biotechnol. 35 (2017) 728–742.

[3] L. Syedmoradi, M. Daneshpour, M. Alvandipour, F.A. Gomez, H. Hajghassem, K. Omidfar, Point of care testing: the impact of nanotechnology, Biosens. Bioelectron. 87 (2017) 373.

[4] E.W. Nery, M. Kundys, P.S. Jeleń, M. Jönnsson-Niedziółka, Electrochemical glucose sensing: is there still room for improvement? Anal. Chem. 88 (2016) 11271–11282.

[5] A. Tricoli, N. Nasiri, S. De, Wearable and miniaturized sensor technologies for personalized and preventive medicine, Adv. Funct. Mater. 27 (2017) 1605271.

[6] J. Kim, R. Kumar, A.J. Bandodkar, J. Wang, Advanced materials for printed wearable electrochemical devices: a review, Adv. Electron. Mater. 3 (2017) 1600260.

[7] A.J. Bandodkar, I. Jeerapan, J. Wang, Wearable chemical sensors: present challenges and future prospects, ACS Sens. 1 (2016) 464–482.

[8] Y. Liu, M. Pharr, G.A. Salvatore, Lab-on-Skin: a review of flexible and stretchable electronics for wearable health monitoring, ACS Nano 11 (2017) 9614–9635.

[9] K. Scholten, E. Meng, A review of implantable biosensors for closed-loop glucose control and other drug delivery applications, Int. J. Pharm. 15 (2018) 319–334. Available from: https://doi.org/10.1016/j.ijpharm.2018.02.022.

[10] F. Wu, P. Yu, L. Mao, Bioelectrochemistry for in vivo analysis: interface engineering toward implantable electrochemical biosensors, Curr. Opin. Electrochem. 5 (2017) 152–157.

[11] S.M.R. Islam, D. Kwak, M.D.H. Kabir, M. Hossain, K.S. Kwak, The internet of things for health care: a comprehensive survey, IEEE Access 3 (2015) 678–708.
[12] E.F. Sato, T. Choudhury, T. Nishikawa, M. Inoue, Dynamic aspect of reactive oxygen and nitric oxide in oral cavity, J. Clin. Biochem. Nutr. 42 (2008) 8–13.
[13] H. Nakane, O. Asami, Y. Yamada, T. Harada, N. Matsui, T. Kanno, et al., Salivary chromogranin A as an index of psychosomatic stress response, Biomed. Res. 19 (1998) 401–406.
[14] S. Ishikawa, M. Sugimoto, K. Kitabatake, A. Sugano, M. Nakamura, M. Kaneko, et al., Identification of salivary metabolomic biomarkers for oral cancer screening, Sci. Rep. 6 (2016) 31520.
[15] X. Wang, K.E. Kaczor-Urbanowicz, D.T.W. Wong, Salivary biomarkers in cancer detection, Med. Oncol. 34 (2017) 7.
[16] S. Kumar, S. Padmashree, R. Jayalekshmi, Correlation of salivary glucose, blood glucose and oral candidal carriage in the saliva of type 2 diabetics: a case-control study, Contemp. Clin. Dent. 5 (2014) 312–317.
[17] W. Gao, S. Emaminejad, H.Y.Y. Nyein, S. Challa, K. Chen, A. Peck, et al., Fully integrated wearable sensor arrays for multiplexed in situ perspiration analysis, Nature 529 (2016) 509–514.
[18] S. Emaminejad, W. Gao, E. Wu, Z.A. Davies, H.Y.Y. Nyein, S. Challa, et al., Autonomous sweat extraction and analysis applied to cystic fibrosis and glucose monitoring using a fully integrated wearable platform, Proc. Natl. Acad. Sci. USA 114 (2017) 4625–4630.
[19] Y. Epstein, W.O. Roberts, The pathopysiology of heat stroke: an integrative view of the final common pathway, Scand. J. Med. Sci. Sports 21 (2011) 742–748.
[20] M.M. Raiszadeh, M.M. Ross, P.S. Russo, M.A. Schaepper, W. Zhou, J. Deng, et al., Proteomic analysis of eccrine sweat: implications for the discovery of schizophrenia biomarker proteins, J. Proteome Res. 11 (2012) 2127–2139.
[21] S. Rieg, H. Steffen, S. Seeber, A. Humeny, H. Kalbacher, K. Dietz, et al., Deficiency of dermcidin-derived antimicrobial peptides in sweat of patients with atopic dermatitis correlates with an impaired innate defense of human skin in vivo, J. Immunol. 174 (2005) 8003–8010.
[22] J. Kim, M. Kim, M.S. Lee, K. Kim, S. Ji, Y.T. Kim, et al., Wearable smart sensor systems integrated on soft contact lenses for wireless ocular diagnostics, Nat. Commun. 8 (2017) 14997.
[23] K. Mitsubayashi, T. Arakawa, Cavitas sensors: contact lens type sensors and mouthguard sensors, Electroanalysis 28 (2016) 1170–1187.
[24] T. Arakawa, Y. Kuroki, H. Nitta, P. Chouhan, K. Toma, S. Sawada, et al., Mouthguard biosensor with telemetry system for monitoring of saliva glucose: a novel cavitas sensor, Biosens. Bioelectron. 84 (2016) 106–111.
[25] J. Kim, G. Valdés-Ramírez, A.J. Bandodkar, W. Jia, A.G. Martinez, A. Ramírez, et al., Non-invasive mouthguard biosensor for continuous salivary monitoring of metabolites, Analyst 139 (2014) 1632–1636.
[26] P. Kassal, J. Kim, R. Kumar, W.R. de Araujo, I.M. Steinberg, M.D. Steinberg, et al., Smart bandage with wireless connectivity for uric acid biosensing as an indicator of wound status, Electrochem. Commun. 56 (2015) 6–10.
[27] T.R. Dargaville, B.L. Farrugia, J.A. Broadbent, S. Pace, Z. Upton, N.H. Voelcker, Sensors and imaging for wound healing: a review, Biosens. Bioelectron. 41 (2013) 30–42.
[28] J. Kim, I. Jeerapan, S. Imani, T.N. Cho, A. Bandodkar, S. Cinti, et al., Noninvasive alcohol monitoring using a wearable tattoo-based iontophoretic-biosensing system, ACS Sens. 1 (2016) 1011–1019.

[29] A. Koh, D. Kang, Y. Xue, S. Lee, R. Pielak, J. Kim, et al., A soft, wearable microfluidic device for the capture, storage, and colorimetric sensing of sweat, Sci. Transl. Med. 8 (2016) 366ra165.
[30] H. Lee, C. Song, Y.S. Hong, M.S. Kim, H.R. Cho, T. Kang, et al., Wearable/disposable sweat-based glucose monitoring device with multistage transdermal drug delivery module, Sci. Adv. 3 (2017) e1601314.
[31] E.V. Karpova, E.V. Shcherbacheva, A.A. Galushin, D.V. Vokhmyanina, E.E. Karyakina, A.A. Karyakin, Noninvasive diabetes monitoring through continuous analysis of sweat using flow-through glucose biosensor, Anal. Chem. 91 (2019) 3778–3783.
[32] S. Wang, Y. Wu, Y. Gu, T. Li, H. Luo, L.H. Li, et al., Wearable sweatband sensor platform based on gold nanodendrite array as efficient solid contact of ion-selective electrode, Anal. Chem. 89 (2017) 10224–10231.
[33] A. Martín, J. Lim, J.F. Kumiawan, J.R. Sempionatto, J.R. Moreto, G. Tang, et al., Epidermal microfluidic electrochemical detection system: Enhanced sweat sampling and metabolite detection, ACS Sens. 2 (2017) 1860–1868.
[34] S. Imani, A.J. Bandodkar, A.M.V. Mohan, R. Kumar, S. Yu, J. Wang, et al., A wearable chemical-electrophysiological hybrid biosensing system for real-time health and fitness monitoring, Nat. Commun. 7 (2016) 11650.
[35] M. Mizukami, S. Oku, S.I. Cho, M. Tatetsu, M. Abiko, M. Mamada, et al., A solution-processed organic thin-film transistor backplane for flexible multiphoton emission organic light-emitting diode displays, IEEE Electron. Dev. Lett. 36 (2015) 841–843.
[36] G.S. Ryu, J.S. Kim, S.H. Jeong, C.K. Song, A printed OTFT- backplane for AMOLED display, Org. Electron. 14 (2013) 1218–1224.
[37] A. Yamamura, H. Matsui, M. Uno, N. Isahaya, Y. Tanaka, M. Kudo, et al., Painting integrated complementary logic circuits for single-crystal organic transistors: A demonstration of a digital wireless communication sensing tag, Adv. Electron. Mater. 3 (2017) 1600456.
[38] M. Jung, J. Kim, J. Noh, N. Lim, C. Lim, G. Lee, et al., All-printed and roll-to-roll-printable 13.56-MHz-operated 1-bit RF tag on plastic foils, IEEE Trans. Electron. Dev. 57 (2010) 571–580.
[39] R. Ganesan, J. Krumm, S. Pankalla, K. Ludwig, M. Glesner, Design of an organic electronic label on a flexible substrate for temperature sensing, Proc. Eur. Solid-State Circuit Conf. 423–426, 2013.
[40] X. Wang, O. Larsson, D. Platt, S. Nordlinder, I. Engquist, M. Berggren, et al., An all-printed wireless humidity sensor label, Sens. Actuators B: Chem. 166–167 (2012) 556–561.
[41] X. Strakosas, B. Wei, D.C. Martin, R.M. Owens, Biofunctionalization of polydioxythiophene derivatives for biomedical applications, J. Mater. Chem. B 4 (2016) 4952–4968.
[42] K. Izumi, Y. Yoshida, S. Tokito, Soft blanket gravure printing technology for finely patterned conductive layers on three-dimensional or curved surfaces, Jpn. J. Appl. Phys. 56 (2017). 05EA03.
[43] A. Miyamoto, S. Lee, N.F. Cooray, S. Lee, M. Mori, N. Matsuhisa, et al., Inflammation-free, gas-permeable, lightweight, stretchable on-skin electronics with nanomeshes, Nat. Nanotech. 12 (2017) 907–913.
[44] K. Hayasaka, H. Matsui, Y. Takeda, R. Shiwaku, Y. Tanaka, T. Shiba, et al., Compact organic complementary D-type flip-flop circuits fabricated with inkjet printing, Adv. Electron. Mater. 3 (2017) 1700208.
[45] Y. Yoshida, H. Wada, K. Izumi, S. Tokito, Highly conductive metal interconnects on three-dimensional objects fabricated with omnidirectional ink jet printing technology, Jpn. J. Appl. Phys. 56 (2017). 05EA01.

[46] D. Shiokawa, K. Izumi, R. Sugano, T. Sekine, T. Minami, D. Kumaki, et al., Development of a silver nanoparticle ink toward fine line patterning using gravure offset printing, Jpn. J. Appl. Phys. 56 (2017). 05EA04.
[47] K. Izumi, Y. Yoshida, S. Tokito, Improved fine layer patterning using soft blanket gravure printing technology, Flexible and Printed Electronics, vol. 3, 2018, pp. 15011–15018.
[48] K. Izumi, Y. Yoshida, S. Tokito, Real-time observation of the reception of silver ink in soft blanket gravure printing, IEEE-ICEP, 16932399, 2017.
[49] H. Nakamura, Y. Yoshida, K. Izumi, Simultaneous printing of interconnects with different line width using soft blanket gravure printing, IEEE-ICEP, 16932476, 2017.
[50] K. Izumi, Y. Yoshida, S. Tokito, Novel soft blanket gravure printing technology with an improved ink transfer process, Flexible and Printed Electronics vol. 2 (2017) 24003–24009.
[51] Y. Takeda, Y. Yoshimura, R. Shiwaku, K. Hayasaka, T. Sekine, T. Okamoto, et al., Organic complementary inverter circuits fabricated with reverse offset printing, Adv. Electron. Mater. 2017 (4) (2018) 1700313.
[52] K. Fukuda, Y. Yoshimura, T. Okamoto, Y. Takeda, D. Kumaki, Y. Katayama, et al., Reverse-offset printing optimized for scalable organic thin-film transistors with submicrometer channel lengths, Adv. Electron. Mater. 1 (2015) 1500145.
[53] S.V. Murphy, A. Atala, 3D bioprinting of tissues and organs, Nat. Biotechnol. 32 (2014) 773–785.
[54] C. Ru, J. Luo, S. Xie, Y. Sun, A review of non-contact micro- and nano-printing technologies, J. Micromech. Microeng. 24 (2014) 053001.
[55] T. Minami, Y. Sasaki, T. Minamiki, S. Wakida, R. Kurita, O. Niwa, et al., Selective nitrate detection by an enzymatic sensor based on an extended-gate type organic field-effect transistor, Biosens. Bioelectron. 81 (2016) 87–91.
[56] T. Minami, T. Sato, T. Minamiki, S. Tokito, An extended-gate type organic FET based biosensor for detecting biogenic amines in aqueous solution, Anal. Sci. 31 (2015) 721.
[57] T. Minami, T. Sato, T. Minamiki, K. Fukuda, D. Kumaki, S. Tokito, A novel OFET-based biosensor for the selective and sensitive detection of lactate levels, Biosens. Bioelectron. 74 (2015) 45–48.
[58] T. Minamiki, T. Minami, Y. Sasaki, S. Wakida, R. Kurita, O. Niwa, et al., Label-free detection of human glycoprotein (CgA) using an extended-gated organic transistor-based immunosensor, Sensors 16 (2016) 2033.
[59] T. Minamiki, T. Minami, Y. Sasaki, R. Kurita, O. Niwa, S. Wakida, et al., An organic field-effect transistor with an extended-gate electrode capable of detecting human immunoglobulin A, Anal. Sci. 31 (2015) 725.
[60] T. Minamiki, T. Minami, R. Kurita, O. Niwa, S. Wakida, K. Fukuda, et al., A label-free immunosensor for IgG based on an extended-gate type organic field effect transistor, Materials 7 (2014) 6843–6852.
[61] T. Minamiki, T. Minami, R. Kurita, O. Niwa, S. Wakida, K. Fukuda, et al., Accurate and reproducible detection of proteins in water using an extended-gate type organic transistor biosensor, Appl. Phys. Lett. 104 (2014) 243703.
[62] K. Fukuda, T. Minamiki, T. Minami, M. Watanabe, T. Fukuda, D. Kumaki, et al., Printed organic transistors with uniform electrical performance and their application to amplifiers in biosensors, Adv. Electron. Mater. 1 (2015) 1400052.
[63] R. Shiwaku, H. Matsui, K. Nagamine, M. Uematsu, T. Mano, Y. Murayama, et al., A printed organic amplification system for wearable potentiometric electrochemical sensors, Sci. Rep. 8 (2018) 3922.

[64] H. Furusawa, Y. Ichimura, S. Harada, M. Uematsu, S. Xue, K. Nagamine, et al., Electric charge detection of sparse organic acid molecules using an organic field-effect transistor (OFET)-based sensor, Bull. Chem. Soc. Jpn. 91 (2018) 1020–1025.
[65] Y. Sasaki, T. Minami, T. Minamiki, S. Tokito, An organic transistor-based electrical assay for copper (II) in water, Electrochemistry 85 (2017) 775–778.
[66] T. Minamiki, Y. Sasaki, S. Tokito, T. Minami, Label-free direct electrical detection of a histidine-rich protein with sub-femtomolar sensitivity using an organic field-effect transistor, ChemistryOpen 6 (2017) 472–475.
[67] Y. Sasaki, T. Minamiki, S. Tokito, T. Minami, A molecular self-assembled colourimetric chemosensor array for simultaneous detection of metal ions in water, Chem. Commun. 53 (2017) 6561–6564.
[68] T. Minami, T. Minamiki, S. Tokito, Detection of mercury (II) ion in water using an organic field-effect transistor with a cysteine-immobilized gold electrode, Jpn. J. Appl. Phys. 55 (2016). 04E02.
[69] T. Minami, T. Minamiki, S. Tokito, Electric detection of phosphate anions in water by an extended-gate type organic field-effect transistor functionalized by a zinc(II)-dipicolylamine derivative, Chem. Lett. 45 (2016) 371–373.
[70] T. Minamiki, T. Minami, P. Koutnik, P. Anzenbachen Jr, S. Tokito, Antibody- and label-free phosphoprotein sensor device based on an organic transistor, Anal. Chem. 88 (2016) 1092–1095.
[71] T. Minami, Y. Sasaki, T. Minamiki, P. Koutnik, P. Anzenbacher Jr, S. Tokito, A mercury (II) ion sensor device based on an organic field effect transistor with an extended-gate modified by dipicolylamine, Chem. Commun. 51 (2015) 17666.
[72] T. Minami, T. Minamiki, S. Tokito, An anion sensor based on an organic field effect transistor, Chem. Commun. 51 (2015) 9491–9494.
[73] T. Minami, T. Minamiki, K. Fukuda, D. Kumaki, S. Tokito, Cysteine detection in water using an organic field-effect transistor, Jpn. J. Appl. Phys. 54 (2015) 04DK01.
[74] T. Minami, T. Minamiki, Y. Hashima, D. Yokoyama, T. Sekine, K. Fukuda, et al., An extended-gate type organic field effect transistor functionalised by phenylboronic acid for saccharide detection in water, Chem. Commun. 50 (2014) 15613–15615.
[75] M.J. Esplandiú, H. Hagenström, D.M. Kolb, Functionalized self-assembled alkanethiol monolayers on Au(111) electrodes: 1. surface structure and electrochemistry, Langmuir 17 (2001) 828–838.
[76] Z. Sonner, E. Wilder, J. Heikenfeld, G. Kasting, F. Beyette, D. Swaile, et al., The microfluidics of the eccrine sweat gland, including biomarker partitioning, transport, and biosensing implications, Biomicrofluidics 9 (2015) 031301.
[77] J. Choi, Y. Xue, W. Xia, T.R. Ray, J.T. Reeder, A.J. Bandodkar, et al., Soft, skin-mounted microfluidic systems for measuring secretory fluidic pressures generated at the surface of the skin by eccrine sweat glands, Lab Chip 17 (2017) 2572–2580.
[78] S. Xu, Y. Zhang, L. Jia, K.E. Mathewson , K.I. Jang, J. Kim, et al., Soft microfluidic assemblies of sensors, circuits, and radios for the skin, Science 344 (2014) 70–74.
[79] V. Beni, D. Nilsson, P. Arven, P. Norberg, G. Gustafsson, A.P.F. Turner, Printed electrochemical instruments for biosensors, ECS J. Solid State Sci. Technol. 4 (2015) S3001–S3005.
[80] D.P. Rose, M.E. Ratterman, D.K. Griffin, L. Hou, N. Kelley-Loughnane, R.R. Naik, et al., Adhesive RFID sensor patch for monitoring of sweat electrolytes, IEEE Trans. Biomed. Eng. 62 (2015) 1457–1465.

Self-monitoring of fat metabolism using portable/wearable acetone analyzers

20

Satoshi Hiyama
Research Laboratories, NTT DOCOMO, Inc., Yokosuka, Japan

20.1 Introduction

In April 2017, NTT DOCOMO announced the Medium Term Strategy 2020 "Declaration beyond" with a focus on value generation through cocreation with diverse partners and solutions to social issues through services [1]. Since these social issues include those in the medical and health care fields, R&D at NTT DOCOMO aims to solve issues arising in the various stages of life such as lifestyle-related diseases and to contribute to healthy and long lives.

NTT DOCOMO has undertaken the visualization of fat metabolism with the aim of alleviating and preventing obesity, which can lead to all sorts of diseases and raise the risk of acquiring lifestyle-related diseases. Specifically, we have developed breath analyzers [2,3] and arm-wearable monitors [4,5] for measuring acetone: a marker of fat metabolism. Biological gases such as acetone contained in breath or emitted from the skin's surface provide abundant biological data on metabolic processes, reflecting individual differences without the pain of drawing blood, and the use of such devices requires no special qualifications in collecting or analyzing samples. Therefore, there are high expectations for applying biological gases to self-health management through trouble-free multiitem analysis that can be performed at home or outside. This book chapter provides overviews of these newly developed devices and assumed applications.

20.2 Portable breath acetone analyzer

More than 200 compounds have been identified in human breath, some of which have been correlated to various diseases such as metabolic disorders and lung or gastrointestinal diseases. Thus, breath analysis has received increasing attention in recent years for noninvasive clinical diagnoses [6–8]. It is well known that the concentrations of endogenous compounds found in human breath, such as inorganic

Chemical, Gas, and Biosensors for the Internet of Things and Related Applications.
DOI: https://doi.org/10.1016/B978-0-12-815409-0.00021-8

gases (e.g., nitric oxide and carbon dioxide) and volatile organic compounds (e.g., acetone, ethanol, isoprene, and ammonia), are in the range of parts per million (ppm) to parts per trillion (ppt) and their compositions vary widely from person to person [7].

Acetone is a metabolite derived from fat-burning, produced in the blood, that is, expelled through alveoli of the lungs during exhalation [6]. Thus, there is a strong correlation between the concentration of breath acetone and that of blood acetone [9]. Considering that breath acetone could be a good indicator for monitoring fat metabolism, extensive previous works have studied the relationships between breath acetone and diabetes [10–12], ketogenic meals [13–15], exercise [16,17], sleep [18], age and gender [19], dieting [3,20], and blood glucose (BG) and hemoglobin A_{1c} (HbA_{1c}) [10,11]. To detect low concentrations of breath acetone accurately, typically, gas chromatography (GC)-based methods and/or mass spectrometry (MS)-based methods have been used [12–14,16–20]. However, current GC- and MS-based methods are still not practically suitable for point-of-care instrumentation for diet-conscious people who wish to monitor their own fat metabolism at home or outside, because these methods often require large instruments and skilled operators. As obesity increases the risk of lifestyle-related diseases, enabling users to measure breath acetone concentration by themselves and to monitor the state of fat-burning could play a pivotal role in daily diet management and diabetes diagnosis [3,20–22].

20.2.1 Prototyped analyzer

We aimed to prototype a portable and easy-to-use breath acetone analyzer [2,3] that does not rely on GC- and MS-based methods (Fig. 20.1). Our prototype consists of a pressure sensor to detect the blowing breath and two types of semiconductor-based gas sensors with no gas separation columns to calculate the breath acetone concentration from the output signals collected by the two sensors. Our prototype is easy to use, simply involving blowing into the prototype for six seconds. The measured results are automatically sent to an Android-based smartphone or tablet via

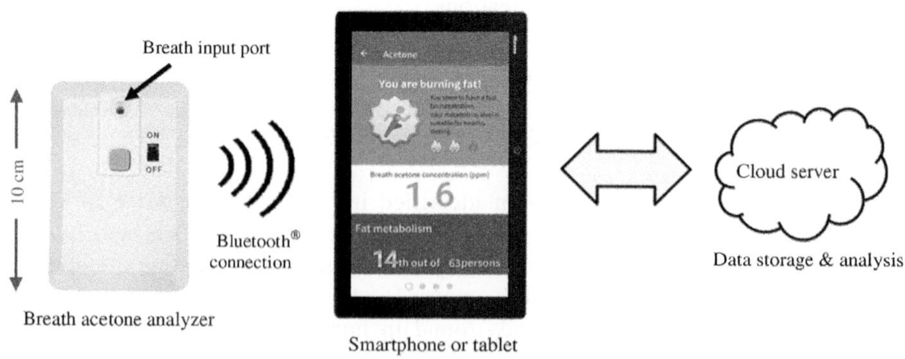

Figure 20.1 Overview of our prototyped system (portable breath acetone analyzer).

either a wireless Bluetooth connection or a wired audio cable and are displayed a few seconds later. Thus, our prototype requires, at most, only 10 seconds for a single measurement. The prototyped analyzer was $65 \times 100 \times 25$ mm in chassis size and 125 g in weight, including two AA batteries. The measured results are also automatically sent to a cloud server from the smartphone or tablet to be stored and analyzed as big data. This configuration could work as an example of an Internet of Things (IoT) system.

Our prototype was designed to detect breath acetone ranging from 0.2 to 50 ppm and to display the measured results with a resolution of 0.1 ppm. Our design is feasible because breath acetone concentration typically varies from 0.2 to 2.4 ppm in fasting healthy adults [19] to below 10 ppm in diabetics [11]. It is known that semiconductor-based gas sensors are compact, highly sensitive, cheap, and maintenance free, and so they are often used in commercially available mouth odor and breath alcohol checkers. However, the concentration of breath acetone is one or two orders of magnitude lower than mouth odor or alcohol concentration, meaning that the effects of interference from other types of gaseous constituents must be taken into account when detecting breath acetone. To meet this requirement, our prototype calculates the breath acetone concentration using the signals from two sensors with different sensitivity characteristics; the first sensor was developed by FIS Inc. (Osaka, Japan) and was made of platinum-doped tungsten oxide that has particularly high sensitivity to acetone, whereas the second sensor was commercially available and was made of tin oxide (SB-30, FIS Inc.) that has almost equal sensitivity to both acetone and interference gases such as hydrogen and ethanol. Both sensors were operated at 400°C and required at least two minutes to stabilize their sensitivities after their power supplies were started. We used a coil heater small enough to fit within a micrometer-scaled ellipsoidal space (ca. $300 \times 300 \times 500$ μm), and thus the housing of each sensor was not hot during heating. Our prototype was able to run at least 10 hours under a continuous running test during which the heaters and Bluetooth modules were continuously on.

Considering that breath hydrogen, ethanol, and relative humidity may affect sensor sensitivity, we prepared simulated breath that contained known concentrations of acetone and/or ethanol, with background hydrogen (50 ppm) and humidity (saturated water vapor at 37°C), to determine calibration curves in advance of actual breath tests. The calibration curves obtained were stored within the circuit of our prototype for the calculation of breath acetone concentration. Which calibration curves to use were systematically determined, depending on the output resistance signals from the two sensors for input breath. This enables breath acetone to be calculated while taking into account the presence of ethanol, hydrogen, and humidity.

To investigate the accuracy of our prototype, the concentrations of breath acetone obtained from our prototype and from conventional GC were compared [3]. Fig. 20.2 shows the scatter plots of breath acetone concentrations obtained from 17 healthy adult subjects (11 men and 6 women volunteers, ranging from 21 to 70 years old). The plots show that the concentrations of breath acetone obtained from our prototype and from conventional GC have a strong correlation ($R = 0.95$, $P < .001$); demonstrating that our prototype was used as a practical breath acetone

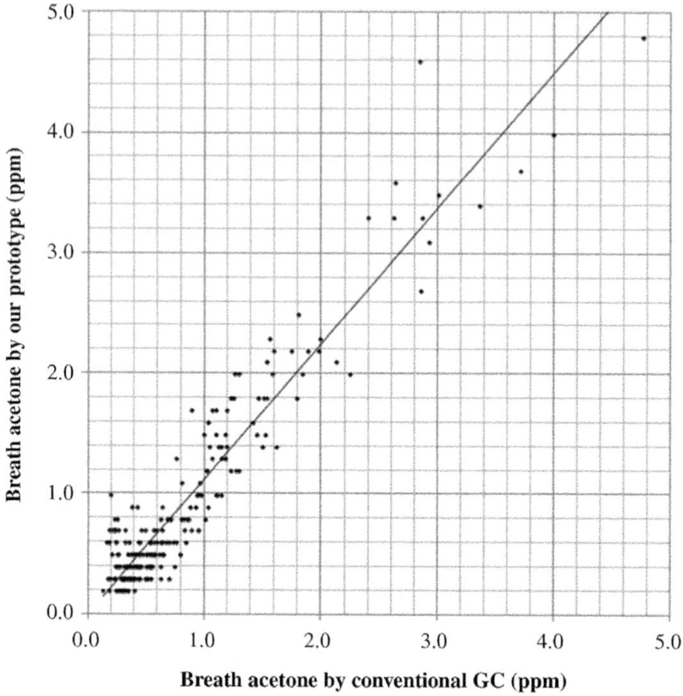

Figure 20.2 Scatter plots of breath acetone concentrations obtained from our prototype and from a conventional GC.
Source: Copyright IOP Publishing. Reproduced with permission from T. Toyooka, S. Hiyama, Y. Yamada, J. Breath Res. 7 (2013) 036005. All rights reserved.

checker with a reasonable range of measurement deviations. By comparing with the results obtained from the conventional GC, we calculated the standard error of the results obtained from our prototype and found that it was ± 0.1 ppm. This indicates that our prototype cannot distinguish between concentration differences below 0.2 ppm. The errors may have arisen from the uncontrolled flow rate and disturbing breath constituents of each individual, which were not taken into account in our calibration curves. Although an improvement in sensor sensitivity may be required in future work, we believe that our prototype is still useful because it is portable and easy-to-use. All of the volunteer subjects, including persons who were not familiar with electronic devices, were able to use our prototype and were able to measure their breath acetone concentrations by themselves. Furthermore, unlike the typical GC- and MS-based methods, our prototype can measure the concentrations of breath acetone within 10 seconds; indicating that we can monitor changes in breath acetone even if the time between measurements is very short (e.g., measurements every minute during exercise).

Currently NTT DOCOMO has started to sell the portable breath acetone analyzer for corporate customers such as health food product companies.

20.2.2 Applicability to diet support

To investigate the applicability to diet support, 17 healthy adult subjects (11 men and 6 women volunteers, ranging from 21 to 70 years old) were divided into three groups, according to the enforcement or otherwise of a controlled caloric intake and/or light exercise [3]. Loss of body fat from day 1 and breath acetone concentrations were measured in the morning before breakfast throughout the experiments, over 14 days (Fig. 20.3). We found that subjects who kept to their regular life

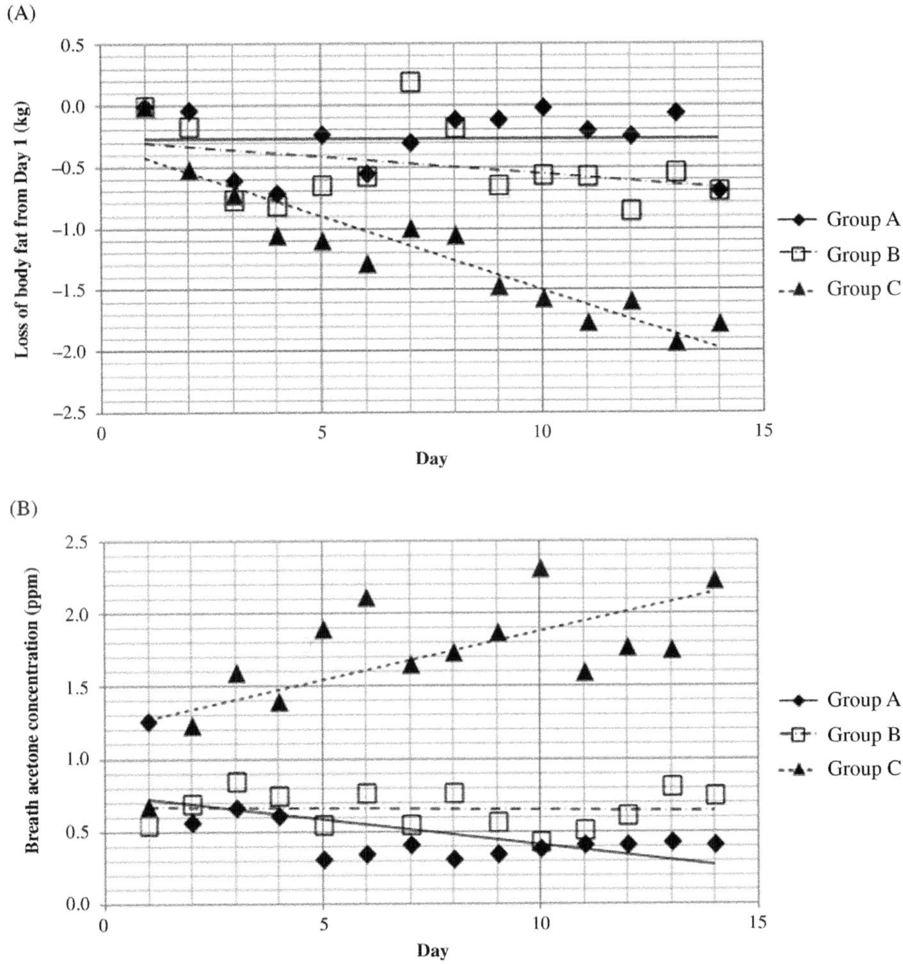

Figure 20.3 Monitoring of fat-burning throughout the experiments over 14 days. (A) Average amount of loss of body fat from day 1. (B) Average concentrations of breath acetone during each day. Solid and dashed lines represent simple trend lines that are intended to show whether the plots are in uptrend or downtrend. All of the lines are linearly fitted. Source: Copyright IOP Publishing. Reproduced with permission from T. Toyooka, S. Hiyama, Y. Yamada, J. Breath Res. 7 (2013) 036005. All rights reserved.

(group A) and subjects who followed enforced light exercise (group B) were not able to lose significant fat, and their breath acetone concentrations remained approximately constant. Although subjects in group B definitely performed light exercises, such as jogging or fast walking for 30–60 minutes per day throughout the experiments, the effect of exercise on fat-burning was limited. This is probably because the subjects were able to decide their walking speed and distance on their own, and thus, their exercise intensity was too weak and was far away from the anaerobic threshold in which exercise intensity is suitable for losing fat. Alternatively, considering that the fat of the subjects in group B decreased slightly over time and previous works that found an increase in breath acetone concentration during exercise [16,17], we should have measured them during/after the light exercise in addition to the early morning measurements. In contrast, subjects who followed enforced controlled caloric intake and light exercise (group C) were able to lose significant fat, and their breath acetone concentrations were increased significantly. Taking into account that their exercise intensity was almost the same as that of group B, we can suppose that controlled caloric intake was very effective for fat-loss, resulting in an increased concentration of breath acetone. These results indicate that breath acetone could be a good indicator for monitoring fat-burning and support the results obtained from previous work aimed at daily dietary management [20].

20.2.3 Applicability to diabetes care at home

Cells from people with diabetes cannot metabolize sugars normally and thus they metabolize fat even when sufficient glucose remains in the blood. Breath acetone could also be an indicator of diabetes control [23–25] because uncontrolled patients tend to exhale high breath acetone concentrations and their breath smells of acetone [26,27]. Although breath acetone has been extensively investigated from the point of view of type of diabetes [23,24,28], BG level [11,24], and the situation such as exercise [29], or pregnancy [30], it has not yet been investigated from the therapeutic method or medication point of view in diabetes care.

Seventy-seven subjects with diabetes (37 men and 40 women aged 32–95 years) and 11 nondiabetic but ill subjects (4 men and 7 women aged 65–90 years) were recruited to investigate the breath acetone from the therapeutic method and medication point of view of diabetes care. Analysis of breath acetone using our portable device and blood tests to obtain the BG and HbA_{1c} level of each subject were conducted before breakfast during the medical examinations at a clinic. We first explored whether breath acetone analysis was applicable for noninvasive self-monitoring of BG levels in patients with diabetes. Fig. 20.4 shows the relationship between breath acetone concentration of (A) either BG or HbA_{1c} levels of all the diabetes subjects and (B) diabetes subjects using dietary therapy and those without diabetes. The results showed no correlations between breath acetone concentrations and either BG ($R = 0.18$) or HbA_{1c} levels ($R = 0.14$) of all those with diabetes, while correlations were found between breath acetone concentrations and either BG ($R = 0.57$) or HbA_{1c} levels ($R = 0.72$) of those with diabetes using dietary therapy. These results indicate that some medications may affect fat metabolism and cause

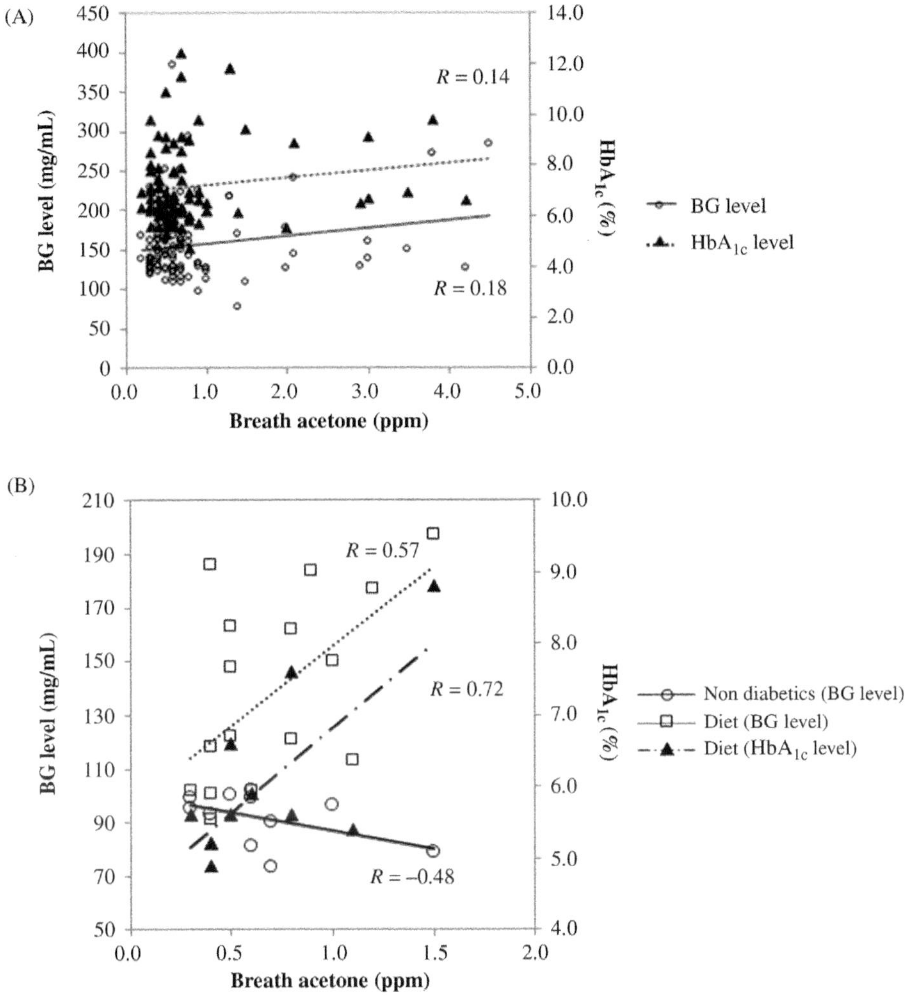

Figure 20.4 (A) Scatter plots between breath acetone concentrations and either blood glucose (BG) or HbA_{1c} levels of all diabetes subjects. (B) Scatter plots between breath acetone concentrations and either blood glucose (BG) or HbA_{1c} levels of diabetes subjects on dietary therapy and nondiabetes subjects.

the disruption of significant correlation between breath acetone and BG level. Nevertheless, it is notable that BG and HbA_{1c} levels of diabetes subjects on dietary therapy can be roughly estimated from the breath acetone concentration. Our results also found that there was a negative correlation between breath acetone concentrations and BG levels in nondiabetes subjects ($R = -0.48$), while a positive correlation was observed between breath acetone concentrations and BG levels of diabetes subjects using dietary therapy ($R = 0.57$). These results reflect the difference in the glucose-metabolizing capability of cells between those with/without diabetes.

To investigate the impact of oral agents and/or insulin therapy on breath acetone concentration, we grouped the subjects with diabetes under oral agent therapy and/or insulin therapy into five categories according to their prescribed medications. Then, their average breath acetone concentrations were compared with those of diabetes subjects under dietary therapy (Fig. 20.5). The five types of medication are as follows: pioglitazone, which enhances insulin sensitivity; glimepiride or alogliptin benzoate, which are insulin secretagogues; phosphate hydrate, which causes antihyperglycemic action; voglibose, which delays absorption of glucose; and insulin. We found that subjects who took pioglitazone had a low concentration of breath acetone, while subjects who took phosphate hydrate or insulin had a high concentration of breath acetone compared with subjects under dietary therapy. These results indicate that some antidiabetic medications affect fat metabolism and breath acetone concentration. If the extent of the disruption can be estimated and the breath acetone concentrations are corrected, the BG and HbA_{1c} levels of diabetes patients using oral agents or insulin therapy could be noninvasively estimated.

Our data suggest that a portable and easy-to-use breath acetone analyzer could noninvasively be used to roughly estimate the BG and HbA_{1c} levels of subjects with diabetes under dietary therapy at home. This study is remarkable in showing

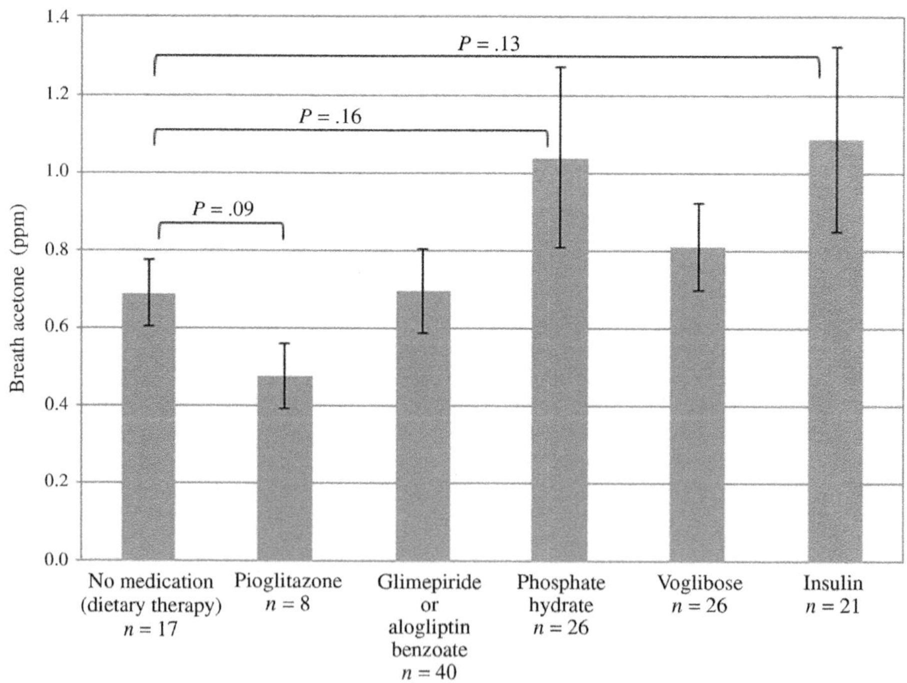

Figure 20.5 Average breath acetone concentrations of diabetes subjects with and without medications.

the possibility of alleviating blood sampling and introducing a method for noninvasive self-monitoring of BG levels for diabetes.

20.2.4 Applicability to "Health Kiosk"

A "health kiosk" is a piece of equipment that enables users to check for health problems by activating a variety of sensors and health management devices installed in the kiosk while following directions on a screen [31]. This kiosk, which was developed and manufactured by Nelsite Inc. under the guidance of the Experimental Center for Social System Technologies (Fukuoka Industry, Science & Technology Foundation) and System LSI Research Center, Kyushu University, is capable of performing more than 14 types of self-health examinations including height, weight, blood pressure, body fat percentage, body temperature, pulse, visual acuity, hearing acuity, lung capacity, glaucoma, cataracts, electrocardiogram, mental health, and dementia (Fig. 20.6). After a personal authentication process by a smart card, the kiosk displays the results of each examination on the screen for the user to view while also storing them on a network server. The user can review these stored results later on a personal computer or mobile terminal.

Our portable breath acetone analyzer is easy to operate and has a compact and light configuration. These features make for easy installation in a health kiosk enabling the provision of a function for examining fat metabolism. The addition of

① Display for guidance and glaucoma, cataract, mental health and dementia tests

② Body height sensor

③ Headphones for audiometric test

④ IC card reader

⑤ Remote controller for eyesight test

⑥ Body temperature sensor

⑦ Spirometer

⑧ Blood pressure, pulse and electrocardiographic monitor

⑨ Breath analyzer

⑩ Peephole for eyesight test

⑪ Weight scale

Figure 20.6 Overview of a health kiosk booth.

this breath analyzer has made this health kiosk the world's most advanced self-health examination device for determining the presence of metabolic disorders brought on by diabetes, eating disorders, and excessive diets.

This health kiosk is expected to be useful in determining the need for consultation at a clinic or hospital, for managing company workers' health and productivity, and for early detection of diseases. It should be particularly useful in regions having a shortage of physicians and/or on remote islands. Going forward, the plan is to conduct usage experiments both inside and outside Japan targeting the company workers and the residents of such areas with the aim of assessing the effectiveness of this health kiosk. We would also like to see this health kiosk installed in companies, public facilities, drug stores, and other establishments outside the home to make it easy for people to perform self-health examinations.

20.3 Wearable skin acetone analyzer

Many compounds, including acetone, are emitted from human skin [32,33], and breath and skin acetone concentrations are correlated [34]. Compared with breath acetone, skin acetone is advantageous for analysis because it is continuously emitted, and analysis requires no active action on the part of the user, unlike the deep exhalation required for breath analysis. Skin analysis is also more precise than breath analysis as it excludes factors such as the flow rate of exhalation.

A wearable analyzer for skin acetone would provide a powerful tool for preventing and alleviating lifestyle-related diseases. However, development of such devices is challenging. This is because skin acetone is typically emitted at concentrations of only several tens of parts per billion (ppb), which is too low to be detected by small commercially available sensors. For example, semiconductor-based gas sensors have limits of detection (LOD) around 200 ppb [3]. In contrast, existing high-sensitivity methods for skin acetone analysis use large apparatus, such as GCs or liquid chromatographs. To develop a wearable device, either the size of the high-sensitivity devices or the sensitivity requirements of the small devices needs to be reduced.

20.3.1 Skin acetone concentrator

To reduce the detector sensitivity requirements, we proposed skin acetone concentrator that concentrates acetone emitted from skin [5]. Zeolite was selected for concentrating skin acetone, which was then detected using a semiconductor-based gas sensor. Zeolites are microporous materials that adsorb gaseous molecules and are widely used as adsorbents, ion exchangers, and catalysts [35]. More than 200 structure types of zeolites have been reported, and we found that a specific zeolite could be targeted to efficient acetone adsorption/desorption agent by the selection of appropriate structure type and its hydrophobicity. The zeolite 390HUA (Tosoh Corp., Tokyo, Japan) that has FAU structure (with 7.4 Å pores) and has high hydrophobicity (500 in SiO_2/Al_2O_3 ratio) was the best to concentrate skin acetone [5].

To investigate the feasibility of our proposed method, skin acetone was measured using the semiconductor-based acetone sensor with and without the concentrator (390HUA zeolite) under simulated real life conditions [5]. To achieve this, the closed space was intentionally collapsed in a cyclic manner during the 20 minutes gas collection (Fig. 20.7). The sensitivity of the acetone sensor without the concentrator drastically decreased from 1.00 to 0.83 within the first 5 minutes from the start of the measurement because of the increased concentration of skin acetone in the closed space. It then plateaued at around 0.83 because of the intentional and cyclic collapsing of the closed space. Such collapsing is very likely in daily life, where some activities are accompanied by twisting and elastic movement of the skin surface. After the 20 minutes collection time, the sensitivity increased back to 1.00 because the closed space was opened and the enclosed gas was completely exchanged with clean air. The sensitivity is indicated by the ratio of the electrical resistance of the gas sensor in air (R_{air}) to the electrical resistance of the gas sensor in the target gas (R). We set the LOD to $R/R_{air} = 0.8$ to avoid false detection and to conduct reliable gas sensing. Considering that the sensitivities of the acetone sensor without the concentrator were always $R/R_{air} > 0.8$, we concluded that the acetone sensor alone would not be able to detect skin acetone under the simulated conditions. In comparison, the sensitivities of the acetone sensor with the concentrator were all much less than 0.8 when the adsorbed skin acetone was desorbed every 20 minutes by flash heating the concentrator. Therefore,

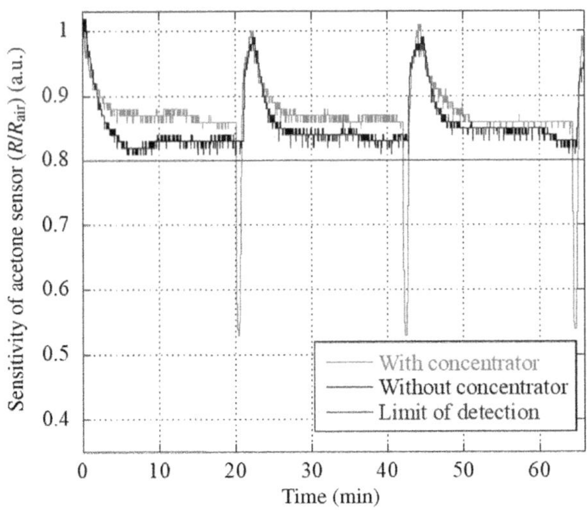

Figure 20.7 Continuous monitoring of the sensitivities of the semiconductor-based acetone sensor with and without the concentrator made from 390HUA zeolite under simulated real conditions.
Source: Reprinted with permission from Y. Yamada, S. Hiyama, T. Toyooka, S. Takeuchi, K. Itabashi, T. Okubo and et al., Anal. Chem. 87 (2015) 7588. Copyright 2015 American Chemical Society.

skin acetone could be easily detected under the simulated conditions with the concentrator made from zeolite. We also found that almost the same peak sensitivities of the acetone sensor with the concentrator (0.53 ± 0.01) were obtained from the three measurements. These results indicate that our proposed method is feasible and shows potential for implementation in wearable devices.

20.3.2 Prototyped analyzer

We prototyped a wearable acetone analyzer that implemented an acetone sensor with the 390HUA concentrator [4]. Our prototype is easy to use; the user simply wears the prototype and the measured results are automatically sent to an Android-based smartphone via a wireless Bluetooth connection. The device has a weight of 54 g and is 40×78 mm in size—smaller around than a credit card—and 13 mm in thickness, which can easily fit into a breast pocket (Fig. 20.8).

To investigate the accuracy of our prototype, we conducted the performance evaluation. Since skin acetone is a constituent of biological gas that originates in the blood and can therefore be measured from various parts of the body as well as the arms, we measured skin acetone from the palms of multiple subjects and compared measurements taken with the device with measurements taken using conventional GC. This test showed that the measurements taken by the device we developed had a strong positive correlation with those of the conventional GC ($R = 0.96$) (Fig. 20.9), confirming that our prototype can be used as a practical skin acetone checker with a reasonable range of measurement deviations [4]. The performance of this prototype remains close to conventional GC, but we have succeeded in drastically reducing size and weight.

20.3.3 Assumed usage scenario

Our device could be used to offer services to users who are concerned about their diet or users who are troubled by metabolic syndromes, and could be used at work and at home on a daily basis. Specifically, this opens the potential to provide dietary support programs that include appropriate timing for meals and exercise, recommendations for menus and the size of meals, and exercise amount and load tailored to the characteristics of the user's metabolism. Fig. 20.10 shows the results

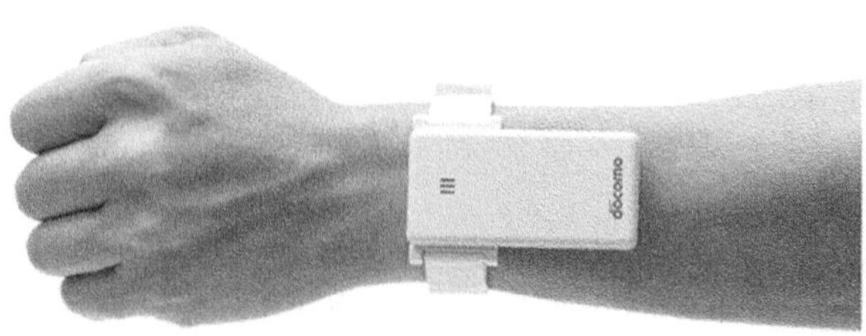

Figure 20.8 Photograph of our prototyped wearable skin acetone analyzer.

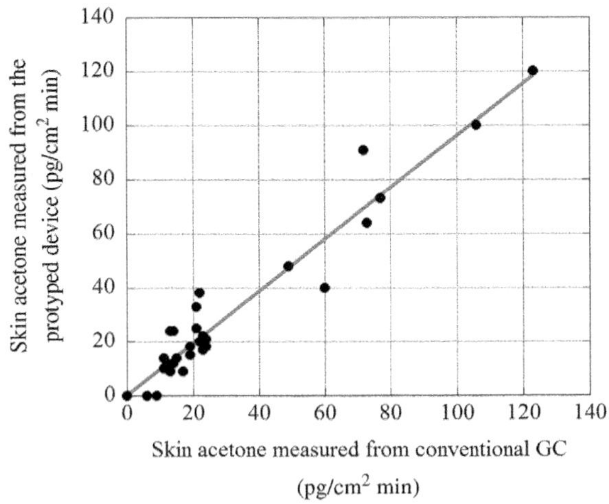

Figure 20.9 Accuracy of our prototyped skin acetone analyzer.

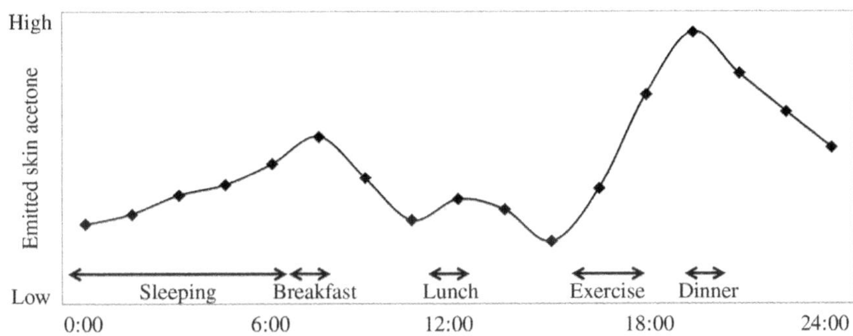

Figure 20.10 Example of skin acetone emission within a day.

of measurement taken by the device being worn for 1 day. It is clear that the amount of emitted acetone from the skin changes throughout the day. For example, when the amount of emitted skin acetone is low before lunch, body fat is not being burnt much, suggesting that there is a high amount of sugar remaining in the body. In this case, as eating can lead to weight gain at this time, the user is advised to eat only a light snack and avoid large amounts of carbohydrates. In contrast, when the amount of emitted skin acetone is markedly high, the user is advised to beware of potentially excessive dieting with unreasonable dietary restriction. In a different usage scenario, the user can compare the amount of emitted skin acetone before and after exercise to determine whether the exercise burnt fat effectively. Fig. 20.10 shows a successful exercise session; however, if there is no marked change in the amount of emitted skin acetone before and after exercising, either the load was too low or the session was too short. Thus, advice can be given to adjust the amount of time and load for exercise in stages.

20.4 Conclusions

In this book chapter, we overviewed a portable breath acetone analyzer and a wearable skin acetone analyzer and introduced their assumed applications in self-monitoring of fat metabolism. By making these devices easy to use on a routine basis, we expect them to be useful in maintaining and improving a person's health and in the prevention and early detection of disease.

In Japan, the difference between average lifespan and healthy lifespan is 9.13 years for men and 12.68 years for women as of 2010, which indicates that there is an unhealthy period that limits everyday life lasting from about 9 to 13 years. In addition, a comparison of data between 2001 and 2010 reveals that average lifespan in Japan increased by 1.48 years for men and 1.37 years for women while healthy lifespan increased by only 1.02 years for men and 0.97 years for women. That is to say, the increase in healthy lifespan was smaller than the increase in average lifespan. Both of these differences are expected to escalate in the years to come, meaning that the unhealthy period that consumes considerable healthcare and nursing-care benefits will be increasing. This is another reason why extending the healthy lifespan is necessary.

NTT DOCOMO seeks to contribute to solutions for social issues through various R&D initiatives in the medical and health care fields such as extending the healthy lifespan to shorten its difference with the average lifespan.

References

[1] NTT DOCOMO Press Release, Apr. 27 (2017). https://www.nttdocomo.co.jp/english/info/media_center/pr/2017/0427_00.html

[2] Y. Yamada, S. Hiyama, Breath acetone analyzer to achieve "biochip mobile terminal", NTT DOCOMO Tech. J. 14 (2012) 51.

[3] T. Toyooka, S. Hiyama, Y. Yamada, A prototype portable breath acetone analyzer for monitoring fat loss, J. Breath Res. 7 (2013) 036005.

[4] Y. Yamada, S. Hiyama, T. Toyooka, Wearable skin acetone analyzer and its applications in health management, NTT DOCOMO Tech. J. 17 (2015) 77.

[5] Y. Yamada, S. Hiyama, T. Toyooka, S. Takeuchi, K. Itabashi, T. Okubo, et al., Ultratrace measurement of acetone from skin using zeolite: toward development of a wearable monitor of fat metabolism, Anal. Chem. 87 (2015) 7588.

[6] A. Manolis, The diagnostic potential of breath analysis, Clin. Chem. 29 (1983) 5.

[7] W. Cao, Y. Duan, Breath analysis: potential for clinical diagnosis and exposure assessment, Clin. Chem. 52 (2006) 800.

[8] R.M.S. Thorn, J. Greenman, Microbial volatile compounds in health and disease conditions, J. Breath Res. 6 (2012) 024001.

[9] O.B. Crofford, R.E. Mallard, R.E. Winston, N.L. Rogers, J.C. Jackson, U. Keller, Acetone in breath and blood, Trans. Am. Clin. Climatol. Assoc. 88 (1977) 128.

[10] C. Wang, A.B. Surampudi, An acetone breath analyzer using cavity ringdown spectroscopy: an initial test with human subjects under various situations, Meas. Sci. Technol. 19 (2008) 105604.

[11] C. Wang, A. Mbi, M. Shepherd, A study on breath acetone in diabetic patients using a cavity ringdown breath analyzer: exploring correlations of breath acetone with blood glucose and glycohemoglobin A1C, IEEE Sens. J. 10 (2010) 54.

[12] M. Storer, et al., Measurement of breath acetone concentrations by selected ion flow tube mass spectrometry in type 2 diabetes, J. Breath Res. 5 (2011) 046011.

[13] K. Musa-Veloso, S.S. Likhodii, S.C. Cunnane, Breath acetone is a reliable indicator of ketosis in adults consuming ketogenic meals, Am. J. Clin. Nutr. 76 (2002) 65.

[14] P. Spanel, K. Dryahina, A. Rejskova, T.W.E. Chippendale, D. Smith, Breath acetone concentration; biological variability and the influence of diet, Physiol. Meas. 32 (2011) N23.

[15] B.J. Stubbs, P.J. Cox, R.D. Evans, P. Santer, J.J. Miller, O.K. Faull, et al., On the metabolism of exogenous ketones in humans, Front. Physiol. 8 (2017) 848.

[16] J. King, et al., Isoprene and acetone concentration profiles during exercise on an ergometer, J. Breath Res. 3 (2009) 027006.

[17] H. Sasaki, S. Ishikawa, H. Ueda, Y. Kimura, Response of acetone in expired air during graded and prolonged exercise, Adv. Exerc. Sports Physiol. 16 (2011) 97.

[18] J. King, A. Kupferthaler, B. Frauscher, H. Hackner, K. Unterkofler, G. Teschl, et al., Measurement of endogenous acetone and isoprene in exhaled breath during sleep, Physiol. Meas. 33 (2012) 413.

[19] K. Schwarz, et al., Breath acetone-aspects of normal physiology related to age and gender as determined in a PTR-MS study, J. Breath Res. 3 (2009) 027003.

[20] S.K. Kundu, J.A. Bruzek, R. Nair, A.M. Judilla, Breath acetone analyzer: diagnostic tool to monitor dietary fat loss, Clin. Chem. 39 (1993) 87.

[21] M. Righettoni, A. Tricoli, S.E. Pratsinis, Si:WO(3) Sensors for highly selective detection of acetone for easy diagnosis of diabetes by breath analysis, Anal. Chem. 82 (2010) 3581.

[22] M. Righettoni, A. Tricoli, Toward portable breath acetone analysis for diabetes detection, J. Breath Res. 5 (2011) 037109.

[23] A. Reyes-Reyes, R.C. Horsten, H.P. Urbach, N. Bhattacharya, Study of the exhaled acetone in type 1 diabetes using quantum cascade laser spectroscopy, Anal. Chem. 87 (2015) 507.

[24] T.P. Blaikie, J.A. Edge, G. Hancock, D. Lunn, C. Megson, R. Peverall, et al., Comparison of breath gases, including acetone, with blood glucose and blood ketones in children and adolescents with type 1 diabetes, J. Breath Res. 8 (2014) 046010.

[25] F.J. Pasquel, G.E. Umpierrez, Hyperosmolar hyperglycemic state: a historic review of the clinical presentation, diagnosis, and treatment, Diabetes Care 37 (2014) 3124.

[26] Z. Wang, C. Wang, Is breath acetone a biomarker of diabetes? A historical review on breath acetone measurements, J. Breath Res. 7 (2013) 037109.

[27] G. Boden, M. Laakso, Lipids and glucose in type 2 diabetes: what is the cause and effect? Diabetes Care 27 (2004) 2253.

[28] N. Tanda, Y. Hinokio, J. Washio, N. Takahashi, T. Koseki, Analysis of ketone bodies in exhaled breath and blood of ten healthy Japanese at OGTT using a portable gas chromatograph, J. Breath Res. 8 (2014) 046008.

[29] D. Samudrala, G. Lammers, J. Mandon, L. Blanchet, T.H. Schreuder, M.T. Hopman, et al., Breath acetone to monitor life style interventions in field conditions: an exploratory study, Obesity (Silver Spring) 22 (2014) 980.

[30] S. Halbritter, M. Fedrigo, V. Höllriegl, W. Szymczak, J.M. Maier, A.G. Ziegler, et al., Human breath gas analysis in the screening of gestational diabetes mellitus, Diabetes Technol. Ther. 14 (2012) 917.

[31] Y. Yamada, S. Hiyama, Application of biological gas analysis toward self-health management, NTT DOCOMO Tech. J. 18 (2017) 4.
[32] P. Mochalski, J. King, K. Unterkofler, H. Hinterhuber, A. Amann, Emission rates of selected volatile organic compounds from skin of healthy volunteers, J. Chromatogr. B 959 (2014) 62.
[33] L. Dormont, J.M. Bessière, A. Cohuet, Human skin volatiles: a review, J. Chem. Ecol. 39 (2013) 569.
[34] C. Turner, B. Parekh, C. Walton, P. Spanel, D. Smith, M. Evans, An exploratory comparative study of volatile compounds in exhaled breath and emitted by skin using selected ion flow tube mass spectrometry, Rapid Commun. Mass Spectrom. 22 (2008) 526.
[35] W.J. Roth, P. Nachtigall, R.E. Morris, J. Čejka, Two-dimensional zeolites: current status and perspectives, Chem. Rev. 114 (2014) 4807.

Air pollution monitoring network of $PM_{2.5}$, NO_2 and radiation of ^{137}Cs

21

Yasuko Yamada Maruo
Tohoku Institute of Technology, Sendai, Japan

21.1 Introduction

In the field of air quality monitoring, many small sensors have been developed for indoor or outdoor measurements of pollutants (carbon nanotube sensors [1], metal oxide semiconductor sensors [2], screen printed electrochemical sensors [3,4], three-dimensional graphene sensors [5], and environmental sensors on plastic foil [6]). Since a number of these sensors can operate with wireless networks [7–11], a new application and service have been proposed for collecting and processing air quality by connecting small sensors via wireless networks [12–15]; for example, in the field of moisture sensing, a soil moisture sensor that can use electric power generated by a natural magnetic field, which allows for continuous operation. Using the information measured by the soil moisture sensor, a feedback system can relay data to the water supply system [16]. In the field of atmospheric pollution, many small sensors have been developed to detect fine particulate matters ($PM_{2.5}$) [17,18], nitrogen dioxide (NO_2) [4,19–25], formaldehyde [26–35] and volatile organic compounds (VOCs) [36–40]. However, there are few sensors that work well in outdoor environments with varying temperature and humidity.

The World Health Organization estimated that 6.5 million people are dying annually from air pollution [41], and the European Study of Cohorts for Air Pollution Effects (ESCAPE) project reported a significant increase in the hazard ratio for $PM_{2.5}$ of 1.07 (95%) per 5 $\mu g/m^3$ [42]. Town-scale pollution mapping would increase risk awareness in the local community because $PM_{2.5}$ and NO_2 concentration are known to be very localized. Regarding radiation monitoring, demand for residential and personal detection devices is increasing due to the accident at the Fukushima nuclear power plant. Several radiation monitoring networks were constructed after the accident [43,44], and according to that information, it is clear that the level of radiation varies by state or region and the local environment. If many people possessed small sensors and each person transmits concentration information together with location information via wireless networking infrastructure, mapping and time change of concentration distribution can be obtained in real time at a microresidential scale, where many people live and lead active lives.

Chemical, Gas, and Biosensors for the Internet of Things and Related Applications.
DOI: https://doi.org/10.1016/B978-0-12-815409-0.00022-X

In addition, personal exposure to atmospheric pollutants and radiation can be estimated, and the relationship between pollutant levels and morbidity is considered. A life pattern is also proposed for reducing individual exposure to atmospheric pollutants and radiation using this real-time data. In addition, a disaster prevention system can be constructed by processing the data in a diffusion simulation when harmful substances are released in high concentrations during an accident. In this chapter, air quality monitoring and its applications to $PM_{2.5}$, NO_2, and radiation of ^{137}Cs are discussed with particular reference to robust and high-function sensors that are reliable in outdoor environments with varying temperature and humidity.

21.2 $PM_{2.5}$ monitoring system

21.2.1 *Introduction*

Particulate air pollution is a complex mixture of small and large particles of varying origin and chemical compositions. Particles with diameters ranging from 2.5 to 100 μm usually comprise dust from agriculture, construction, road traffic, plant pollens, and other natural sources. Smaller particles with diameters less than 2.5 μm are typically associated with industry and the combustion of fossil fuels. These particles include soot from vehicle exhaust that often coats surfaces with various chemical contaminants or metals. The largest sources of fine particles are coal-fired power plants, fuel combustion in industry, and combustion engine vehicle exhaust [45]. Ambient air concentrations of $PM_{2.5}$ represent a significant threat to human health and the environment, and epidemiological studies have distinguished the outdoor and nonoutdoor components of $PM_{2.5}$ and the outdoor component was found to be most strongly associated with health risks [46–49]. As $PM_{2.5}$ is associated with more adverse health effects than larger particles, accurate measurement and monitoring of this contaminant is essential to increasing the understanding of $PM_{2.5}$ [50]. A combination of small contaminant sensors connected via wireless networks represents a promising method for evaluating the concentrations in microresidential areas and for evaluating personal exposure levels. In this section, three methods are described: (1) concentration measurement and density mapping along urban roads using a small $PM_{2.5}$ sensor carried by individuals, (2) $PM_{2.5}$ multipoint measurements in residential areas using small $PM_{2.5}$ sensors fixed at six different locations, and (3) measurement of $PM_{2.5}$ advection using three small $PM_{2.5}$ sensors.

21.3 Monitoring device (small $PM_{2.5}$ sensor)

The $PM_{2.5}$ measurements were conducted using small $PM_{2.5}$ sensors that employ the light scattering method [51]. The sensor has a laser light emitting diode (LED), photodiode (PD), fan, amplifier, and universal serial bus encoder. Atmospheric particles were introduced by a fan, flowed along a path that was designed to classify particle sizes, and scattered light was measured. The data contained in the transmission signal was then converted into concentrations of $PM_{2.5}$ and PM_{10}, and the measurements were performed one per second. Principles of measurement and photographs of the small $PM_{2.5}$ sensor are provided in Fig. 21.1. The sensor connected to the tablet terminal received power from the tablet device and was controlled by the operation software installed in the tablet terminal. Measurement data was displayed on the tablet terminal as shown in Fig. 21.1, then recorded in the tablet terminal with a time stamp. The data also can be transmitted in real time along with the position information via wireless networks. In the case of performance estimation in outdoor measurements, the small $PM_{2.5}$ sensor was powered by solar cells and used a 3G/4G network [52]. The combined sensor was installed in close proximity to the publicly accessible $PM_{2.5}$ measurement instrument that uses the beta-ray absorption method, and the data from both sources were compared. The relationship between the two data sets is presented in Fig. 21.2. The temperature and humidity ranges in the measurement period were 0°C–25°C and 50%–90%, respectively. As shown in Fig. 21.2, the slope of the approximate line and the correlation coefficient were 0.86 and 0.84, respectively. Considering these results, reliable outdoor $PM_{2.5}$ monitoring using the small $PM_{2.5}$ sensor is operationally viable. The deviation of measurement data was also examined by comparing the output of six $PM_{2.5}$ sensors under the same conditions. The obtained data had error range in ± 8%, therefore we can construct the mapping of $PM_{2.5}$ concentrations using several strategically placed $PM_{2.5}$ sensors.

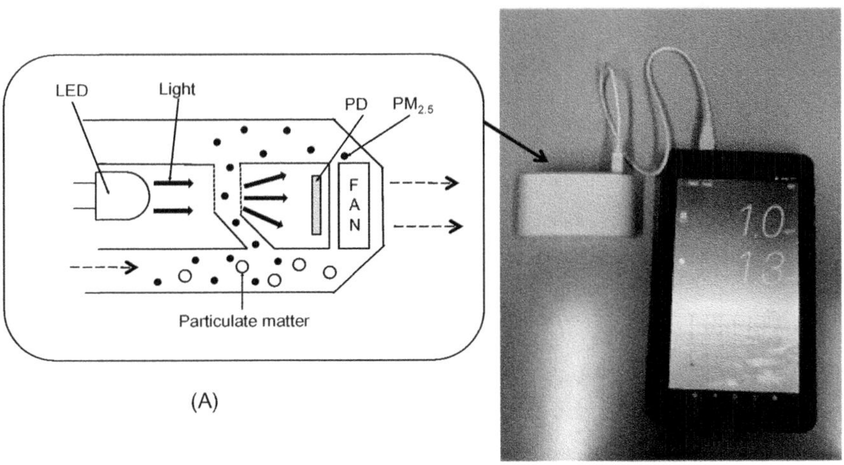

(B)

Figure 21.1 Principle of measurement and photograph of the small $PM_{2.5}$ sensor. (A) PM2.5 sensor module circuit schematic, (B) photograph of the small $PM_{2.5}$ sensor connected to the tablet terminal.

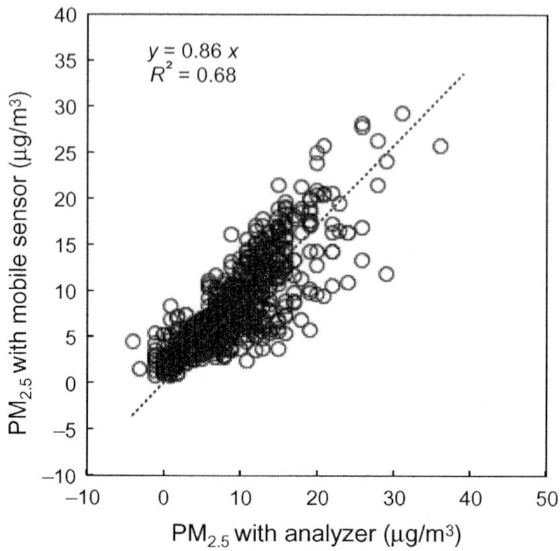

Figure 21.2 Relationship between $PM_{2.5}$ concentrations of $PM_{2.5}$ measurement instrument and of mobile $PM_{2.5}$ sensor at outdoor condition.

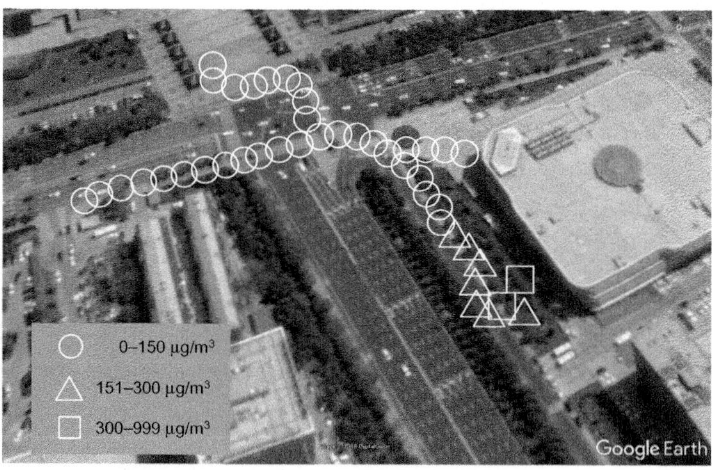

Figure 21.3 $PM_{2.5}$ concentration mapping in Weihai, China.

21.4 Mobile sensing of outside $PM_{2.5}$

Fig. 21.3 shows pollution mapping results in Weihai, China that used small $PM_{2.5}$ sensors. The data was collected by walking with mobile $PM_{2.5}$ sensors. The $PM_{2.5}$ concentrations were high at intersections and roadsides with high traffic volume. However, there were several high concentration areas that could not be predicted from the traffic volume. There was a high concentration level (200–300 $\mu g/m^3$)

near grill restaurants and areas that permit smoking. Pollution mapping that used mobile $PM_{2.5}$ sensors while walking has been conducted in Paris, France; Ikebukuro, Japan; and Seoul, South Korea, and the same trend was observed. Since one sensor was used in these studies, comparison of the $PM_{2.5}$ concentrations in different locations at the same time was not collected. It will become possible to perform real-time mapping of $PM_{2.5}$ concentrations once a critical mass of people carry the $PM_{2.5}$ sensor and send their data wirelessly. It was also possible to evaluate the personal exposure to $PM_{2.5}$ using the collected data.

21.5 Measurement at several points

Ambient air quality measurements of $PM_{2.5}$ were performed at the Sendai Port area, Miyagi, Japan using six mobile $PM_{2.5}$ sensors. The surface area and the data collection points are depicted in Fig. 21.4A. The Sendai Port area is a residential–industrial area with a highway traffic junction with a traffic density of approximately 3700 vehicles per hour [53] and a coal-fired power plant serving as the primary sources of $PM_{2.5}$ emissions. Data from the six $PM_{2.5}$ sensors was collected on November 8, 2017 and is summarized in Fig. 21.4B. Average concentrations were 24, 26, 21, 22, and 25 $\mu g/m^3$ at data point No. 2, No. 3, No. 4, No. 5, and No. 6, respectively. The average concentration values of $PM_{2.5}$ at each site were relatively similar, while the average concentrations were affected by wind direction (15 $\mu g/m^3$ for NW and 20 $\mu g/m^3$ for SE) [54]. The results indicated that $PM_{2.5}$ concentrations were quite similar in the microarea, with the exception of concentrations detected at the roadside. It was assumed that the effect of emissions from vehicles on the highway and the nearby coal-fired power plant were similar in a 1-km^2 area. The monitoring results from 14:30 to 14:45 are presented in Fig. 21.4C. The time series of the $PM_{2.5}$ concentrations detected at each data collection point revealed average concentrations that were virtually identical. However, as shown in Fig. 21.4C, each point displayed different patterns in contaminant concentration variability, with time transitions being shifted by several minutes. These time-shift patterns indicate advection of $PM_{2.5}$ by wind, and this tendency could be successfully monitored using small $PM_{2.5}$ sensors installed every several hundred meters to measure $PM_{2.5}$ concentrations once per second. Therefore, a monitoring network constructed by installing these small sensors at distance intervals of several hundred meters makes it possible to predict sources and apply them to disaster prevention systems at the time of an accident.

21.6 NO_2 monitoring system

21.6.1 Introduction

NO_2 is a major air pollutant that is subject to environmental regulation in many countries. In the outdoor environment, NO_2 is mainly emitted from vehicles and

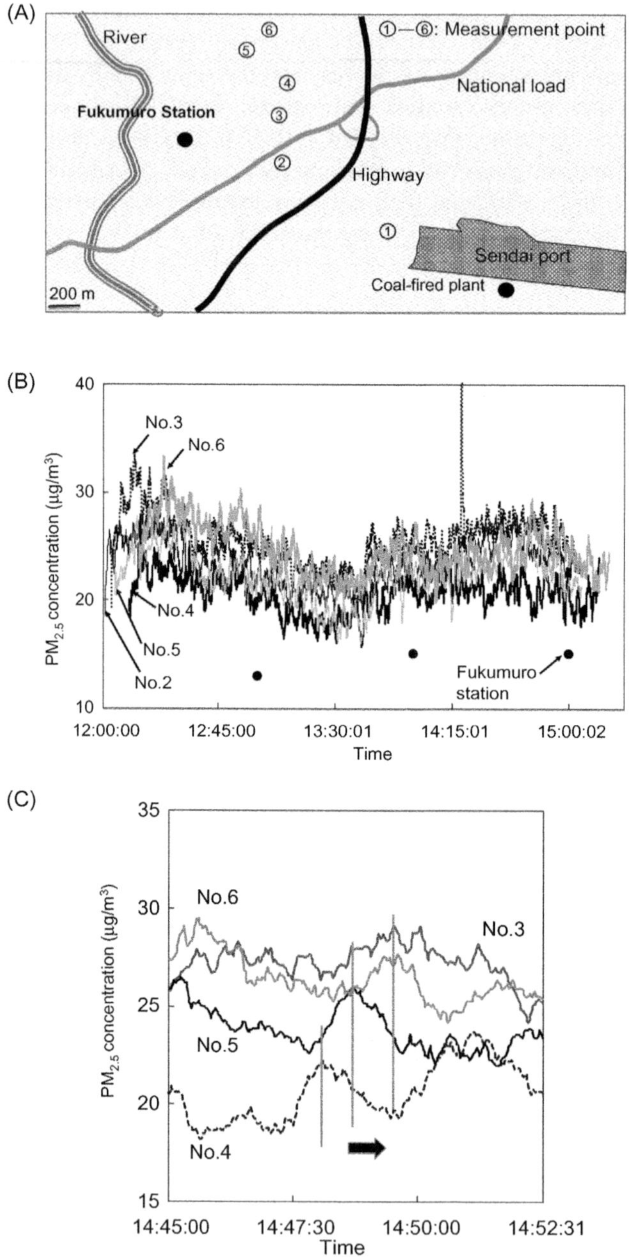

Figure 21.4 Ambient $PM_{2.5}$ measurements in Sendai port area.
(A) Surface area of ambient $PM_{2.5}$ measurements and the locations of the 6 $PM_{2.5}$ monitoring sites in Sendai port area. (B) Time series plots of $PM_{2.5}$ concentrations at No. 2–No. 6 sites, and hourly average $PM_{2.5}$ concentrations at Fukumuro Station. (C) Time series plots of $PM_{2.5}$ concentrations at No. 3–No. 6 sites.

factories [54]. Its distribution is often highly localized, and high NO_2 concentrations are detected at roadsides and in close proximity to factories. NO_2 clearly has an adverse effect on human lungs [55–58], and the US EPA strengthened the national ambient air quality standard for NO_2 (1 hour at a level of 100 ppb) [59]. There are several forecasting methods for NO_2. However, because the grid scale used in the simulation was large (1–10 km), and the forecast concentrations were defined by a single value per grid, localized NO_2 concentrations could not be forecasted properly. Networks for sensing NO_2 can provide the information of the localized NO_2 concentration, though NO_2 monitoring remains difficult due to the lack of small and inexpensive monitoring devices with high sensitivity. To realize an effective NO_2 monitoring network, a highly sensitive NO_2 monitoring device is needed, with detection limits in the range of a few ppb to 100 ppb with high time resolution. Extensive research has been conducted in the field of many NO_2 sensing systems, such as sensors for specific chemical compounds [60], semiconductors [25], and passive samplers [61]. However, some have insufficient sensitivity for monitoring urban atmospheres, while others have a response time that is insufficient for monitoring the changes at the roadside over time. In this section, a small NO_2 sensing device with sufficient sensitivity and sufficient time resolution is described [62], and a constructed NO_2 mobile sensing network is also described [63]. In addition, its application in route-finding systems for people who commute by bicycle is also described.

21.7 NO_2 monitoring device

The sensor element used 4-nm-diameter porous glass as a substrate and the high sensitivity was achieved by the large surface area of the porous glass. The reaction between NO_2 and diazo coupling reagents in the porous glass was used to detect NO_2 [64]. An azo compound with an absorption peak at 525 nm formed as a product of the reaction between the diazo coupling reagents and the NO_2 in the atmosphere, and the absorbance change values was converted into NO_2 concentration data. The sensor response was examined in an artificial environment, and it was found that 95% of the absorbance change occurred within 5 minutes in the porous glass, and then it was limited to measuring the NO_2 concentration at a high time resolution of about 10 minutes. A portable NO_2 sensing device that a viable mobile sensing network could deploy is presented in Fig. 21.5. Each side of the device was equipped with a pair of louvers to allow air ventilation and to prevent outside light from entering the device. The outputs of the photodiodes were converted from analog to digital through an A-D converter, and the NO_2 concentration was calculated using digital signals. With dimensions of $100 \times 100 \times 40$ mm, and dry-cell battery for power, the device was compact and self-contained. Fig. 21.6 shows the results of outdoor measurements taken with a time interval of 10 minutes [62]. The correlation coefficient was 0.63, which indicated good agreement, even in an outdoor environment. An environmental monitoring terminal unit containing the NO_2 sensor

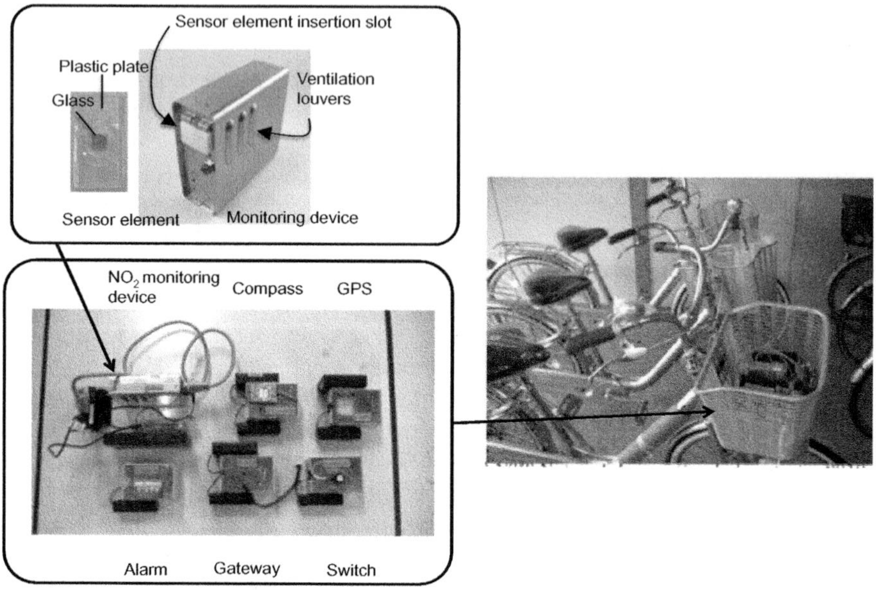

Figure 21.5 Portable NO_2 sensing device and environmental monitoring terminal unit for bicycle.

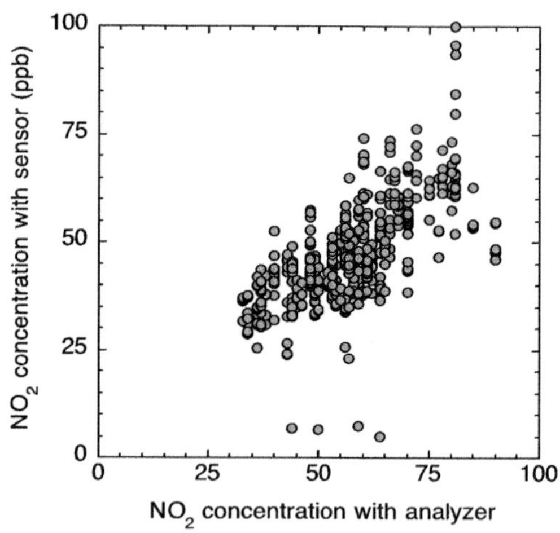

Figure 21.6 Relationship between NO_2 concentrations of commercial analyzer and that of mobile NO_2 sensor at outdoor condition.

device, a switch, a compass, a Global Positioning System (GPS) device, and a wireless gateway based on Mote, was developed as shown in Fig. 21.5. The terminal unit was designed to be mounted on a bicycle. The compass and the GPS device were used to obtain the position of the bicycle, and its direction of travel. The switch was used for the navigation system, and the terminal unit was connected to a wide-area ubiquitous wireless network through a wireless gateway. The field trial was limited to two terminal units.

21.8 Mobile sensing of outside NO_2

The demonstration experiment was carried out in certain areas of Katsushika, Tokyo, Japan. The topography in the study is depicted in Fig. 21.7, and was characterized as a flat, riverside area containing one main road with very high traffic density (56,000 vehicles a day [65]). To understand area variations in NO_2, two bicycles (bicycle I and bicycle II) were individually equipped with a monitoring terminal unit. The NO_2 data was acquired every 10 minutes. Bicycle I began its route at point B, then moved along the load during heavy traffic to intersection A. The route is depicted as a solid line and was completed from 14:10 to 15:40, while NO_2 concentrations and GPS data was transmitted thorough the network at 10-minute intervals. Bicycle II began its route at point B, then moved along the street bordering the park. The route is depicted as a dotted line, and NO_2 concentrations and GPS data was transmitted through the network at the same 10-minute intervals. For the purposes of this experiment, transmitted NO_2 concentrations were considered to be the average value over 10 minutes. The time series for NO_2 concentration data transmitted by both bicycles are shown in Fig. 21.7. The NO_2 concentrations from bicycle I remained consistently high, while NO_2 concentrations from bicycle II were initially high, then decreased as the bicycle's distance from the road with heavy traffic increased and finally increased again during the approach to the intersection.

A real-time data sharing application was conceptualized based on the collection and network sharing air quality data among cyclists to identify routes with lower levels of air pollution. A map displaying the latest NO_2 concentrations could be exchanged among cyclists via mobile phones. In a well-organized route navigation system, a cyclist would activate an onboard terminal unit with an alarm LED to alert the rider to the presence of elevated NO_2 concentrations in the bicycle's direction of travel. For example, when a cyclist wants to get from point A to point B and travels along a high-traffic road, a red LED would alert the cyclist to the conditions. Then, when the cyclist nears a street that borders a park, the red LED would be deactivated. A visual depiction of such a mobile phone application with an outline of the service is shown in Fig. 21.8.

21.9 Radiation of ^{137}Cs monitoring system

21.9.1 Introduction

The Daiichi Nuclear Power Plant accident in Fukushima has stimulated interest and demand from ordinary people for personal radiation sensors. The public agencies

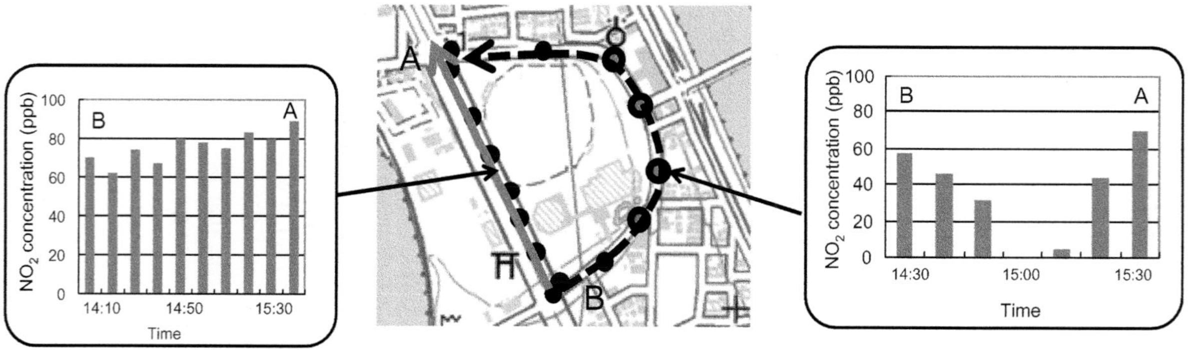

Figure 21.7 The time series of the NO_2 concentration of both bicycles and an application image on the mobile phone.

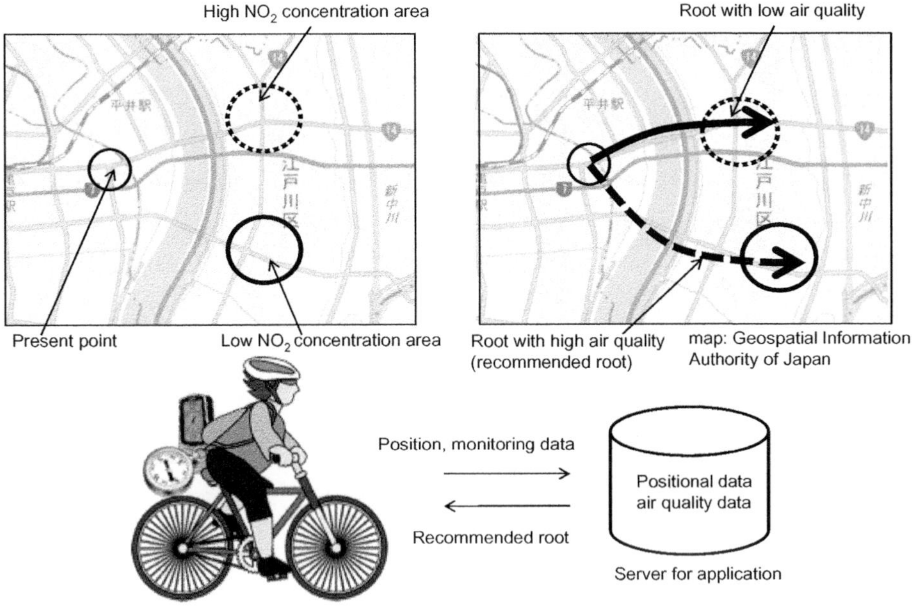

Figure 21.8 Application image and outline of the service.

have been collecting radiation data at fixed monitoring stations and open-sourced the radiation mapping data in large areas, so people could better understand the geographical trends of radiation levels from those high-risk sites [66]. However, there is also a rising commercial demand for personal sensors to measure radiation levels in and around homes, as well as exposure levels away from home. Unfortunately, conventional radiation sensing instruments are too expensive for the general public, and current models are typically difficult to obtain and use. The following sections describe a simplified method for measure personal radiation levels with small and inexpensive mobile radiation sensor and their application in trial monitoring conducted in Fukushima, Japan.

21.10 Radiation of ^{137}Cs monitoring device

A PIN photodiode is capable of detecting various forms of nuclear radiation at its depletion layer, and its effectiveness has been shown using a general PIN photodiode and charge amplifier. Inexpensive mobile devices that can accurately detect radiation have been developed to address the rising public demand for an affordable personal radiation detector. The detector used a combination of a PIN photodiode detector connected to a smartphone via a microphone cable. The detector circuit design was optimized for simplicity and low cost, whereas the smartphone software application was tasked with handling the complex processing required.

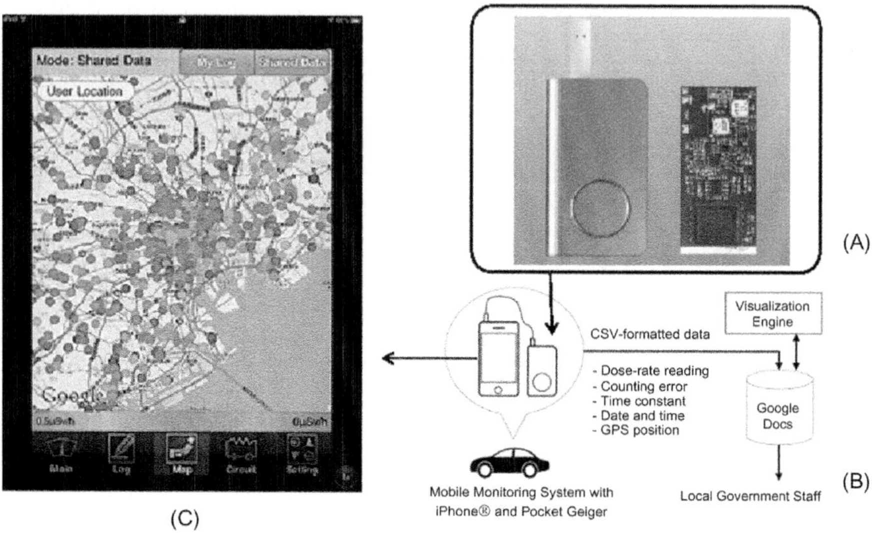

Figure 21.9 Application image and outline of the service: (A) photograph of the detector, (B) system used in the test to collect, (C) heat map of the radiation levels measured around the power plants and Kanto area to Tohoku area.

Furthermore, the device also used the GPS and networking capabilities of the smartphone for logging and data sharing. A photograph of the detector is shown in Fig. 21.9A [67]. According to the results of performance estimating, the detector displayed a 5-decade linear response for ^{137}Cs, and a measurement range of approximately 0.05 μSv/h to 10 mSv/h and from 0.01 cpm to 300 kcpm. This range covers most radiation levels measured in Japan.

21.11 Field test in Fukushima and other areas

The field test was conducted in the evacuation zone around the Fukushima Daiichi Nuclear Power Plants, and at the roadside from the Kanto to Tohoku areas. Fig. 21.9B shows the system used to collect and visualize dose-rate data in the evacuation zone [67]. In the tests, the detector was set on the dashboard in a car. Fig. 21.9C shows a heat map of the radiation levels measured around the power plants and along the road from Kanto area to Tohoku area. As shown in Fig. 21.9, the level tended to be higher nearer the plants, but still differed, within a particular zone. For example, the reading range was from 2 to 20 μSv/h in an area 2–4 km away from the accident site, and higher dose rates were recorded to the north. This wide variation in radioactivity readings resulted from various natural environmental conditions, such as weather, vertical intervals, and vegetation.

Small radiation sensors that collect and transmit contamination concentration data together with location information over wireless networks are a promising method of monitoring an individual's dose and exposure, and are capable of contributing to radiation safety information infrastructure by mapping radiation in different areas, which could be utilized as a useful basis for radiation research [68].

Acknowledgment

Thanks to Dr. Y. Ishigaki and Prof. N. Matsumoto for providing personal monitoring data of $PM_{2.5}$, and radiation.

References

[1] J. Li, Y. Lu, Q. Ye, M. Cinke, J. Han, M. Meyyappan, Carbon nanotube sensors for gas and organic vapor detection, Nano Lett. 3 (2003) 929–933.
[2] G.F. Fine, L.M. Cavanagh, A. Afonja, R. Binions, Metal oxide semi-conductor gas sensors in environmental monitoring, Sensors 10 (2010) 5469–5502.
[3] A. Hayat, J.L. Marty, Disposable screen printed electrochemical sensors: tools for environmental monitoring, Sensors 14 (2014) 10432–10453.
[4] M.I. Mead, O.A.M. Popoola, G.B. Stewart, P. Landshoff, M. Calleja, M. Hayes, et al., The use of electrochemical sensors for monitoring urban air quality in low-cost, high-density networks, Atmos. Environ. 70 (2013) 186–203.
[5] F. Yavari, Z. Chen, N. Abhay, V. Thomas, W. Ren, H.-M. Cheng, et al., High sensitivity gas detection using a macroscopic three-dimensional graphene foam network, Sci. Rep. 1 (2011). article number 166.
[6] D. Briand, A. Oprea, J. Courbat, N. Barsan, Making environmental sensors on plastic foil, Mater. Today 14 (2011) 416–423.
[7] P. Kumar, L. Morawska, C. Martani, G. Biskos, M. Neophytou, S.D. Sabatino, et al., The rise of low-cost sensing for managing air pollution in cities, Environ. Int. 75 (2015) 199–205.
[8] K.K. Khedo, R. Perseedoss, A. Mungur, A wireless sensor network air pollution monitoring system, Int. J. Wireless Mobile Network 2 (2002) 31–45.
[9] D. Hasenfratz, O. Saukh, S. Sturzengger, L. Thiele, Participatory air pollution monitoring using smartphones, 2nd Inter. Workshop on mobile sensing, April 16-20, 2012, Beijing, China.
[10] S. Devarakonda, P. Sevusu, H. Liu, R. Liu, L. Iftode, B. Nath, Real-time air quality monitoring through mobile sensing in metropolitan areas, The 2nd ACM SIGKDD Inter. Workshop on urban computing, article No. 15, August 11, 2013, Chicago, USA.
[11] V. Jelicic, M. Magno, D. Brunelli, Context-adaptive multimodal wireless sensor network for energy-efficient gas monitoring, IEEE Sens. J. 13 (2013) 328–338.
[12] A.R. A-Ali, I. Zualkernan, F. Aloul, A mobile GPRS-sensors array for air pollution monitoring, IEEE Sens. J. 10 (2010) 1666–1671.
[13] J.P. Langrish, X. Li, S. Wang, M.M.Y. Lee, G.D. Barnes, M.R. Miller, et al., Reducing personal exposure to particulate air pollution improves cardiovascular health in patients with coronary heart disease, Environ. Health Perspect. 120 (2012) 367–372.
[14] A. Mainwaring, J. Polastre, R. Szewczyk, D. Culler, J. Anderson, Wireless sensor networks for habitat monitoring, The 1st ACM inter. workshop on wireless sensor network and appl., 88–97, Septem. 28, 2002, Atlanta, USA.

[15] D. Hasenfratz, O. Saukh, C. Walser, C. Hueglin, M. Fierz, T. Arn, et al., Deriving high-resolution urban air pollution maps using mobile sensor nodes, Pervasive Mobile Computing 16 (B) (2015) 268–285.

[16] Y. Kawahara, H. Lee, M.M. Tentzeris, SenSprout: inkjet-printed soil moisture and leaf wetness sensor, The 2012 ACM Conf. on Ubiquitous Computing, 545, Pittsburgh, USA, 2012.

[17] S. Hankey, J.D. Marshall, Land use regression models of on-road particulate air pollution (particle number, black carbon, $PM_{2.5}$, particle size) using mobile monitoring, Environ. Sci. Technol. 49 (2015) 9194–9202.

[18] D.M. Holstius, A. Pillarisetti, K.R. Smith, E. Seto, Field calibrations of a low-cost aerosol sensor at a regulatory monitoring site in California, Atmos. Meas. Tech. 7 (2014) 1121–1131.

[19] S. Ji, H. Wang, T. Wang, D. Yan, A high-performance room-temperature NO_2 sensor based on an ultrathin heterojunction film, Adv. Mater. 25 (2013) 1755–1760.

[20] L. You, Y.F. Sun, J. Ma, Y. Guan, J.M. Sun, Y. Du, et al., Highly sensitive NO_2 sensor based on square-like tungsten oxide prepared with hydrothermal treatment, Sens. Actuators B 157 (2011) 401–407.

[21] G. Lu, J. Xu, J. Sun, Y. Yu, Y. Zhang, F. Liu, UV-enhanced room temperature NO_2 sensor using ZnO nanorods modified with SnO_2 nanoparticles, Sens. Actuators B 162 (2012) 82–88.

[22] X. Liang, S. Yang, J. Li, H. Zhang, Q. Diao, W. Zhao, et al., Mixed-potential-type zirconis-based NO_2 sensor with high-performance three-phase boundary, Sens. Actuators B 158 (2011) 1–8.

[23] S.T. Navale, A.T. Mane, M.A. Chougule, R.D. Sakhare, S.R. Nalage, V.B. Patil, Highly selective and sensitive room temperature NO_2 gas sensor based on polypyrrole thin films, Synth. Metals 189 (2014) 94–99.

[24] G. Ko, H.-Y. Kim, J. Ahn, Y.-M. Park, K.-Y. Lee, J. Kim, Graphene-based nitrogen dioxide gas sensors, Curr. Appl. Phys. 10 (2010) 1002–1004.

[25] P. Wu, J.H. Sun, Y.Y. Huang, G.F. Gu, D.G. Tong, Solution plasma synthesized nickel oxide nanoflowers: an effective NO_2 sensor, Mater. Lett. 82 (2012) 191–194.

[26] N. Han, X. Wu, D. Zhang, G. Shen, H. Liu, Y. Chen, CdO activated Sn-doped ZnO for highly sensitive, selective and stable formaldehyde sensor, Sens. Actuators B 152 (2011) 324–329.

[27] Q. Ma, H. Cui, X. Su, Highly sensitive gaseous formaldehyde sensor with CdTe quantum dots multilayer films, Biosens. Bioelectron. 25 (2009) 839–844.

[28] X. Wang, F. Cui, J. Lin, B. Ding, J. Yu, S.S. Al-Deyab, Functionalized nanoporous TiO_2 fibers on quartz crystal microbalance platform for formaldehyde sensor, Sens. Actuators B 171–172 (2012) 658–665.

[29] C. Dong, X. Liu, B. Han, S. Deng, X. Xiao, Y. Wang, Nanoqueous synthesis of Ag-functionalized In_2O_3/ZnO nanocomposites for highly sensitive formaldehyde sensor, Sens. Actuators B 224 (2016) 193–200.

[30] Z. Ye, H. Tai, T. Xie, Z. Yuan, C. Liu, Y. Jiang, Room temperature formaldehyde sensor with enhanced performance based on reduced graphene oxide/titanium dioxide, Sens. Actuators B 223 (2016) 149–156.

[31] Y. Suzuki, N. Nakano, K. Suzuki, Development of an optical formaldehyde sensor based on the use of immobilized pararosaniline, Environ. Sci. Technol. 37 (2003) 5695–5700.

[32] Q. Meng, T. Han, G. Wang, N. Zheng, C. Cao, S. Xie, Preparation of a natural dye doped Ormosil coating for the detection of formaldehyde in the optical gas sensor, Sens. Actuators B 196 (2014) 238–244.

[33] C.-Y. Lee, C.-M. Chiang, Y.-H. Wang, R.-H. Ma, A self-heating gas sensor with integrated NiO thin-film for formaldehyde detection, Sens. Actuators B 122 (2007) 503–510.
[34] P.-R. Chung, C.-T. Tzeng, M.-T. Ke, C.-Y. Lee, Formaldehyde gas sensors: a review, Sensors 13 (2013) 4468–4484.
[35] Y.Y. Maruo, J. Nakamura, M. Uchiyama, Development of formaldehyde sensing element using porous glass impregnated with β-diketone, Talanta 74 (2008) 1141–1147.
[36] C. Tasaltin, F. Basarir, Preparation of flexible VOC sensor based on carbon nanotubes and gold nanoparticles, Sens. Actuators B 194 (2014) 173–179.
[37] Y. Sakai, M. Kadosaki, I. Matsubara, T. Itoh, Preparation of total VOC sensor with sensor-response stability for humidity by noble metal addition to SnO_2, J. Ceram. Soc. Jpn. 117 (2009) 1297–1301.
[38] K. Kanda, T. Maekawa, Development of a WO_3 thick-film-based sensor for the detection of VOC, Sens. Actuators B 108 (2005) 97–101.
[39] C. Chen, K.D. Campbell, I. Negi, R.A. Iglesias, P. Owens, N. Tao, et al., A new sensor for the assessment of personal exposure to volatile organic compounds, Atmos. Environ. 54 (2012) 679–687.
[40] N.V. Quy, T.M. Hung, T.Q. Thong, L.A. Tuan, T.Q. Huy, N.D. Hoa, Novel synthesis of highly ordered mesoporous Fe_2O_3/SiO_2 nanocomposites for a room temperature VOC sensor, Curr. Appl. Phys. 13 (2013) 1581–1588.
[41] United Nations Report. http://www.un.org/apps/news/story.asp?NewsID = 55138#.WXWnFYjyhaQ (24.07.17).
[42] R. Beelen, D. Raaschou-Nielsen, M. Stafoggia, Z.J. Andersen, G. Weinmayr, B. Hoffmann, et al., Effects of long-term exposure to air pollution on natural-cause mortality: an analysis of 22 European cohorts within the multicentre ESCAPE project, The lancet 383 (2014) 785–795.
[43] Radmonitor311 Project. https://sites.google.com/site/radmonitor311/ (2018.05.25).
[44] SAFECAST Project. https://blog.safecast.org/ja/about/ (2018.05.25).
[45] R. Kumar, A.E. Joseph, Air pollution concentrations of $PM_{2.5}$, PM_{10}, and NO_2 at ambient and kerbsite and their correlation in metro city-Mumbai, Environ. Monit. Assess. 119 (2006) 191–199.
[46] C.R. Pope, R. Burnett, E. Calle, D. Krewski, K. Ito, G. Thuston, Lung cancer, cardiopulmonary mortality, and long-term exposure to fine particulate air pollution, J. Am. Med. Assoc. 287 (2002) 1132–1141.
[47] S.T. Ebels, E.W. Wilson, M. Brauer, Exposure to ambient and nonambient components of particulate matter: a comparison of health effects, Epidemiology 16 (2005) 396–405.
[48] R.W. Allen, T. Mar, J. Koening, L.J. Liu, T. Gould, C. Simpson, et al., Changes in lung function and airway inflammation among asthmatic children residing in a woodsmoke-impacted urban area, Inhal. Toxicol. 20 (2008) 423–433.
[49] J. Wichmann, T. Lind, M.A.-M. Nilsson, T. Bellander, $PM_{2.5}$, soot and NO_2 indoor-outdoor relationships at homes, pre-schools and schools in Stockholm, Sweden, Atmos. Environ. 44 (2010) 4536–4544.
[50] W.E. Wilson, H.H. Suh, Fine particles and coarse particles: concentration relationships relevant to epidemiological studies, J. Air Waste Manage. Assoc. 47 (1997) 1238–1249.
[51] Yaguchi denshi. https://www.yaguchidenshi.jp/development (19.05.18).
[52] Y. Matsumoto, Y. Ishigaki, K. Tanaka, IEEE Inter. Conf. on Sensing Tech. (2015). Available from: https://doi.org/10.1109/ICSensT.2015.7438498.
[53] Ministry of Land, Infrastructure and Transport, Japan. http://www.mlit.go.jp/road/census/h27/data/pdf/kasyo04.pdf (19.05.18).

[54] Z.H. Shon, K.H. Kim, S.K. Song, Long-term trend in NO_2 and Nox levels and their emission ratio in relation to road traffic activities in East Asia, Atmos. Environ. 45 (2011) 3120–3131.

[55] P. Sicard, O. Lesne, N. Alexandre, A. Mangin, R. Collomp, Air quality trends and potential health effects-development of an aggregate risk index, Atmos. Environ. 45 (2011) 1145–1153.

[56] A. Chaloulakou, I. Mavroidis, I. Gavriil, Compliance with the annual NO_2 air quality standard in Athens. Required NOx levels and expected health implications, Atmos. Environ. 42 (2008) 454–465.

[57] A.F. Somoano, M. Estarlich, F. Ballester, R.F. Patier, A.A. Alfaro, M. Dolores, et al., Outdoor NO_2 and benzene exposure in the INMA (environment and children) Asturias cohort(Spain), Atmos. Environ. 45 (2011) 813–823.

[58] J.F. Mejia, S.L. Choy, K. Mengersen, L. Morawska, Methodology for assessing exposure and impacts of air pollutants in school children: data collection, analysis and health effects-a literature review, Atmos. Environ. 45 (2011) 1145–1153.

[59] US Environmental Protection Agency. https://www.epa.gov/no2-pollution (25.05.18).

[60] T.X. Fu, Novel NO_2 gas sensor based on Cr(III) complex thin film, Electroanalysis 20 (2008) 68–74.

[61] S. Vardoulakis, J. Lumbreras, E. Solazzo, Comparative evaluation of nitrogen oxides and ozone passive diffusion tubes for exposure studies, Atmos. Environ. 43 (2009) 2509–2517.

[62] Y.Y. Maruo, M. Nakamura, Y. Higashijima, Y. Kikuya, M. Nakamura, Development of highly sensitive nitrogen dioxide monitoring device and its application to wide-area ubiquitous network, Sens. Actuators B 173 (2012) 191–196.

[63] T. Nakamura, Y. Kikuya, Y. Arakawa, M. Nakamura, Y. Higashijima, Y.Y. Maruo, et al., Personal of the web framework for ubiquitous sensor network and its application using NOx sensor mounted on bicycle, IEEE/IPSJ 12th Inter. Symp. Appl. Inter. (SAINT) (2012). Available from: https://doi.org/10.1109/SAINT.2012.20.

[64] T. Tanaka, A. Gulleux, T. Ohyama, Y.Y. Maruo, T. Hayashi, Coloration reaction between NO_2 and organic compounds in porous glass for cumulative gas sensor, Sens. Actuators B 47 (1998) 65–69.

[65] Ministry of Land, Infrastructure and Transport, Japan. http://www.ktr.mlit.go.jp/honkyoku/road/census/h18/tokyo.htm (14.05.12).

[66] Nuclear Regulation Authority, Japan. http://radioactivity.nsr.go.jp/map/ja/ (10.05.18).

[67] Y. Ishigaki, Y. Matsumoto, R. Ichimiya, K. Tanaka, Development of mobile radiation monitoring system utilizing smartphone and its field tests in Fukushima, IEEE Sens. J. 13 (2013) 3520–3526.

[68] E. Han, C. Rott, S.-W. Hong, Intent to use a smartphone application for radiation monitoring in correlation with anxiety about exposure to radiation, recognition of risks, and attitudes toward the use of radiation, J. Rad. Prot. Res. 42 (2017) 205–211.

Wireless sensor network with various sensors

22

Junichi Kodate
NTT Device Innovation Center, NTT Corporation, Atsugi, Japan

22.1 Sensing system with network

Data sensing systems are widely used for many applications in science, engineering, manufacturing, medical usage, home automation, vehicle electronics, and so on. Most simple sensing systems are made with a sensor module and a data convertor and display as shown in Fig. 22.1. The sensor module consists of a sensor, preamplifier, and peripheral parts for data interface. The sensor module converts a physical or chemical quantity to electrical signals. The data converter has many types of components such as an analog level shifter, analog filter, an amplifier, A/D convertor, digital filter, and signal processor. The components and circuit of the data convertor is determined from the requirement of the application or services. The data display also has many variations such as an analog meter, a digital meter, a graphical display. An output device such as a USB memory or an SD card is an optional component. The output device enables to store the sensor data and to bring it to other apparatus. Combining sensor data with data computing and analysis, data sensing systems are made with a sensor and data convertor module, and a personal computer or a laptop connected by peripheral cables or data bus interface (Fig. 22.2A). If you need more powerful computing and analysis resources, you had better to use network resources or cloud computing services. The configuration of a sensing system with network resources such as remote access computers or cloud servers is shown in Fig. 22.2B. The sensing system with network technology has been commonly used in many applications.

22.2 Wireless sensor network as a sensing system

Wireless sensor networks are widely used in many applications such as a factory automation, home automation, environment sensing, and vital sensing [1–4]. For example, the sensor system in a factory gathers various data from sensors that are attached to manufacturing lines and utilities. Monitoring sensors such as vibration or temperature sensors are attached to the motors and manufacturing machines for monitoring the status of the machines. The sensing system with the monitoring

Chemical, Gas, and Biosensors for the Internet of Things and Related Applications.
DOI: https://doi.org/10.1016/B978-0-12-815409-0.00023-1

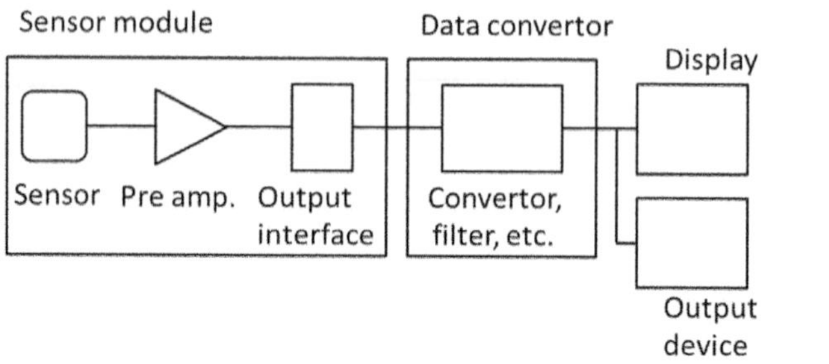

Figure 22.1 Simple sensing system with a sensor module and a data convertor.

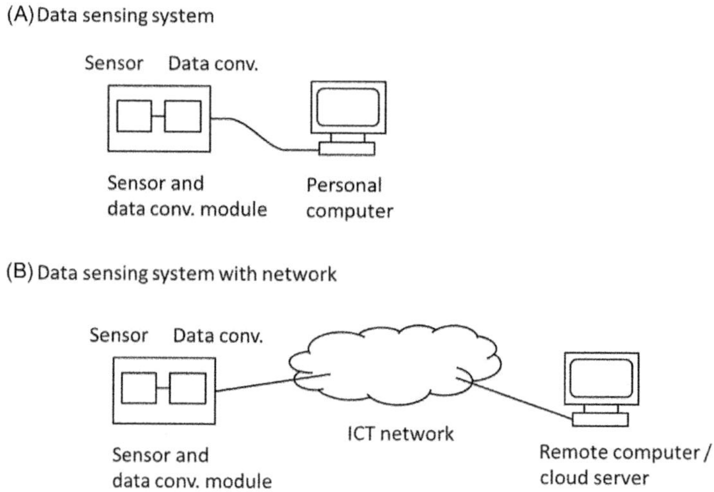

Figure 22.2 Sensing system with data computing and analysis, using (A) a personal computer, or (B) network resources.

sensors gathers sensing data from many machines and apparatus in the factory and detects the trouble and failure of the machines. The environment sensing system monitors various environmental parameters and works for temperature monitoring, traffic monitoring, NOx/SOx monitoring, and so on. An example of environmental sensing is weather monitoring. The weather sensors at numerous places in a wide area detect temperature, humidity, wind speed, and rain conditions in each place. The weather monitoring system gathers many data from the weather sensors in the desired region or country and analyzes the data for weather forecasting. There are many kinds of sensors used in these systems such as temperature, pressure, object detection, physical force, and chemical analysis sensors. These sensor systems are categorized as sensing systems for things and objects around us, and have been already used in many fields as Internet of Things [5] applications and services.

On the other hand, vital sensing, that is, a sensing method and system for human vital signals, enables health check and medical applications. For example, heart rate

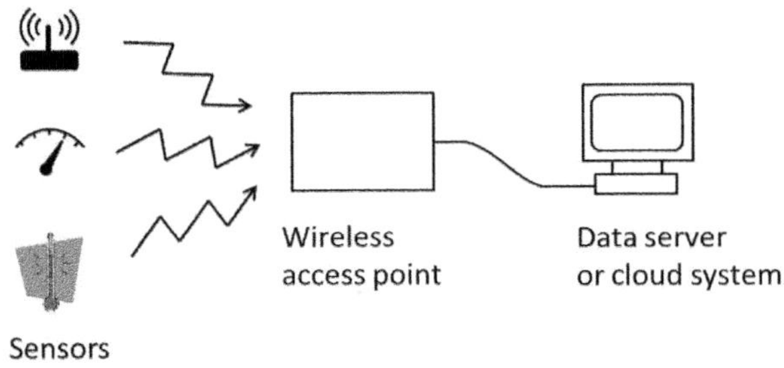

Figure 22.3 Wireless vital sensing system with various types of sensors.

and body temperature are the barometer vital conditions in personal activity. The other sensors such as a blood sensor or an electrocardiograph are for medical usage. It also has possibility for other types of services, such as personal vital monitoring, health monitoring of aged persons, business workers health check, heart-rate monitoring for fitness, and sports usage. As shown in Fig. 22.3, the vital sensing system consists of various types of sensors, a data logging server, and cable or wireless connections to transfer the sensor data to the server, and other peripheral apparatus. The sensors for the vital sensing system are a heat-rate, electrocardiograph, blood pressure, blood component, motion detection sensor, and so on. Recently, a wireless-connected vital sensing system has been used for healthcare, fitness, and sports usage. These vital sensing applications have the potential for new "life assisted" applications and services.

For both Internet of Things and life assisted services, networking of many sensors and connecting the sensors to a server or cloud system are important. Especially, a wireless network that connects many and various sensors to a network system should satisfy the following requirements:

1. The wireless sensor system can connect many and various types of sensors.
2. The system should work with several types of sensor data in one networking system.
3. The system should work with various and multiwireless standards/regulations.
4. The data from sensors are stored in the system with the information of date/time, places, and sensing objects in proper format.

The wireless sensor network with many and various sensors enables us to create new applications and services.

22.3 Wireless sensing system for health condition monitoring with a wearable and flexible sensor

In this section, a wireless sensing system for health condition monitoring is described as an example. The health condition monitoring system consists of a wearable and flexible sensor and a wireless data-transmitting module.

22.3.1 Wearable and flexible electrode with a conductive fiber

As an electrode of biomedical sensor, there has been developed a flexible conductive fiber fabricated by coating fiber material such as silk with a conductive polymer (poly(3,4-ethylenedioxythiroohene)-poly(styrenesulfonate): PEDOT-PSS) [6–8]. The conductive fiber is a functional material capable of measuring biomedical signals to obtain a person's electrocardiogram or electromyogram. A wearable electrode with the conductive fiber is a cloth-based electrode material with elasticity. When embedded in an inner T-shirt, it can be used to easily obtain long-term recordings of the user's heart rate and electrocardiogram in a variety of everyday scenarios. Changes in a person's mental state can be estimated by analyzing fluctuations in that person's heart rate over time. This feature enables continuous and long-term recording of a person's heart rate fluctuation for wide variety of life style scenarios that include rapid and intense motion in sports. For example, an inner T-shirt embedded with a wearable electrode can be monitored the heart rate and condition of athlete's during competition. An inner T-shirt with a wearable electrode can obtain electrocardiogram waveforms in addition to heart rate fluctuation. Moreover, electrocardiogram waveform can be transmitted to a smartphone via wireless means such as Bluetooth when the T-shirt is affixed with an electrocardiogram transmitter, as shown in Fig. 22.4. The electrode arrangement in this T-shirt can obtain a waveform similar to that of CC5 leads in a Holter monitor. In addition to sharp electrocardiogram QRS waves, preceding and succeeding P and T waves can be clearly measured.

22.3.2 Wireless data-transmitting module with many sensors

For the wireless data-transmitting module with many sensors, there has been the issue of latency due to transmission and processing as well as synchronizing when sensor data was being used by multiple users and sensors. When controlling

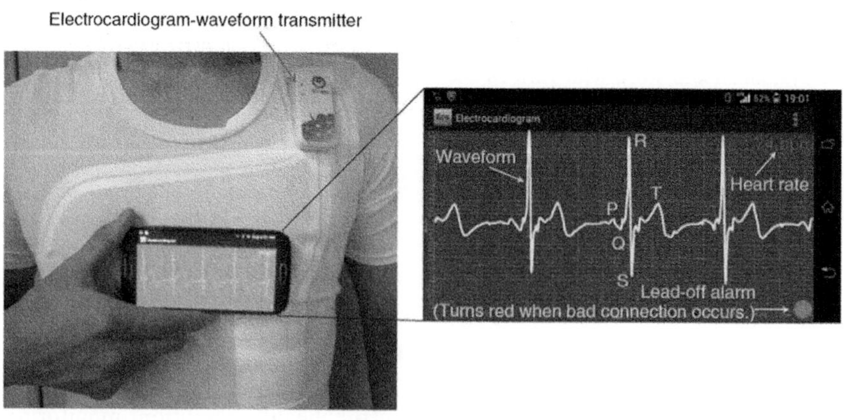

Figure 22.4 Health condition monitoring system with a wearable sensor and a wireless data transmitter.

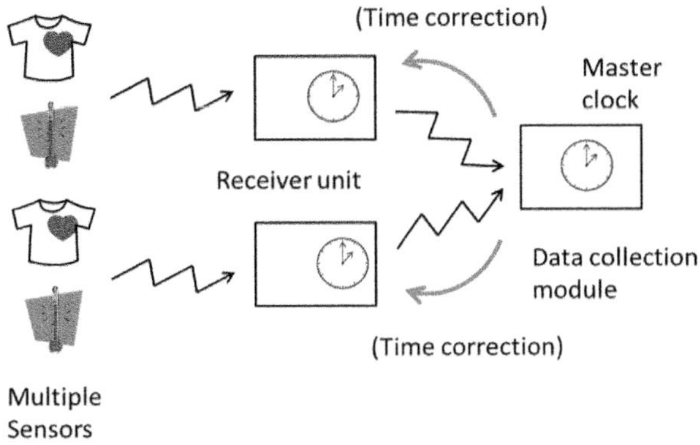

Figure 22.5 Time synchronization mechanism for a vital sensing system.

multiple data with synchronizing multiple sensors, the time management of the sensors and network system becomes important. Sensor time management is the technology that collects information over various networks using protocol conversion while synchronizing wearable devices. With the time management technology, it is possible to unify a sensing time and data from various types of sensors with synchronization and time stamp correction.

We will explain the characteristics and mechanism of time synchronization with Fig. 22.5 using sports as an example. The athlete has sensors attached to their body for measuring electromyogram, electrocardiogram, and temperature. The sensors send measured data to a data collection module every 50 msec. If the synchronization of each sensor is off and the time is delayed while this data is being sent the collected data contains noise because of time stamp errors. The data correction module has a master time and the receiver unit that is connected the sensors references the master time and performs time correction. This method enables synchronization without relying on a specific wireless standard such as Bluetooth, Zig Bee, and Wi-Fi standards. Additionally, since this sensor time management technology can be applied to a multiple user system, the correct time stamp for big data analysis is added to the sensor data from multiple users. This technology is expected to be useful not only in sports applications but also in medical care such as in monitoring patients in home health care and at hospitals. There has been reported the application of a wireless monitoring system with a wearable and flexible sensor for rehabilitation medicine [9].

References

[1] B. Warneke, M. Last, B. Liebowtiz, K.S.J. Pister, Smart dust: communicating with a cubic-milimeter computer, IEEE Comput. (2001) 44–48.

[2] R. Min, M. Bhardwaj, S. Cho, N. Ickes, E. Shih, A. Sinha, et al., Energy-centric enabling technologies for wireless sensor networks, IEEE Wireless Commun. (2002) 28–38.

[3] T.G. Zimmerman, Personal Area Networks (PAN): Near-Field Intra-Body Communication, M.S. thesis, MIT Media Laboratory, 1995.
[4] T. Tanaka, S. Yamashita, K. Aiki, H. Kuriyama, K. Yano, Life Microscope: Continuous Daily-Activity Recording System with Tiny Wireless Sensor, in: 5th Int. Conf. on Networked Sensing Systems, June 2008.
[5] K. Ashton, That 'Internet of Things' Thing, RFID Journal, June 22, 2009. <https://www.rfidjournal.com/articles/view?4986/>, (accessed 13.02.19).
[6] S. Tsukada, H. Nakashima, K. Torimitsu, Conductive polymer combined silk fiber bundle for bioelectrical signal recording, PLoS One 7 (4) (2012) e33689.
[7] S. Tsukada, N. Kasai, R. Kawano, K. Takagahara, K. Fujii, K. Sumitomo, Electrocardiogram monitoring simply by wearing a shirt – for medical, Healthcare, Sports, Entertainment, NTT Tech. Rev. 12 (4) (2014).
[8] K. Takagahara, K. Ono, N. Oda, T. Teshigawara, 'hitoe' - a wearable sensor developed through cross-industrial collaboration, NTT Tech. Rev. 12 (9) (2014).
[9] T. Ogasawara, K. Matsunaga, H. Ito, M. Mukaino, Application for rehabilitation medicine using wearable textile 'hitoe', NTT Tech. Rev. 16 (9) (2018).

Data analysis targeting healthcare-support applications using Internet-of-Things sensors

23

Akihiro Chiba, Kana Eguchi and Hisashi Kurasawa
NTT Service Evolution Laboratories, Yokosuka, Japan

23.1 Motivation for data analysis

An increasing amount of medical and health data is currently produced by a variety of devices in daily life. When people go to clinics, diagnosis-related data (e.g., results of laboratory testing) and treatment-related data (e.g., prescriptions) are generated and stored in electronic health records (EHRs). When a person wears wearable devices such as smartwatches, he/she is able to measure his/her health data (e.g., heart rate) for self-monitoring, and these data can be stored as personal health records (PHRs). The development of network infrastructure and computing technology as well as legislation regarding personal information enable us to easily gather and use a huge amount of medical and health data.

However, medical and health data cannot be of any value without a strategy. Since we usually gather medical and health data for understanding specific things and events, each piece of data represents only one facet of that specific thing or event. We therefore should try to gather all the data related to that target thing or event to appropriately understand what we are interested in. Furthermore, unless we successfully collect enough quantity or quality of data for the target thing or event, that raw data cannot become valuable; therefore we need to find patterns and facts behind the data for understanding the targeted thing or event. For instance, we should at least investigate the distribution of collected data value when we determine whether a person's medical condition is abnormal through laboratory testing.

Data analysis is a strategic means to handle such conditions by conducting appropriate data collection and data processing, which enables us to find something new. The data, information, knowledge, and wisdom (DIKW) hierarchy shown in Fig. 23.1 is a well-known guideline of data analysis.

In short, extracting information from data is the objective in the first step of data analysis, and finding wisdom, or even accumulating it, is the objective in the last step.

Chemical, Gas, and Biosensors for the Internet of Things and Related Applications.
DOI: https://doi.org/10.1016/B978-0-12-815409-0.00024-3

Figure 23.1 DIKW hierarchy.

In this article, we introduce a general procedure for data analysis as an example in health care, which might become the reader's initial knowledge of data analysis. We explain the entire procedure of data analysis, then cite the example of analyzing a heartbeat.

23.2 Procedure of data analysis

Data analysis consists of the following six steps: analysis design, data collection, data cleansing, feature extraction, learning, and evaluation, as shown in Fig. 23.2 bottom to top. The first step (i.e., analysis design) involves determining the data-analysis design including data processing by translating an expected situation and all related information into a mathematical representation that can be processed on computers. The second step (i.e., data collection) involves gathering all the data related to the expected situation. The third step (i.e., data cleansing) involves refining the quality of raw data through data cleansing. The fourth step (i.e., feature extraction) involves calculating a set of fundamental features from the data to appropriately analyze the characteristics of data and enable faster more accurate learning. The fifth step (i.e., learning) involves finding patterns between features and target status by tuning parameters of the analysis model. In the final step (evaluation), we evaluate the performance of data analysis from the viewpoint of whether the results meet the initial expectation.

Note that the more accurate the processing in one step, the more precise the results we are able to obtain in subsequent steps. In other words, if a fatal processing fault occurs in one step, there is no recovery from it regardless of the effort put in in the subsequent steps. Therefore it is important to pursue data analysis in each step from analysis design to evaluation.

23.2.1 Analysis design

In this step, the objective of one's data analysis must be defined with the expected output form, and an appropriate analysis method for it must be selected. The output form is, of course, limited to what is representable on computers. We explain typical data formats and relevant analysis methods in the rest of this subsection.

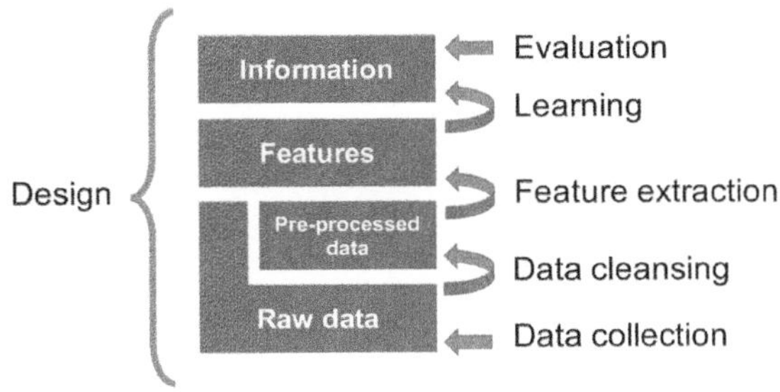

Figure 23.2 Procedure of data analysis.

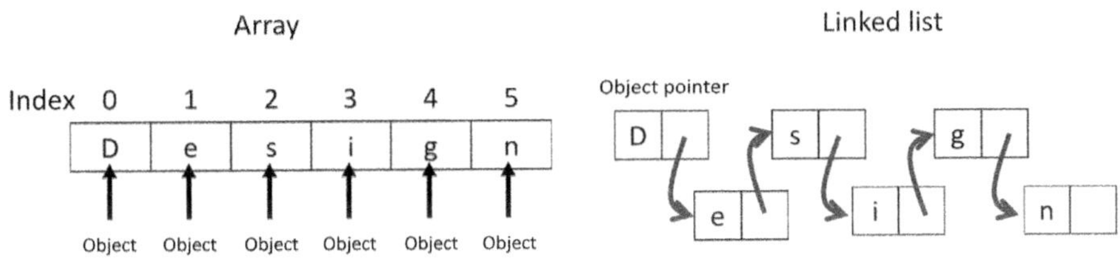

Figure 23.3 Data structure.

23.2.1.1 Fundamental data format on computer

A fundamental data format on computers is data structure [1]. An array and linked list are typical examples, as shown in Fig. 23.3. An array is an indexed set of objects and can be represented as a string, set, vector, or matrix. A linked list consists of ordered objects and can be represented as a sequence, tree, or graph. All information handled on a computer should be described by a combination of these data structures.

23.2.1.2 Example of data format: cluster

A cluster is a format used for representing a group of similar objects, as shown in Fig. 23.4. A clustering algorithm is used for forming clusters. We should select a cluster format and clustering algorithm when, for example, we have many medical images and need to categorize them. This algorithm receives a set of objects, such as vectors and matrices as input, and measures the similarities among objects by using a distance function, then forms clusters based on the calculated similarities.

The k-means algorithm, which is one of the most common clustering algorithms, partitions objects into k clusters, in which each object is linked to the nearest cluster. Many other clustering algorithms have been proposed and satisfy a particular structure of clusters. BIRCH constructs a hierarchical structure of clusters [2], and DBSCAN builds arbitrarily shaped clusters [3].

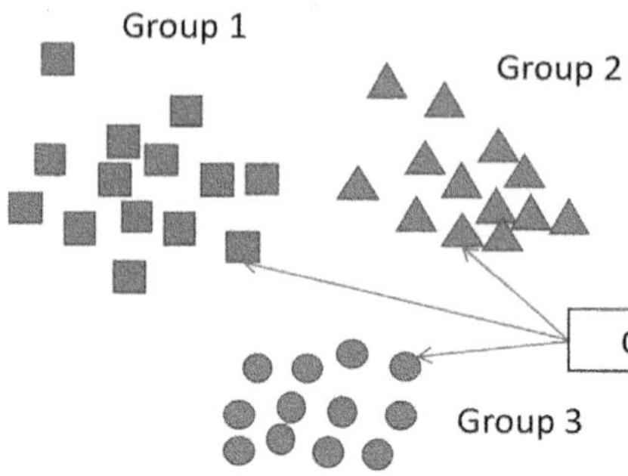

Figure 23.4 Cluster.

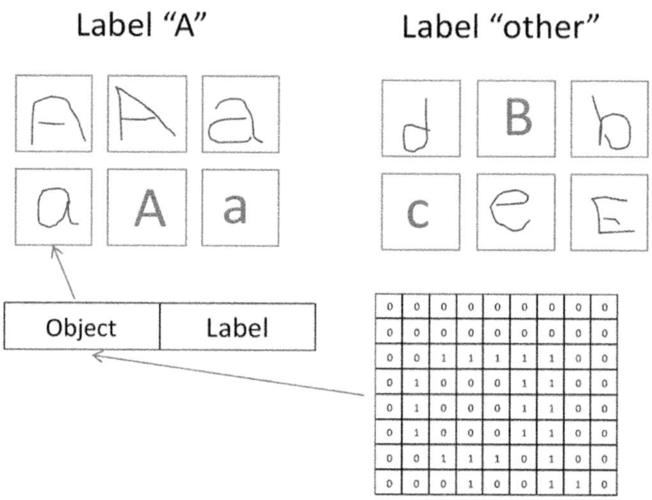

Figure 23.5 Label.

23.2.1.3 Example of data format: label

A label is a format used for representing attributions, as shown in Fig. 23.5. A classification algorithm is used for identifying the label of an object. We should select this format when we, for example, have many pairs of a label and an image and need to estimate the label of an unknown image. This algorithm receives a set of such pairs and creates a rule about the probability of an object attributed to each label. After tuning the rule, this algorithm can select the label relevant to a new object with high accuracy.

There are two types of common classification algorithms. One type uses numerical formulas as the rule, and the other uses a set of rules. Examples of the first type are logistic regression, which uses the logit function [4]; support vector machine, which uses a hyperplane [5]; and deep learning, which uses neural networks that

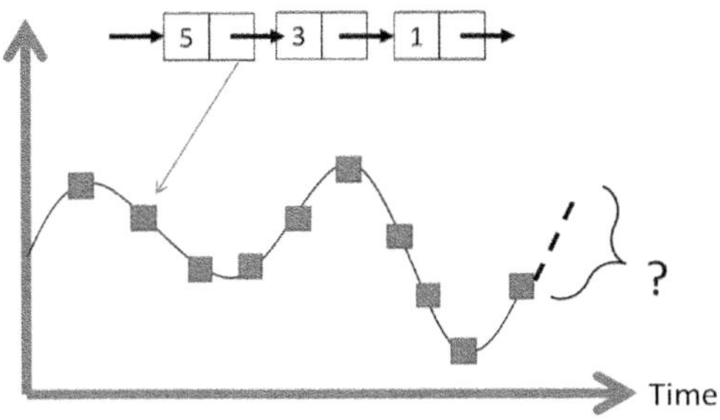

Figure 23.6 Value.

consist of linear, nonlinear, and activation functions [6]. One example of the second type is random forest, which uses a decision tree [7].

23.2.1.4 Example of data format: value

Value is a format used for representing objectively observable conditions of data, as shown in Fig. 23.6. A regression algorithm is used for estimating the numeric value of an arbitrary condition. We should select this format when we have, for example, many pairs of acceleration data and number of steps and need to estimate the number of steps by acceleration data that we separately measure afterward. This algorithm receives a set of such pairs as input and tunes the parameters of a given numerical formula to estimate the value for an unknown condition.

23.2.1.5 "Tips" in analysis design

Classification of unsupervised learning and supervised learning are helpful when searching appropriate data formats and analysis methods relevant to the information, unsupervised learning is used when we have no sample of the information and the information format related clustering algorithm. Supervised learning, on the other hand, is used when we have a sample of the information and the classification and regression algorithms related to the format of information. We should choose the appropriate data format as well as analysis method for a specific purpose, even when the latest or most common methods might seem to be the most interesting.

23.2.2 Data collection

Once the analysis design is determined, we move on to data collection. There are two strategies in data collection: one using existing data, such as EHRs or open data, and the other obtaining new data from participants through biological experiments or questionnaires for data analysis.

Although collecting existing data takes less time than measuring new data, it is not necessarily easy to do; one has to search for target data from a huge storage of EHRs and extract them from for data analysis. This procedure requires special skills including data-manipulation language such as structured query language, so one might have to ask for help from the EHR system engineers.

On the other hand, collecting new medical or health data is more expensive and time consuming than collecting existing data because special instruments for specific purposes as well as consumable supplies (e.g., disposable electrodes) used for such experiments are expensive and may result in few measured data due to the lack of preparations. This restriction may prevent the collecting of a huge amount of data.

Not enough quantity of data is a serious problem in data analysis and machine learning because there is a relationship between the amount of data and analysis accuracy or machine learning models in general. In particular, the latest technology of deep neural networks requires huge data sets. Therefore it might be helpful to consider using existing data at first and later consider whether to collect new data through experiments. When it is too difficult to collect enough quantity of data, one should step back to analyze the design, as described above, and consider an analysis method that is effective even with a small amount of data.

Note that medical and health data require careful data handling because the data may contain sensitive personal information. Therefore one must obtain an informed consent agreement from patients or experiment participants. Furthermore, it may require secure data storage or even external media or file servers with encryption and access limitation. One should generally pass the review from the ethics committee for the entire data-collection procedure beforehand. Drawing out a scheme of the data analysis including background, objective, and assumptions might be helpful to prepare for such a review.

23.2.3 Data cleansing

Ideally, we should collect "good enough" data and analyze them to clarify their characteristics. We use the term "good enough" to mean satisfying the following four Cs of data-quality analysis [8]: complete, coherent, correct, and accountable.

✓ *Complete*: Is everything here that's supposed to be here?
✓ *Coherent*: Does all of the data add up?
✓ *Correct*: Are these, in fact, right values?
✓ *Accountable*: Can we trace the data?

However, when collecting biosignal data in the field, it is difficult to satisfy all four Cs. This is why data cleansing is needed in data analysis. That is, data cleansing modifies raw data as effectively as possible to approach the ideal status from the perspective of the four Cs. The following are example causes of false data from the perspective of three of the four Cs, complete, coherent, and correct, collected in the field.

✓ Complete
- Example cause: data loss
- Example solution: missing data complement

✓ Coherent
  - Example cause: discordance of measurement units, individual difference in measured data
  - Example solution: unification of measurement units, data standardization

✓ Correct
  - Example cause: false data (e.g., mistyped data in EHRs, mismeasured data)
  - Example solution: data exclusion, data replacement

One of the most important issues in data cleansing is what assumption we use as "good enough" data. In particular, regarding missing data complement, data exclusion, and data replacement, the processing method varies depending on the assumption for ideal data; therefore we should determine how to conduct data cleansing in detail by comparing nearly ideal data to false data and verifying whether data cleansing appropriately handles the target data without degrading them based on the wrong assumption.

Although some processes conducted in data cleansing will not necessarily ensure data reversibility, it is important to clarify the validity of the target data and the analysis method when evaluating data analysis. Therefore we should ensure data accountability, the last of the essential four Cs, to enable follow-up verification of data cleansing from the following perspectives; validity of assumption, performance of method used, and influence on subsequent analyses.

We should remember that data cleansing is not a perfect method and has technical limitations. Although data cleansing is able to make raw data approach ideal data to some extent, it may fail to conserve the data's original features when there is a large amount of the data set as the target of data cleansing (e.g., data complement bias). Furthermore, it is sometimes technically impossible to solve certain systematic errors occurring in data collection, such as measurement faults of wearable Internet-of-Things devices including measurement electrode misplacement during electrocardiogram (ECG) measurement (i.e., misplaced positive to negative and vice versa). For these reasons, we repeat that it is important to collect adequate data during data collection to prevent such systematic errors under appropriate experimental conditions.

23.2.4 Feature extraction

We calculate certain features as feature extraction to understand the characteristics of the target data by using certain criteria. In this section, we briefly explain the knowledge indispensable for feature extraction, that is, type of data, feature examples, missing data handling, and tips in feature extraction.

23.2.4.1 Type of data

Data are roughly divided into qualitative and quantitative variables, and each variable is divided into two scales: nominal and ordinal for qualitative variables, and interval and ratio for quantitative variables.

Qualitative variables are for distinguishing the classification or types of data and cannot conduct any calculation. The nominal scale qualitative variables only involves data labels (e.g., sex, blood type, telephone number), and the ordinal scale qualitative variables have meaningful order, and enable comparison of each ordinal scale qualitative variable according to its value (e.g., rank, customer satisfaction).

Quantitative variables have meaningful numerical values and are calculable such as by addition or subtraction. Interval scale quantitative variables have meaningful order and intervals (e.g., date, time, temperature), and ratio scale quantitative variables have meaningful ratios as well as order and intervals (e.g., height, weight). The main differences between interval and ratio scales are the origin of the data (i.e., absolute zero that stands for "nothing"), which enables the calculation of ratios; the interval scale only allows addition and subtraction, whereas the ratio scale is used for four arithmetic operations: addition, subtraction, multiplication, and division.

Since data analysis, including machine learning, generally handles only quantitative variables, qualitative variables should be converted to quantitative variables by using dummy variables. There are roughly two types of qualitative variable conversion methods; one involves not increasing the dimensionality of the data such as label encoding, and the other involves increasing it such as one-hot encoding. Note that we should use data dimensionality reduction methods including feature hashing when we use the latter methods to handle large categories, which may become large and sparse features.

23.2.4.2 Example of features

Typical examples of features calculated from individual quantitative variables are fundamental statistics such as average, variance, and standard deviation. We might also define new features based on the target physiological response. As one example of the certain features including aforementioned two kinds of features, we show some features used in heart rate variability (HRV) analysis as shown in Table 23.1 [9]. Several time domain measures of HRV, such as mean R−R intervals (RRIs) and standard deviation of normal-to-normal R waves (SDNN), are calculated from fundamental statistics, whereas the other time measures of HRV, including pNN50, and frequency domain measures of HRV, such as the ratio of low-frequency components to high-frequency components (LF/HF), are calculated using a specific calculation method that is defined based on the physiological reaction of the heart or circulatory system.

Note that when we analyze two quantitative variables, certain statistic features focusing on the relationship of two variables are also usable such as sum of products of deviations, covariance, and correlation coefficient. When we analyze a combination of qualitative and quantitative variables, one solution may be that we first classify the target quantitative variable in accordance with the qualitative variable and later focus on the characteristics of the quantitative variable in each class.

Table 23.1 Examples of HRV features

	Variable	Units	Description
Time domain measures of HRV	Mean RRI	ms	Average of RRIs
	SDNN	ms	Standard deviation of all NN intervals
	RMSSD	ms	The square root of the mean of the sum of the squares of differences between adjacent NN intervals
	pNN50	%	The proportion derived by diving NN50 by the total number of all NN intervals
Frequency domain measures of HRV	LF	ms^2	Power in low-frequency range (frequency range 0.04–0.15 Hz)
	HF	ms^2	Power in high-frequency range (frequency range 0.15–0.40 Hz)
	LF/HF		Ratio LF (ms^2)/HF (ms^2)

23.2.4.3 Missing data handling

There are three types of missing data in accordance with their generation mechanisms: missing completely at random (MCAR), missing at random (MAR), and missing not at random (MNAR) [10]. As the name suggests, MCAR means missing data generated completely at random, while MAR means certain missing data generated by other factors (e.g., certain censored data due to patients changing hospitals). MNAR, which is also known as nonignorable missing, is neither MCAR nor MAR, and their missing reason is related to the missing itself.

There are two types of missing data processes: alternative data imputation and data deletion. Regarding data imputation, there are also two types of data imputation: data imputation using a fixed value (e.g., average, median, and mode) and using an estimated value calculated from a certain analysis method such as regression analysis or full information maximum likelihood (FIML). Regarding data deletion, listwise and pairwise deletions are the most traditional and representative data deletion methods. However, since with both methods it is assumed that the target data are MCAR, they may fail to appropriately delete missing data when the data are MAR or MNAR. When the target data are regarded as MAR, we should use multiple imputation (MI) or FIML, which can be used to calculate nonbiased estimated values [11], instead of listwise or pairwise deletion. Note that data deletion methods sometimes weaken statistical power, even when the target data are regarded as MCAR because they delete certain data containing missing values. On the other hand, since data imputation, such as MI or FIML, uses all data except for missing values, its statistical becomes power comparably high (in other words, its statistical power becomes low when the target data contain many missing values). Therefore even when the target data are regarded as MCAR, data imputation, such as MI or FIML, may be used to handle missing values.

When we analyze the target data regarded as MNAR, we should internalize the mechanism of missing values to data analysis instead of data imputation, which cannot generate nonbiased estimated values in MNAR data. Some MNAR data may be considered as MAR by integrating appropriate variables called auxiliary variables into data analysis, which does not affect the target data analysis [10].

23.2.4.4 Tips for feature extraction

Standardization: The characteristics of biological data may vary from person to person and from time to time even in the same person. This difference may prevent appropriate comparison among different conditions. One method of solving this problem is standardization with respect to each variable. We again emphasize that the assumption of "good enough" data is important in this standardization; although there are different standardization methods, each one standardizes the target data based on the assumed "good enough" data. Therefore it is important to consider adequate characteristics of ideal data before conducting standardization.

Curse of dimensionality: It is not necessarily better to handle more features. The more features there are, the more possibility we have to face what we call "curse of dimensionality" and might fail to improve generalization performance, resulting in certain models overfitting features and failing to estimate unknown data. Therefore we should conduct feature selection or dimensionality reduction to avoid the curse of dimensionality.

23.2.5 Learning

Leaning is a technique to obtain better information by tuning parameters, such as weights of features, and analysis method inherent hyperparameters; features that would contribute to obtaining better information should have larger weights, whereas hyperparameters should be tuned based on noise and errors in learning.

We should first define an objective function based on the analysis method. Note that we should design an objective function that is equal to the refinement criteria of information with which we are able to obtain the minimization problem. We discuss several examples of objective functions with related algorithms. Clustering algorithms use the total distance between each object and its nearest cluster as their objective function, classification algorithms use the total score of misidentified labels as an objective function, and regression algorithms use the difference between actual and estimated values as their objective function.

We then optimize an objective function under the initial hyperparameters and set weights to each feature. One of the most common optimizations is gradient descent, which calculates the gradient of an objective function at an arbitrary point by repeatedly moving that point in the direction of the negative gradient [4]. To improve calculation speed and memory efficiency, many other methods on gradient descent have been proposed, such as stochastic approximation [12] and momentum [13].

If we cannot obtain sufficient accuracy with the combination of tuned feature weights and initial hyperparameters, we then move on to hyperparameter tuning. Hyperparameters in each analysis method significantly affect information granularity. For instance, many clustering algorithms have hyperparameters on the number of clusters; smaller number of clusters would provide broader clusters. Classification algorithms on the other hand have hyperparameters on the types of labels; we can reduce the score of misidentified labels by removing immoderately rare labels.

When we fail to improve the performance of the model in spite of feature weight optimization and hyperparameter tuning, we should further conduct regularization. Regularization is an additional rule to tune weights of the features against "curse of dimensionality" [4]. When the size of the training data is smaller than the dimension of feature vectors, or there are only biased training data, learning often overfits the weights of features for training data, and generality is lost for newly obtained information. Regularization methods mitigate that overfitting and improve generalizability. One of the most popular regularization methods, for example, L2-norm regularizer, acts as mitigating penalty by imposing large weights to certain features that are frequently observed only in training data.

23.2.6 Evaluation

As the final step in data analysis, evaluation is very important for validating the performance of the learning against the purpose which mentioned in Section 23.2.1. We give several evaluation examples below. Note that evaluation methods should be determined based on the purpose of data analysis.

23.2.6.1 Clustering

To evaluate clustering, we should focus on the degree of separation among clusters. The easiest way is visualization, such as drawing a graph of each cluster. For instance, histograms or box plots are useful for one dimensional data, whereas for two- or three-dimensional data, scatter plots help in the understanding of the structure of obtained clusters. When four or higher dimensional data are evaluated, principal component analysis may have to be introduced to reduce dimensionality. Clusters of course have no labels, so that analysts must pick up several records from clusters to determine whether reasonable clusters can be obtained as shown in Fig. 23.7.

23.2.6.2 Classification

To evaluate classification, we should focus on the coincidence of the predicted labels against the correct labels. When the predicted label is the same as the correct label, it is counted as success. Conversely, when the predicted label is NOT the same as the correct one, it is counted as failure. We then count the following fundamental evaluation factors based on the confusion matrix shown in Fig. 23.8:

- ✓ *True positive (TP)*: the number of positive labels predicted as positive
- ✓ *True negative (TN)*: the number of negative labels predicted as negative
- ✓ *False positive (FP)*: the number of positive labels predicted as negative
- ✓ *False negative (FN)*: the number of negative labels predicted as positive

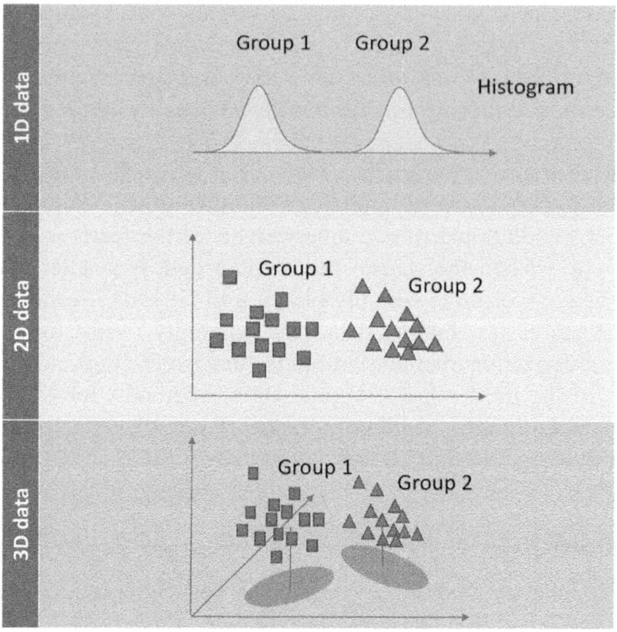

Figure 23.7 Well-separated sample.

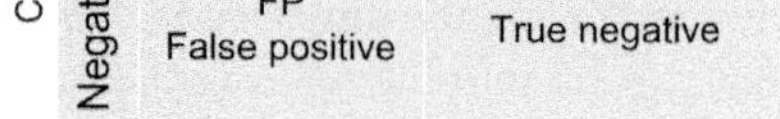

Figure 23.8 Confusion matrix.

These factors enable the calculation of several useful indicators including the following examples.

- ✓ *Accuracy* expressed as (TP + TN)/(TP + TN + FP + FN) is a standard indicator for model performance.
- ✓ *Recall* expressed as TP/(TP + FN) is the ratio of successful positive detections to correct detections, which called "sensitivity" in the medical field.
- ✓ *Precision* expressed as TP/(TP + FP) is the ratio of successful detections to all detections
- ✓ *Specificity* expressed as TN/(FP + TN) is the ratio of successful negative detections to all predicted negatives.
- ✓ *F-measure* is the harmonic mean of recall and precision expressed as 2 × (recall × precision)/(recall + precision)

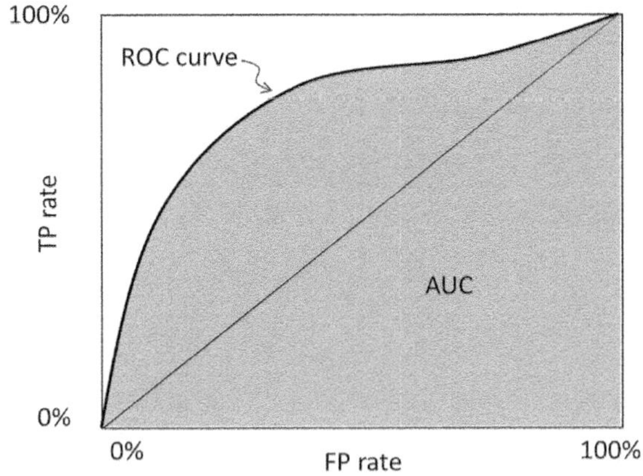

Figure 23.9 ROC curve and AUC.

Note that there is generally a trade-off between FP and TP. Therefore it is better to evaluate the performance of a classifier by plotting a receiver operating characteristic (ROC) curve, which plots precision on the vertical axis and 1-specificity on the horizontal axis [14].

The area under the ROC curve, or area under curve (AUC), as shown in Fig. 23.9, roughly represents the performance of the model; the better model shows the larger AUC in ROC. AUC also means the probability that the classifier ranks a randomly chosen positive instance above a randomly chosen negative one [15]. Therefore when analyzing the certain data set in which the number of each label is imbalanced (e.g., medical and health data), the AUC becomes one of the most useful evaluation methods. As another example of an evaluation tool, there is a recall precision curve, which plots precision on the vertical axis and recall on the horizontal axis, which is frequently used in information retrieval [16].

23.2.6.3 Regression

To evaluate regression, we should use the difference between the model-predicted values and correct values. We give several frequently used indicators below:

- ✓ *Mean absolute error* is the difference between the predicted and correct values.
- ✓ *Root mean squared error* is the root value of the difference between the predicted and correct values.
- ✓ *Correlation coefficient* is the relationship between the predicted and correct values.
- ✓ *Max error* is the maximum error value.

To correctly evaluate error, we should define the range of correct values, then regard the values deviating from that range as errors. For instance, when we estimate body weight, an error of 0.1 kg in adults is still acceptable but is a significant

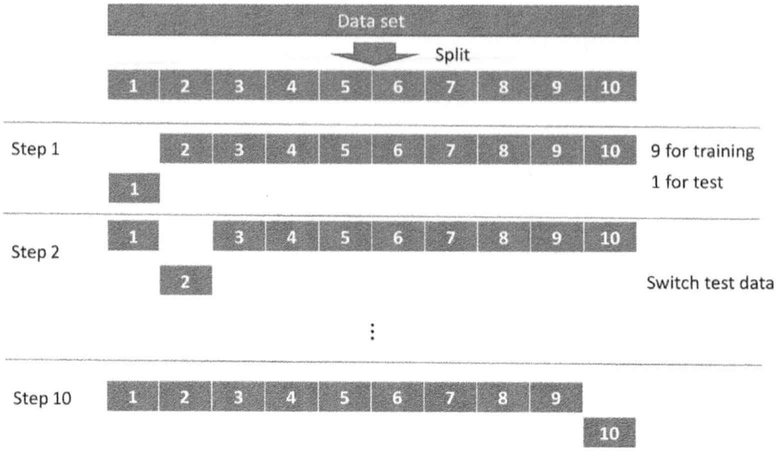

Figure 23.10 Procedure of 10-fold cross-validation.

error in infants. To prevent this problem, the error between the predicted value and correct value is divided by the correct value.

Note that when we evaluate the learned model, we must split the data set into two, one for training and the other for evaluation, to avoid overvaluation due to the overfitting of the model to the training data.

The robustness of the model for unknown data is called *generalization*; we regard a model as overfitting when robustness is low. If we obtained "good enough" results in the test data, which are not used for training, training may be successful. When evaluation results are not "good enough," we move on to the retraining of the model by changing parameters, as mentioned in the learning section, then reevaluate the retrained model (this process is called *parameter tuning*). However, tuning parameters only for one test data cannot improve the performance of the model for the unknown data. Thus we should introduce cross-validation. The easiest way of cross-validation is to split a data set into two; we train the model with one data set, then evaluate with the other, and vice versa. One of the most frequently used *cross-validation* methods is *N*-fold cross-validation. An *N* of 10 (i.e., 10-fold cross-validation) is typically used, in which we split the data set into 10 sets. We train the model with nine of the sets and evaluate it with the remaining set. We then switch the test data to another split part of data, and reserve the remaining nine split parts for the training data of that test data. We repeat this evaluation 10 times in 10-fold cross-validation as shown in Fig. 23.10.

23.2.6.4 For further evaluation

When we evaluate classification, there are several cases of having only imbalanced data, where the ratio of positive labels to negative labels is imbalanced. It becomes difficult to tune parameters when there are only a few labels. To improve the ratio of each label to 50–50, we generally reduce the one with many labels to the same number as that with only a few labels. However, this method cannot be used to solve the data imbalance, especially when the one label stands for an extremely rare disease. Therefore we should use several other methods including partial AUC maximization [17] to solve this problem.

23.3 Example of health data analysis

In this chapter, we show an example of health data analysis for a specific purpose. We assume that the goal of this data analysis is to clarify the fatigue state of drivers with the aim of driver health monitoring to reduce traffic accidents due to overwork [18]. We briefly explain each step according to the previously mentioned procedure as if we were the analysts.

Design: Our prime interest in driver health monitoring is how to estimate or detect dangerous situations caused by drowsiness or fatigue based on vital information obtained from sensors attached to drivers. When we want to detect drowsiness and automatically and safely stop the vehicle, our target problem would become classification, and the output format of data analysis will be a label (i.e., drowsy or not). Another example of the target problem is regression; when we estimate the change in fatigue to appropriately manage the bus operation schedule, the output format of this data analysis will be a continuous value. We describe the latter case as follows. In this case, the fatigue state measured from specific questionnaires or special instruments becomes objective values, whereas vital information or features measured from wearable devices becomes explanatory values.

Data collection: Several types of vital information have been used for driver health monitoring, such as electroencephalograms [19] and electrooculograms [20]. We focused on ECGs as practical biosignals that may be easy to measure from wearable devices, with which we are able to measure the time series fluctuation of RRI (i.e., HRV), to possibly reflect autonomic nervous system activity [9]. Regarding fatigue level, we assume that our target fatigue is measurable as critical fusion frequency (CFF) by using special instruments [21]. We measure the ECGs and CFFs of a driver at the same time. Note that CFFs will become sparse time series data since they cannot be measured while the person is driving.

Data cleansing: ECGs measured from wearable devices sometimes include noise or artifacts due to measurement faults (Fig. 23.11) [22,23]. Therefore RRI, the fundamental feature of an ECG, may also include outliers. Furthermore, we must take data loss, such as packet loss, into account when using wireless wearable devices for data collection. We should reduce the effect of those outliers or data loss by outlier exclusion [24] or data interpolation for data cleansing.

Feature extraction: We are able to calculate the feature value of HRV based on HRV analysis [9]. For instance, we can calculate MEAN and SDNN, which are

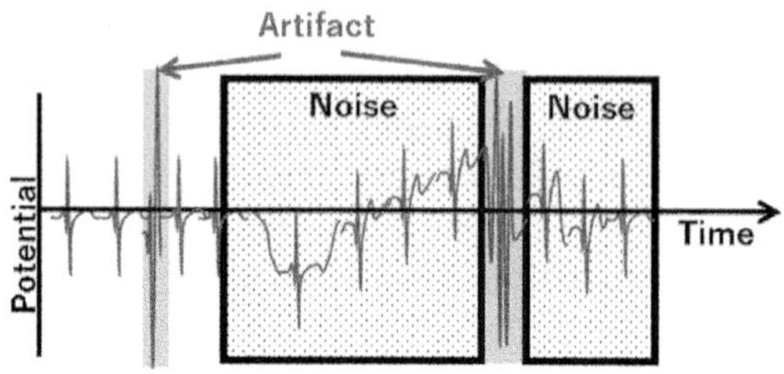

Figure 23.11 Example of measurement faults in ECGs.

fundamental statistics of target RRIs, the pNN50 as a frequently used HRV inherent time feature, or frequency domain measures of HRV such as LF and HF described in Section 23.2.4.2. Note that when using a device other than medical devices such as shirt-type wearable ECG devices, we should also reconsider feature value calculation based on actual data and appropriately change it to reduce the device inherent problems as necessary [23].

Learning: We set a simple model $y = w_1 \cdot \text{MEAN} + w_2 \cdot \text{SDNN} + w_3 \cdot \text{pNN50} + w_4 \cdot \text{LF} + w_5 \cdot \text{HF} + w_0$, where y is an objective value for the fatigue level of drivers, and w_0 to w_5 are regression parameters. We then move on to obtain regression parameters. In this case, a multilinear regression can be used to solve this problem. Note that several types of health data can have nonlinear characteristics or multicollinearity. In such cases, methods including support vector regression [25] or partial least squares regression [26] may be helpful.

Evaluation: We first split a data set into a training data set and test data set. We then create a program code for model training and execute it with the training data set. Finally, we can predict the fatigue level from the HRV features calculated from the measured ECGs. We of course should evaluate the errors between the predicted value and actual measured values (in this example, actual CFFs) and tune the parameters. When the estimation performance of the created model is lower than expected, we should step back to the learning step and reconsider the model including feature addition or change the learning method.

23.4 Conclusion

Recent development of Internet-of-Things devices enables us to measure an extraordinary amount of data. However, just analyzing measured data as it is cannot create productive result; we should analyze it with strategies which suitable for the target objectives. In this article, we explained the process of data analysis consisting of the following six processes: analysis design, data collection, data cleansing, feature extraction, learning, and evaluation. We again repeat that it is important to pursue each process step by step from start to finish with appropriate assumption to obtain more valid analysis results.

References

[1] T.H. Cormen, C.E. Leiserson, R.L. Rivest, C. Stein, Introduction to Algorithms, third ed., The MIT Press, 2009.

[2] T. Zhang, R. Ramakrishnan, M. Livny, BIRCH: an efficient data clustering method for very large databases, SIGMOD Rec. 25 (2) (1996) 103–114.

[3] M. Ester, H.P. Kriegel, J. Sander, X. Xu, A density-based algorithm for discovering clusters a density-based algorithm for discovering clusters in large spatial databases with noise, in: Proceedings of the Second International Conference on Knowledge Discovery and Data Mining (KDD'96), 1996, pp. 226–231.

[4] C.M. Bishop, Pattern Recognition and Machine Learning (Information Science and Statistics), Springer-Verlag, 2006.

[5] B.E. Boser, I.M. Guyon, V.N. Vapnik, A training algorithm for optimal margin classifiers, in: Proceedings of the Fifth Annual Workshop on Computational Learning Theory (COLT '92), 1992, pp. 144–152.

[6] I. Goodfellow, Y. Bengio, A. Courville, Deep Learning, The MIT Press, 2016.

[7] L. Breiman, Random forests, Mach. Learn. 45 (1) (2001) 5–32.

[8] Q.E. McCallum, Bad Data Handbook, O'Reilly Media, 2012.

[9] Task Force of the European Society of Cardiology and the North American Society of Pacing and Electrophysiology, Heart rate variability. Standards of measurement, physiological interpretation, and clinical use, Eur. Heart J. 17 (3) (1996) 354–381.

[10] J.L. Schafer, J.W. Graham, Missing data: our view of the state of the art, Psychol. Methods 7 (2) (2002) 147–177.

[11] D.B. Rubin, Multiple imputation after 18+ years, J. Am. Statist. Assoc. 91 (434) (1996) 473–489.

[12] J. Nocedal, Updating quasi-Newton matrices with limited storage, Math. Comput. 35 (151) (1980) 773–782.

[13] D.E. Rumelhart, G.E. Hinton, R.J. Williams, Learning representations by back-propagating errors, Nature 323 (1986) 533–536.

[14] T. Fawcett, An introduction to ROC analysis, Pattern Recogn. Lett. 27 (8) (2006) 861–874.

[15] A.P. Bradley, The use of the area under the ROC curve in the evaluation of machine learning algorithms, Pattern Recogn. 30 (7) (1997) 1145–1159.

[16] I.H. Witten, E. Frank, Data mining: practical machine learning tools and techniques, Morgan Kaufmann Series in Data Management Systems, second ed., Morgan Kaufmann, 2005.

[17] M.T. Ricamato, F. Tortorella, Partial AUC maximization in a linear combination of dichotomizers, Pattern Recogn. 44 (10) (2011) 2669–2677.

[18] Y. Dong, Z. Hu, K. Uchimura, N. Murayama, Driver inattention monitoring system for intelligent vehicles: a review, IEEE Trans. Intelligent Transport. Syst. 12 (2) (2011) 596–614.

[19] S.K. Lal, A. Craig, P. Boord, L. Kirkup, H. Nguyen, Development of an algorithm for an EEG-based driver fatigue countermeasure, J. Safety Res. 34 (3) (2003) 321–328.

[20] S. Hu, G. Zheng, Driver drowsiness detection with eyelid related parameters by support vector machine, Expert Syst. Appl. 36 (4) (2009) 7651–7658.

[21] E. Simonson, Measurement of fusion frequency of flicker as a test for fatigue of the central nervous system, Ind. Hyg. Toxixol. 23 (1941) 83–89.

[22] G.M. Friesen, T.C. Jannett, M.A. Jadallah, S.L. Yates, S.R. Quint, H.T. Nagle, A comparison of the noise sensitivity of nine QRS detection algorithms, IEEE Trans. Biomed. Eng. 37 (1) (1990) 85–98.
[23] K. Eguchi, R. Aoki, S. Shimauchi, K. Yoshida, T. Yamada, R-R interval outlier processing for heart rate variability analysis using wearable ECG devices, Adv. Biomed. Eng. 7 (2018) 28–38.
[24] K. Eguchi, R. Aoki, K. Yoshida, T. Yamada, Reliability evaluation of R-R interval measurement status for time domain heart rate variability analysis with wearable ECG devices, in: Proceedings of the 39th Annual International Conference on IEEE Engineering in Medicine and Biology Society (EMBC'17), 2017, pp. 1307–1311.
[25] A.J. Smola, B. Scholkopf, A tutorial on support vector regression, Stat. Comput. 14 (3) (2004) 199–222.
[26] P. Geladi, B.R. Kowalski, Partial least-squares regression: a tutorial, Anal. Chim. Acta 185 (1986) 1–17.

Summary and future issues

Throughout the chapters, this book has introduced the growing field of chemical, gas, and biosensors for Internet of Things (IoT) and related applications. All chapters were contributed by the authors, who are active researchers at the leading edge of the field. Recently various sensing devices have been connected to the Internet to monitor vital, clinical, and environmental information. For example, wearable sensors for healthcare applications, which can monitor physical signals including temperature, blood pressure, and electrocardiogram, have been developed for continuous monitoring of health conditions. In contrast, the development of chemical and biochemical sensing devices is not as advanced and these are limited to a few applications, typically glucose monitoring. This book introduced various kinds of chemical and biochemical sensors that will be applied in the future by combination with the Internet after further improvements. The book is divided into three parts.

Part 1 consists of the topics of new sensing materials, detection methods, and devices and their biochemical, clinical, and environmental applications. As new sensing devices and systems, portable urine glucose sensor (Chapter 1), artificial olfactory system (Chapter 3), portable immunoassay devices (Chapter 6), SAW device based immunosensor (Chapter 7), and sensors for pollutant determination (Chapter 10) are illustrated. In contrast, paper based technology (Chapter 2), olfactory receptor based biosensor (Chapter 4), surface modification technology for biosensors (Chapter 5), aptamer based sensors (Chapter 8), and new carbon based chemical sensors (Chapter 9) are introduced as examples of new materials based sensors.

Part 2 consists of flexible, wearable, and mobile sensors and related technology, which are more application-oriented sensing devices compared with those in Part 1. Smart clothing using wearable bioelectrodes that are physical sensors, and which is commercially available, is introduced in Chapter 11. This is a good example for future wearable IoT sensors. As chemical and biochemical sensors, cavitas bio/chemical sensors (Chapter 11) are a good example of wearable sensors. Point of care testing apparatus for immunosensing (Chapter 13), IoT sensors for smart livestock management (Chapter 14), and compact disk-type biosensors (Chapter 15) are introduced as examples of microfluidics based sensors mainly for wet biological or clinical samples. On the other hand, gas phase monitoring becomes more interesting for future clinical diagnosis. A CMOS device for gas sensing (Chapter 16), visualization of odor space and quality (Chapter 17), and bio-sniffer and sniff-cam (Chapter 18) are used for such applications.

Part 3 contains information and network technologies for sensor–IoT applications. Flexible and printed biosensors based on an organic FET device (Chapter 19) are an important technology to develop wearable biosensors with reasonable cost.

Self-monitoring of fat metabolism using a portable/wearable acetone analyzer (Chapter 20) is an example of a wearable clinical gas sensor employing already established sensing materials. Environmental monitoring is another important field for IoT chemical sensors. Air pollution monitoring network of PM2.5, N_2, and radiation of 137Cs (Chapter 21) is an example of such sensors. Wireless sensor network with various sensors (Chapter 22) and data analysis targeting healthcare-support applications using IoT sensors (Chapter 23) are both core technologies to support the above chemical sensors, and are essential for developing IoT sensor system and data analysis obtained from various sensors. At the present stage, most of the chemical sensors are still under development and there is a need to realize more maintenance-free, higher selectivity, longer lifetime (or reusable), and lower cost sensors, which will be solved by employing new technologies including nanomaterials, biotechnology, low cost device fabrication technology, etc. In the 5G Internet society, many more chemical sensors will be connected to the Internet and will gather a huge amount of data, which will be analyzed by employing artificial intelligence technology.

Index

Note: Page numbers followed by "*f*" and "*t*" refer to figures and tables, respectively.

Q

R

S

T

CPI Antony Rowe
Eastbourne, UK
June 23, 2019